A TEXTBOOK OF ALGAE

A TEXTBOOK OF ALGAE

K. S. Bilgrami
D.Phil, D.Sc., FNA
Post-Graduate Department of Botany
Bhagalpur University, Bhagalpur

L.C. Saha
Ph.D.
Post-Graduate Department of Botany
Bhagalpur University, Bhagalpur

CBSPD

CBS Publishers & Distributors Pvt Ltd

New Delhi • Bengaluru • Chennai • Kochi • Kolkata • Lucknow • Mumbai
Hyderabad • Jharkhand • Nagpur • Patna • Pune • Uttarakhand

Disclaimer

Science and technology are constantly changing fields. New research and experience broaden the scope of information and knowledge. The authors have tried their best in giving information available to them while preparing the material for this book. Although all efforts has been made to ensure optimum accuracy of the material, yet it is quite possible some errors might have been left uncorrected. The publisher, the printer and the authors will not be held responsible for any inadvertent errors, omissions or inaccuracies.

A TEXTBOOK OF ALGAE

ISBN: 978-81-239-0049-0

First Edition: 1992
Reprint: 1996, 2001, 2002, 2004, 2006, 2007, 2010, 2012, 2015, 2018, 2020, 2023

Published by **Satish Kumar Jain** and produced by **Varun Jain** for

CBS Publishers & Distributors Pvt Ltd

4819/XI Prahlad Street, 24 Ansari Road, Daryaganj, New Delhi 110 002, India.
Ph: 011-23289259, 23266861, 23266867 Website: www.cbspd.com
Fax: 011-23243014 e-mail: delhi@cbspd.com;

Corporate Office: 204 FIE, Industrial Area, Patparganj, Delhi 110 092
Ph: 011-4934 4934 Fax: 011-4934 4935 e-mail: publishing@cbspd.com; publicity@cbspd.com

Branches

- **Bengaluru:** Seema House 2975, 17th Cross, KR Road, Banasankari 2nd Stage, Bengaluru 560 070, Karnataka, India
  Ph: +91-80-26771678/79 Fax: +91-80-26771680 e-mail: bangalore@cbspd.com
- **Chennai:** 7, Subbaraya Street, Shenoy Nagar, Chennai 600 030, Tamil Nadu, India
  Ph: +91-44-26680620, 26681266 Fax: +91-44-42032115 e-mail: chennai@cbspd.com
- **Kochi:** 42/1325, 1326, Power House Road, Opp KSEB, Power House, Ernakulum Kochi 682 018, Kerala, India
  Ph: +91-484-4059061-65,67 Fax: +91-484-4059065 e-mail: kochi@cbspd.com
- **Kolkata:** 147, Hind Ceramics Compound, 1st Floor, Nilgunj Road, Belghoria, Kolkata-700056, West Bengal, India
  Ph: +033-25633055, 033-25633056 e-mail: kolkata@cbspd.com
- **Lucknow:** Basement, Khushnuma Complex, 7 Meerabai Marg (Behind Jawahar Bhawan),Lucknow-226001, UP, India
  Ph: +0522-4000032 e-mail: tiwari.lucknow@cbspd.com
- **Mumbai:** PWD Shed, Gala no 25/26, Ramchandra Bhatt Marg, Next to JJ Hospital Gate no. 2, Opp. Union Bank of India, Noorbaug, Mumbai-400009, Maharashtra, India
  Ph: 022-66661880/89 e-mail: mumbai@cbspd.com

Representatives

- Hyderabad 0-9885175004
- Jharkhand 0-9811541605
- Nagpur 0-9421945513
- Patna 0-9334159340
- Pune 0-9923910676
- Uttarakhand 0-9716462459

Printed at Sanjay Printer, Sahibabad, UP, India

PREFACE

This text has been written for the university level students. It is essentially based on the syllabus which has recently been prepared and approved by the Curriculum Development Committee of the U.G.C. Care has been taken to include all such important topics which are being taught in different Indian universities. This book will be of special interest to the students preparing for various competitive examinations.

Authors are thankful to Prof. Daniel E. Wujek, Central Michigan Univ., USA for using some of his photographs. We would also like to thank the staff of the Computer Centre, Bhagalpur University for their cooperations in the preparation of the text and Sri Santosh Kumar Thakur for his help in making diagrams.

We realize that there might be many shortcomings in this text and we shall greatly appreciate and welcome suggestions from the teachers and students for improvement in the next eddition.

K.S. Bilgrami
L.C. Saha

CONTENTS

1

INTRODUCTION TO THE ALGAE

DEFINITION

The algae comprise a large heterogenous assemblage of plants which are highly diverse with respect to their organization, dimension, habitat, physiology, biochemistry, reproduction and ecology. The definition of algae given by Fritsch in 1935 still seems to be most comprehensive. According to him "Unless purely artificial limits are drawn, the designation alga must include all holophytic organisms (as well as their numerous colourless derivatives) that fail to reach the level of differentiation characteristic of archegoniate plants". The difficulty of defining an alga still persists as Lewin and Gibbs (1982) state "organisms currently regarded as algae have virtually no common features which would indicate phylogenetic homogeneity".

The algae are chlorophyll bearing simple lower plants which range from unicellular forms to aggregation of cells, filaments, or parenchymatous thalli. Many of the unicellular forms are motile and even most complex multicellular forms show some degree of functions and organization as compared to other groups of plants. The range of morphology is, however, extremely diverse and a number of brown algae (Phaeophyceae) may attain a size comparable to that of a small tree.

Algal reproduction may be accomplished by vegetative, asexual and sexual processes which express considerable variation in different groups. In procaryotic division (Cyanophyta) and in a number of eucaryotic division (e.g., Euglenophyta), sexual reproduction is lacking.

DIFFERENTIATION WITH OTHER PLANTS

A feature which distinguishes algae from other cryptogams is the absence of multicellular wall around the sporangia or gametangia (with the exception of the antheridium in Charophyceae). Sex organs and sporangia of Bryophyta and other plants higher to them in the evolutionary ladder are multicellullar with the outer cells being sterile, whereas, in algae, the sporangia and sex organs are either unicellular, or when multicellular, all the cells are fertile.

Biochemically and physiologically the algae have many common features with other plants. They have the same biochemical pathways, chlorophyll **a** and have carbohydrate and protein as the end products which are comparable to those of higher plants. Due to their small size and easy manipulation in liquid media, they are ideal laboratory tools for physiological studies. The study of algae is known as **algology** or **phycology**. Word phycology is derived from a Greek work **phykos**, which means seaweed.

HABIT AND HABITAT

Algae are ubiquitous, practically occurring in every habitable environment on earth. They are, however, abundant in stagnant waters. The occurrence of algae in habitats like thermal springs, ice, damp soil, rocks, tree trunks, old walls etc. is not uncommon. Free floating assemblage of algae is collectively known as **phytoplankton**.

A number of algae can live inside tissue of other plants and are known as **endophytes. Symbiotic association** i.e., deriving mutual benefit and **epiphytic association** i.e., without causing any harm to other plants are also common.

These photoautotrophic plants are the major primary producers of organic materials and they play a key role as the base of the food chain in aquatic and semiaquatic systems. They constitute a source of food and their chemical extracts are used in the manufacture of synthetic food and other useful products. Sometimes they become 'nuisance' organisms in municipal water and choke the supply system, depending on the degree of eutrophication of reservoirs. Very rarely the algae are the cause of serious human diseases. As the causative organisms of 'red tides', some dinophycean members are source of poisoning. Term **phycotoxin** is collectively used for the toxins secreted by algal forms, especially the blue-greens like *Microcystis aeruginosa, Aphanizomenon flos-aquae* etc.

RANGE OF THALLUS STRUCTURE

The algal plant bodies vary from relatively simple single cells to strikingly complex structures exhibited by the giant kelps. Unicelullar, colonial, filamentous, membranous or foliose and tubular type of algal plant body together with highly complex types that have root like, stemlike and leaf like organs are of frequent occurrence. They however, lack vascular tissues. Various forms of thalli (Fig. 1.1) can be distinguished into following types:

MOTILE HABIT

(a) **Unicellular form :** These are generally spherical, oblong or pear-shaped alga with two flagella in the anterior portion (e.g., *Chlamydomonas*, Fig. 1.1 A).

(b) **Coenobial form :** The definite number of motile cells are embedded in a gelatinous matrix and are held together by cytoplasmic connections (e.g., *Gonium*, Fig. 1.1 B).

(c) **Colonial form :** Indefinite number of motile cells are present in a gelatinous envelope (e.g., *Volvox*, Fig. 1.1 C).

NONMOTILE HABIT (COCCOID HABIT)

(a) **Unicellular form :** The cells are nonflagellate and do not exhibit any movement during major phase of their life cycle (e.g., *Chlorococcum*, Fig. 1.1 D).

(b) **Coenobial form :** Definite number of nonmotile cells also form coenobia. These cells may be arranged in a single layer being closely adpressed to each other (*Scenedesmus*, Fig. 1.1 E) or may form a pentagonal or hexagonal net work (e.g., *Hydrodictyon*, Fig. 1.1 F).

PALMELLOID AND DENDROID HABIT

In this type of habit, motility is lost in the vegetative phase and occurs only during reproduction. In some genera the nonmotile vegetative phase is short lived e.g., *Chlamydomonas* while in a large number of algae this palmelloid habit is almost permanent with the motility being restricted to reproductive cells only.

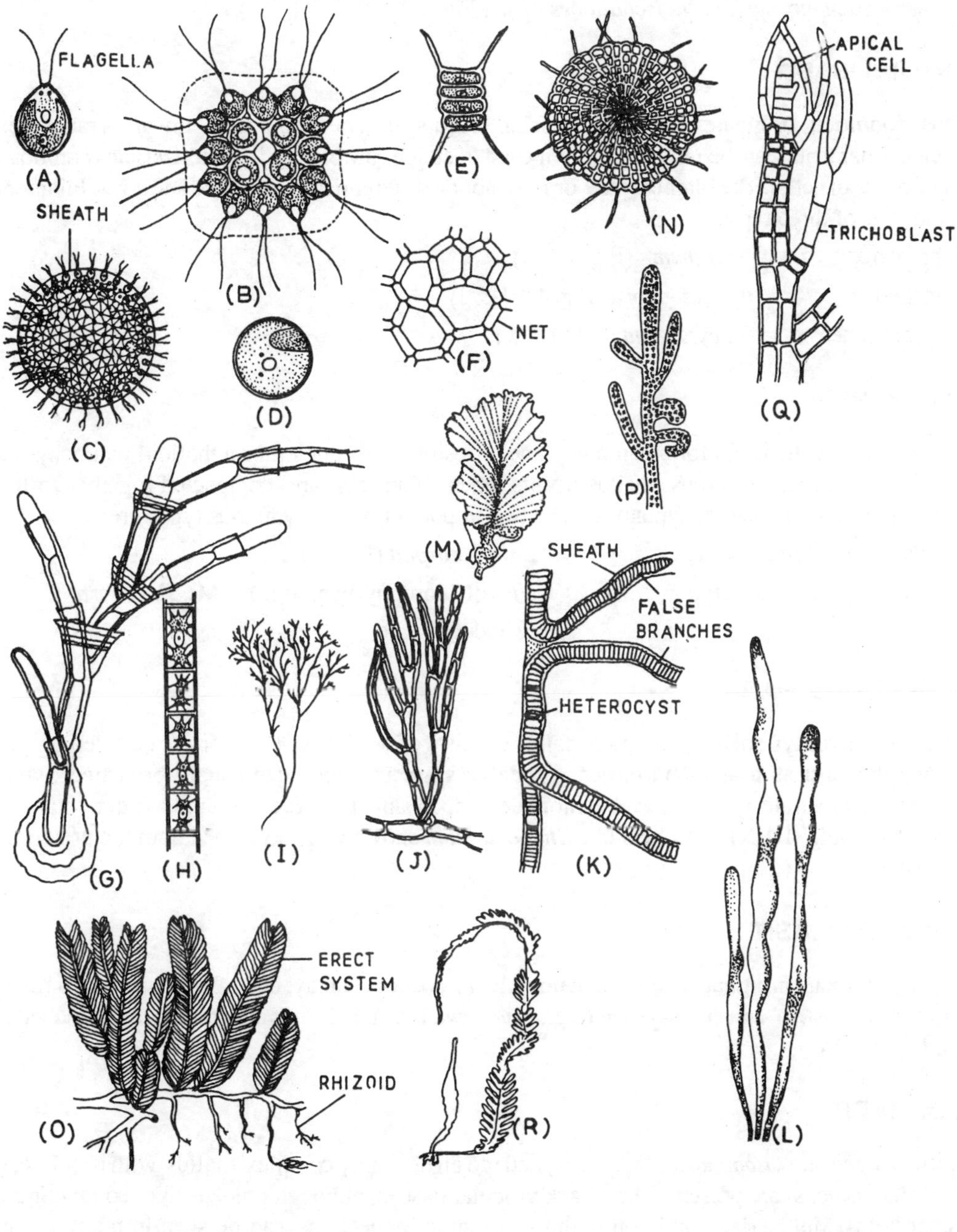

Fig. 1.1. Thallus types in Algae. (A) Unicellular biflagellate *Chlamydomonas*. (B) Coenobial *Gonium*. (C) Motile colonial *Volvox*. (D) Nonmotile unicellular *Chlorococcum*. (E) Nonmotile coenobial *Scenedesmus*. (F) Nonmotile net-like *Hydrodictyon*. (G) Dendroid *Ecballocystopsis*. (H) Unbranched filamentous *Zygnema*. (I,J) Branched filamentous *Cladophora*. (K) False-branched *Scytonema*. (L) Tubular parenchymatous *Enteromorpha intestinalis*. (M) Foliose parenchymatous *Ulva*. (N) Discoid *Coleochaete scutata*. (O) Siphonaceous *Caulerpa*. (P) Siphonaceous simple branched *Vaucheria*. (Q) Polysiphonaceous *Polysiphonia*. (R) Complex thallus of *Laminaria agardhii*.

In dendroid type of thallus mucilage is produced locally, generally at the base of cell which results in cells with basal mucilage stalk (e.g., *Ecballocystopsis*, Fig. 1.1 G).

FILAMENTOUS HABIT

Filamentous condition originates when cell division occurs in one plane with linear arrangement of daughter cells. Filaments can be uniseriate (composed of single row of cells) or multiseriate (composed of more than one row of cells). The filament may or may not be differentiated into base and apex. Filamentous condition may be of three types :

(a) Unbranched - *Zygnema* (Fig. 1.1 H)
(b) Branched - *Cladophora* (Fig. 1.1 I & J)
(c) False branched - *Scytonema* (Fig. 1.1 K)

THALLOID HABIT

When the division of cells is in more than one plane, it results in formation of a thalloid parenchymatous structure. A pseudoparenchymatous thallus occurs when filaments are compacted together to form a structure that appears to be parenchymatous. The two important parenchymatous types are:

(i) Tubular parenchymatous type - *Enteromorpha* (Fig. 1.1 L)
(ii) Foliose parenchymatous type - *Ulva* (Chlorophyceae, Fig. 1.1 M), *Porphyra* (Rhodophyceae)

HETEROTRICHOUS HABIT

This is the most evolved type of habit in which thallus consists of two parts: (1) a prostrate creeping system growing along the substratum and (2) a projecting or erect system composed of usually branched filaments. In many members, prostrate system may be more developed and the erect system is rudimentary (e.g., *Coleochaete scuta*, Fig. 1.1 N). Members like *Stigeoclonium* show well developed both the prostrate and projecting systems.

SIPHONACEOUS HABIT

The thallus is nonseptate multinucleate i.e., **coenocytic.** A coenocyte may be differentiated into rhizoidal system, rhizomatic system and erect system (e.g., *Caulerpa*, Fig. 1.1 O) or it may be simply branched (e.g., *Vaucheria*, Fig. 1.1 P).

COMPLEX HABIT

In many brown algae (e.g., *Laminaria*, Fig. 1.1 k) and red algae highly complex thallus with root like, stem like and leaf like organs are present. These lack vascular tissues, although phloem like conducting cells occur in some kelps. Multiseriate condition with cytoplasmic connections can be seen in many red algae like *Polysiphonia* (Fig.1.1 Q).

Besides these thallus types, many members show **parasitic, endophytic, endozoic, symbiotic** and **epiphytic** adaptations.

REPRODUCTIVE DIVERSITY

Reproductive cells : There are some general terms used in reference to reproductive cells. **Spores** are cells which germinate to new individual without fusion. **Aplanaspores** are nonmotile spores. Aplanospores with greatly thickened cell wall is known as **hypnospores. Gametes** are cells which fuse to form zygote. Motile gametes are **planogametes** and nonmotile gametes are known as **aplanogametes. Autospores** are like parent cell. **Swarmer** is a general term for any motile cell. Vegetative cells which have developed into a spore like stage with very thick walls and abundant food reserves are known as **akinetes.** Unlike aplanospores, akinetes always have additional wall layers around the protoplast fused with the parent wall.

Reproduction : There are three types of reproduction namely vegetative, asexual and sexual (Fig. 1.2). Asexual and sexual reproduction is of wide occurrence in algae. It is the most common method of algal propagation.

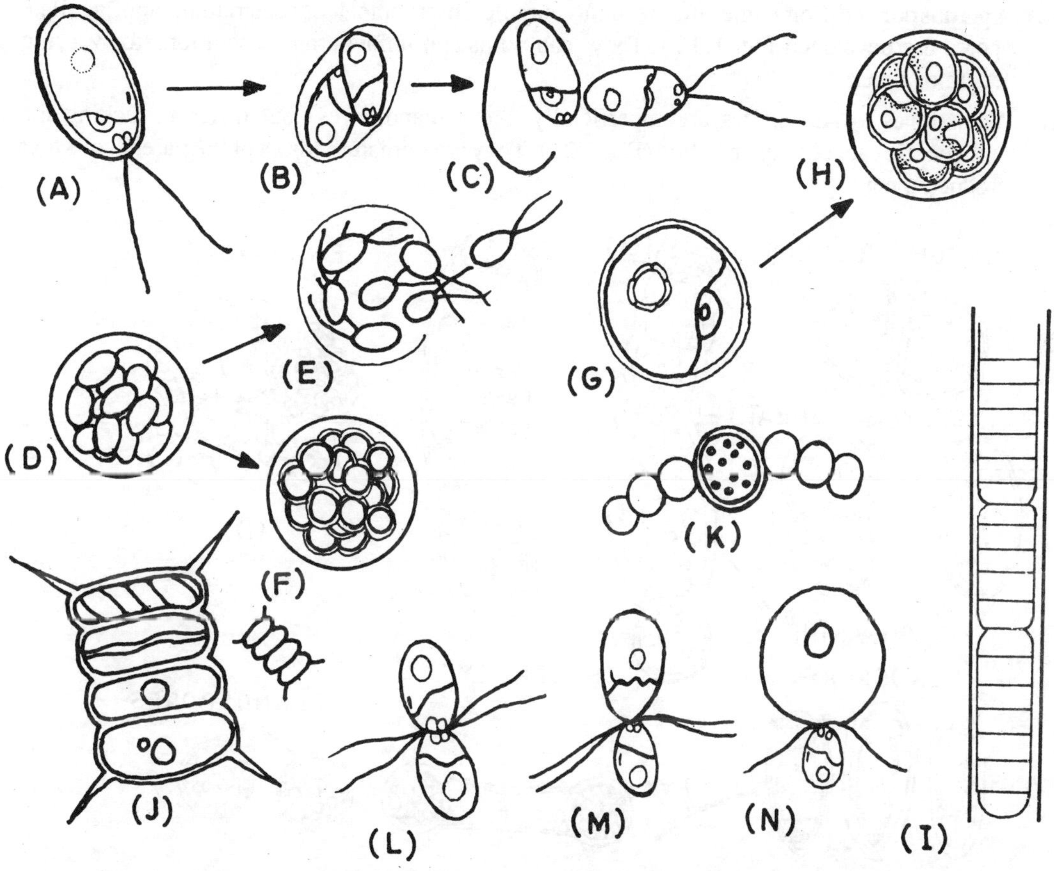

Fig. 1.2. Methods of algal reproduction. (A-C) Binary fission or bipartition. (D-E) Zoospore formation. (F) Aplanospore formation. (G-H) Autospore formation. (I) Fragmentation or formation of hormogonium. (J) Autocolony formation. (K) Akinete formation. (L) Isogamy. (M) Anisogamy. (N) Oogomy.

VEGETABLE : Vegetative reproduction may take place by any of the following two methods :

(1) **Fragmentation** - It may be accidental or by the development of special structures like **separation disc** (e.g., *Oscillatoria*). This process of breakage is common in filamentous forms.

(2) **Cell division** - In unicellular algae, cell division is a common mode of reproduction. This division may be repeated in rapid succession and sometimes known as binary fission.

ASEXUAL : Asexual reproduction may be accomplished by these methods :

(1) **Akinete** - Akinetes are resistant to unfavourable environmental conditions. With the onset of favourable conditions they germinate into new individuals (e.g., *Pithophora*).

(2) **Spores** - There are various kinds of spores (Fig. 1.3). These are :

(a) **Zoospores :** These flagellated spores are either produced inside **zoosporangium** and **aplanosporangium** or directly develop from the vegetative cells. The zoospores may be biflagellate or multiflagellate. In multiflagellate zoospores the flagella may be localised at the apical end or may be distributed throughout the entire body. If zoospores are produced after mitotic cell division, these may be designated as **mitospores** and if they are produced after meiosis then these are known as **meiospores.** Zoospores swarm in water for some time then come to rest, shed their flagella and germinate into new individuals.

(b) **Aplanospores :** Sometimes due to unfavourable environment of substratum, nonflagellate spores are developed(Fig. 1.3 F). They can withstand unfavourable conditions for varying length of time.

(c) **Autospores :** Autospores are superficially like aplanospores but differ in lacking the ontogenetic capacity for motility (Fig. 1.2 J). They are miniafure forms of the parent cells that produce them.

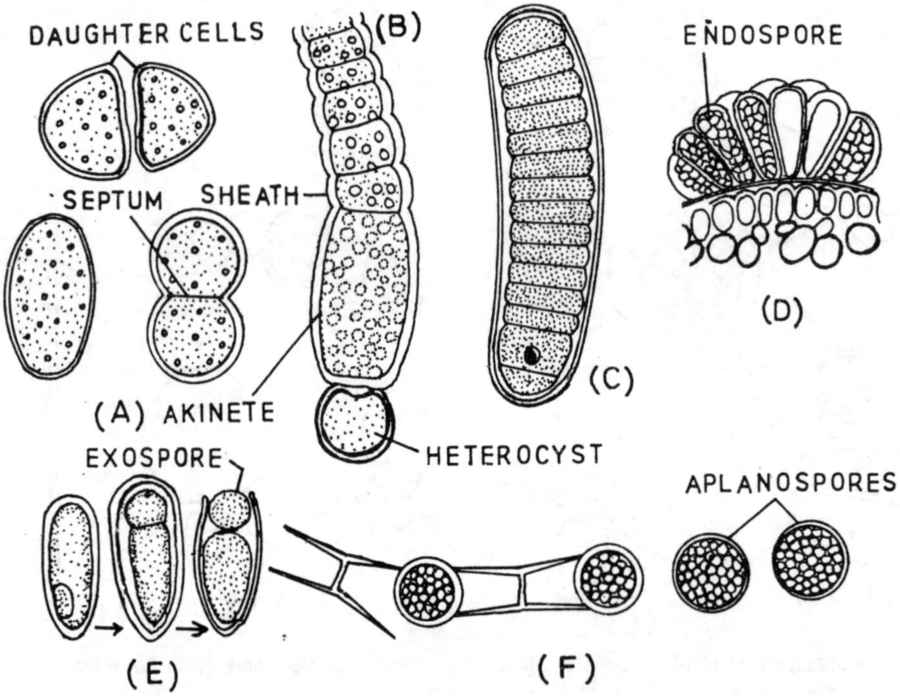

Fig. 1.3. Spore types and asexual reproduction in algae. (A) Simple cell division in unicellular *Synechococcus.* (B) Filament of *Gloeotrichia natans* with heterocyst and akinete. (C) Hormospore of *Westiella.* (D) Endospores of *Dermocarpa prasina.* (E) Exospore formation in *Chamaesiphon incrustans.* (F) Aplanospores of *Microspora willeana.*

(d) **Hormospores :** Hormocysts are the fragments of some filamentous form which develop thick sheath to tide over unfavourable conditions (Fig. 1.3 C).

(e) **Heterocyst :** In many filamentous blue-green algae, certain vegetative cells become larger, transparent and develop thick layers. These are heterocysts whose position may be terminal or intercalary; their number may be one or more. These serve the points of fragmentation of the filaments (Fig. 1.3 B).

(f) **Endospores:** In some unicellular algae, the entire protoplast of the vegetative cell divides to form internally a number of non-flagellate thin walled spores (Fig. 1.3 D). These are endospores which develop into new individuals.

(g) **Exospores:** These thin walled non-flagellate spores are produced externally from the protoplast of vegetative cells (Fig. 1.3 E).

(h) **Other spores** : In red algae, **monospores, tetraspores, paraspores** and **carpospores** are also found. In golden algae **statospores** and **auxospores** are the spores wich produce new individuals.

SEXUAL : Sexual reproduction involves the union of cells **(plasmogamy)**, the union of their nuclei **(karyogamy)**, the association of their chromosomes and genes and meiosis. Sexual reproduction has apparently not yet been confirmed in the Euglenophyta and Cryptophyta but occurs in atleast some members of all the other divisions of algae. In blue -green algae there is the evidence of genetic recombination.

In sexual reproduction, there is the fusion of gametes which form a zygote. In monoecious or homothallic species, gametes of different mating types are formed on the same plant, whereas in heterothallic or dioecious species the gametes of different strains are formed on different plants. Gametes are formed by **gametogenesis**. Depending upon structure and function of the gametes, there are different types of sexual reproduction. These are :

Isogamy : The process of isogamy involves the fusion of motile or nonmotile **isogametes** which are morphologically and physiologically alike. As the two gametes that fuse to form a zygote, look and behave in identical manner, it is not possible to distinguish them as male or female. The gametes are therfore referred to as plus (+) and minus (-). Any vegetative cell of the plant body may produce gamete and thus behave as gametangium (e.g., *Sphaeroplea*) or the gametangia may be morphologically specialized (*Oedogonium*). Morphologically similar gametangia are known as **isogametangia**.

Anisogamy: In anisogamy, the motile or nonmotile gametes are morphologically dissimilar. The larger gamete is the female and the smaller one is the male. The female gamete is usually larger and nonmotile. These structurally different gametes are known as **anisogametes** or **heterogametes**. Sometimes morphologically identical gametes behave differently and thus show **physiological anisogamy**.

Oogamy: In oogamy, there is the fusion of a large nomotile egg or ovum with a smaller motile sperm (except in Rhodophyceae, where the spermatia are non motile). The eggs are formed within an oogonium and the sperm within an antheridium. In the red algae and in several species of the green algae like *Coleochaete* the oogonium bears a protuberance **(trichogyne)** which is the receptive antena for male gametes (Fig. 1.4). The oogonia in red algae are called **carpogonia**. During fertilization, the female gamete remains confined within the oogonium either permanently or until maturity. After maturity antherozoids are liberated out from the antheridium and swim around in water. Ultimately fusion between two gametes take place and **zygote** is formed.

The zygote formed by any of the methods of sexual reproduction has an independent existence for a variable length of time. On germination, the content of zygote divides to form a number of zoospores. These are liberated and after a brief period of swimming they germinate into new plant. More rarely, the zygote germinates directly into the adult plant. As zygote bears diploidic chromosome complement, reduction division (meiotic division) takes place before its germination.

LIFE HISTORY PATTERNS

The sequence of a number of stages or changes in morphological or cytological phases during the growth and development of an alga is its life cycle. On the basis of number or distinct morphological phases, different

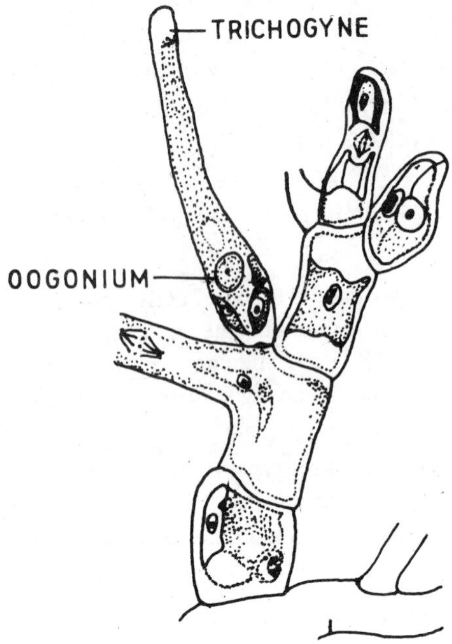

Fig. 1.4. Oogamous reproduction in *Coleochaete* sp. showing oogonium with trichogyne.

kinds of life cycles are **monophasic, diphasic** and **triphasic** or **monomorphic, dimorphic** and **trimorphic** (Drew, 1955) or **monogenetic, digenetic** and **trigenetic** (Feldmann, 1952). On the basis of occurrence of cytolgical phases the different kinds of life cycles are **haplontic, diplontic, haplodiplontic, haplobiontic,** and **diplobiontic** (Fig. 1.5). Haplontic type of life cycle is considered as most primitive, diplontic as advanced and haplodiplontic occupying the intermediate position.

A. **Monogenetic or Monomorphic or Monophasic type** : In this type of life cycle there is only one morphologically distinct generation which may be either sporophytic or gametophytic. Cytologically this monogenetic type may be (i) haplontic or (ii) diplontic.

(1) **Haplontic** : In majority of Chlorophyceae such as *Chlamydomonas*, *Ulothrix*, *Spirogyra*, *Oedogonium* etc. and *Bangia* in Rhodophyceae, this type of life cycle occurs.

In this type of life cycle (Fig. 1.5 A) there is a single thalloid vegetative individual which is haploid in nature containing "n" number of chromosomes. Hence it is called the gametophytic plant. Reproduction by vegetative and asexual means do not play any role in alternation of generation because there is no alternation of haploid phase with diploid and the haploid phase remains throughout and there is no sexual recombination of characters. Towards the end of growing season when unfavourable conditions of the environment persists, the gametophytic plant starts reproducing sexually. Iso - or anisogametes of opposite strains unite to form zygote having diploidic chromosome complement (sporophyte). Since it has to maintain chromosomal race it undergoes meiosis or reduction division which is called as **zygotic meiosis.** As a result 4 haploid spores are produced which give rise to plant proper.

As sporophytic phase in this life cycle occurs at only one step and it is not a prolonged stage, this type of alternation is not a true alternation of generation. Here life cycle is an alternation of prolonged haploid phase and single celled diploid phase. This is supposed to have evolved from a diplohaplontic life cycle by reduction and loss of diploid generation.

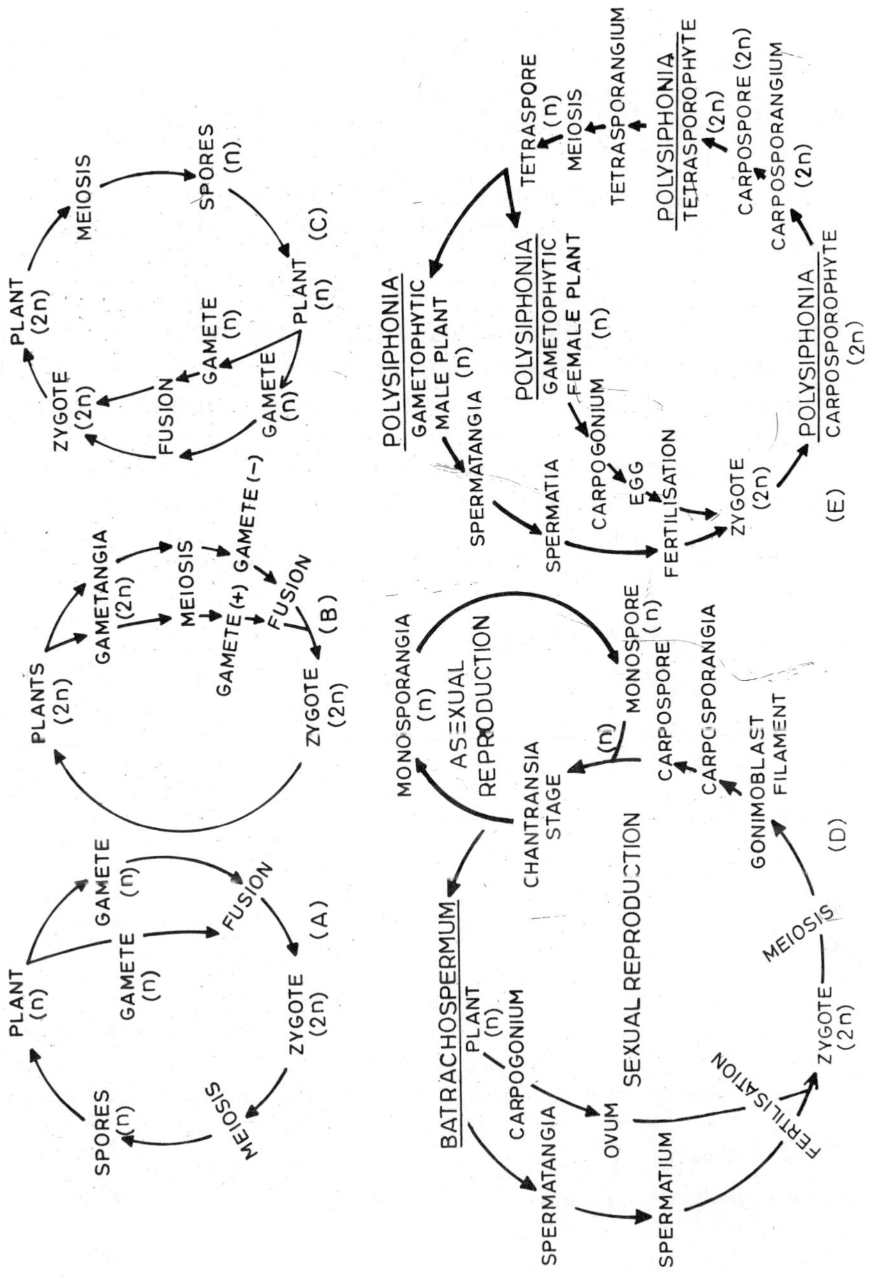

Fig. 1.5. Types of life cycles found in algae. (A) Haplontic life cycle. (B) Diplontic life cycle. (C) Diplohaplontic life cycle. (D) Haplobiontic life cycle. (E) Diplobiontic life cycle.

(2) **Diplontic :** This type of life cycle (Fig. 1.5 B) is exhibited by some of the Siphonales, Siphonocladales of green algae, Fucales in brown algae, diatoms etc.

Here the vegetative plant is diploid. It produces sex organs which give rise to gametes. Reduction division takes place at the time of differentiation of gametes in the sex organs. This type of meiosis is called **gametogenic meiosis**. Haploid gametes fuse together to form a diploid zygotic structure which in turn by repeated mitotic division gives rise to sporophytic plant.

This type of life cycle does not represent true alternation of generation because gametophytic phase is restricted to gametes (n) and remains incospicuous. Many phycologists believe that a diplontic life cycle has evolved from a diplohaplontic cycle by reduction of the gametophytic phase.

B. **Digenetic or Dimorphic or Diphasic type** : In this type of life cycle there are two morphologically distinct generations. Cytologically it may be considered as diplohaplontic life cycle.

Diplohaplontic type : This type of life cycle (Fig. 1.5 C) can be found in Cladophorales, *Ulva* of Chlorophyceae, *Ectocarpus* and *Dictyota* of Phaeophyceae. It differs from the other two types in the respect that here two distinct vegetative individuals alternate to each other one of which is gametophyte retaining "n" number of chromosomes and the other one is sporophyte containing "2n" number of chromosomes. The gametophytic plant is concerned with sexual reproduction whereas the sporophytic plant produces spores through **sporogenic meiosis**. This type of life cycle has two categories :

(1) Isomorphic or homologous diplohaplontic

(2) Heteromorphic or heterologous diplohaplantic

(1) **Isomorphic or homologous diplohaplontic** : When both the gametophytic and sporophytic plant are morphologically identical in diplohaplontic type, it is called isomorphic type. In such life cycle meiosis is delayed so that zygote gives rise to multicellular sporophyte. Examples can be seen in *Ectocarpus, Dictyota* of Phaeophyceae, *Cladophora, Draparnaldiopsis* etc. of Chlorophyceae.

Isomorphic type of alternation of generation is considered to have evolved from haplontic type by sudden mutation in which zygote directly developed into a new vegetative individual by mitosis.

(2) **Heteromorphic or heterologous diplohaplontic** : When two alternating vegetative plants in life cycle are dissimilar in morphology it is known as heteromorphic type. In heteromorphic type of life cycle there are two possibilities- 1. either sporophyte will show more elaborate development than the gametophyte. e.g., Laminariales, Desmarestiales (Phaeophyceae) or 2. the gametophyte will show more developed structure than the sporophyte. e.g., *Cutleria* (Phaeophyceae)

C. **Trigenetic or Trimorphic or Triphasic type** : Three different morphologically distinct generations are found in this type i.e., gametophytic, sporophytic and carposporophytic. On the basis of double occurrence of haploid or diploid phase, it may be haplobiontic or diplobiontic.

(1) **Haplobionitc** : In this life cycle (Fig. 1.5 D), haploid phase occurs twice (gametophyte and carposporophyte) which is represented by many Bangioideae such as *Porphyra* and majority of primitive Florideae (*Nemalion* and *Batrachospermum*).

Like haplontic type of life cycle here zygote also undergoes meiosis to form haploid cells but these haploid cells do not behave as spores. Instead they divide repeatedly to form small somatic individual consisting of number of compact filaments which are haploid carposporophyte. This haploid carposporophyte remains parasitic on the parent gametophyte. The terminal cell of this carposporophyte functions as carposporangia. The protoplast of each carposporangia forms a haploid carpospore which forms haploid gametophytic plant after liberation.

In such life cycle zygote is the only diploid stage and two different haploidic generations (gametophytic and carposporophytic) alternate in which parent gametophyte is free living.

(2) **Diplobiontic:** Here (Fig. 1.5 E) diploid phase occurs twice in the life cycle i.e., sporophyte and carposporophyte. In many Florideae like *Polysiphonia*, the diploid zygote instead of dividing meiotically divides mitotically to form a small compact plantlet, the diploid carposporophyte. It remains attached to the parent gametophytic plant and produces diploid carposporangia. The protoplast of each carposporangium gives rise to a diploid capospore. The germination of carpospores lead to formation of free living tetrasporophyte. This tetrasporophyte produces diploid tertrasporangium which in turn forms haploid tetraspores after reduction division. Tetraspores ultimately form gametophytic plant.

This triphasic life cycle has two free living phases (parent gametophyte and tetrasporophyte). Carposporophyte remains attached to the parent gametophyte as parasite. The two free living generations may be morphologically similar or dissimilar. In case of heteromorphic freeliving generation there are two possibilities - either freeliving gametophyte is more elaborate in development or freeliving sporophyte (tetrasporophyte) is elaborate.

The study of life histories is important in the sense that classification is sometimes based on it. The types of alternation of generations may vary even within the same genus (e.g., *Cladophora*) due to variation in environmental and genetic characters.

PARALLELISM IN EVOLUTION

Most of the morphological trends that are present among the members of the Chlorophyceae occur within the classes Chrysophyceae, Xanthophyceae and others. Thus, motile stages (both unicellular and colonial), coccoid stages, palmelloid forms, filamentous habit (both simple and branched), parenchymatous thalloid structures known in Chrysophyceae, Xanthophyceae and in some other classes are morphologically similar with the thallus development in the Chlorophyceae (Table 1.1). This phenomenon is referred to as **parallel evolution.**

It is widely accepted hypothesis that each phylum or division of algae originated from a unicellular ancestor. Many features indicate that each group is phylogenetically related to others. Because of many features in common, phycologists in the past included Xanthophyceae, Bacillariphyceae and Chrysophyceae in the phylum Chrysophyta. On similar grounds, Dinophyceae and Cryptophyceae were placed in the phylum Pyrrophyta. The parallelism existing within various groups may be due to the plasticity of primitive organisms.

Table 1.1. Parallelism in evolution in different classes of algae

Type of construction	Cyanophyceae	Chlorophyceae	Euglenophyceae	Dinophyceae	Xanthophyceae
Non motile unicell	*Synechocystis*	*Chlorella*	—	*Tetradinium*	*Botrydiopsis*
Motile unicell	—	*Chlamydomonas*	*Phacus*	*Gymnodinium*	*Chloromeson*
Non motile colony	*Microcystis*	*Gloeocystis*	*Colacium*	*Gloeodinium*	*Chlorosaccus*
Motile colony	—	*Volvox*	—	*Polykrikos*	—
Unbranched filament	*Oscillatoria*	*Ulothrix*	—	*Dinothrix*	*Tribonema*
Branched filament	*Nostochopsis*	*Stigeoclonium*	—	*Dinoclonium*	*Monocilia*
Siphonaceous	—	*Dichotomosiphon*	—	—	*Botrydium*
Foliose parenchymatous	—	*Prasiola*	—	—	—

Type of construction	Chrysophyceae	Bacillariophyceae	Rhodophyceae	Phaeophyceae
Non motile unicell	*Chrysocapsa*	*Navicula*	*Porphyridium*	—
Motile unicell	*Chromulina*	—	—	—
Non motile colony	*Chrysocapsa*	*Asterionella*	—	—
Motile colony	*Synura*	—	—	—
Unbranched filament	*Nematochrysis*	*Bacillaria*	—	—
Branched filaments	*Phaeothamnion*	—	*Erythrocladia*	*Ectocarpus*
Siphonaceous	—	—	—	—
Foliose parenchymatous:	—	—	*Porphyra*	*Punctaria*

Besides these common types of plant bodies, many examples are also available which show parallel evoultion in algae.

2

CLASSIFICATION OF THE ALGAE

The classification of plants and animals has occupied man since the time of Aristotle. A Swedish botanist Linnaeus (1754) was the first to use the name Algae for a particular group of plants. However, his group actually consisted of a mixture of algae and Hepaticae. Since then sevaral phycologists classified algae keeping one or more than one principles into consideration. In recent days there has been considerable debate over the major divisions of algae.

PRINCIPLES

Unlike other groups (Bryophyta, Pteridophyta and higher plants) the morphology of mature vegetative stages does not afford a reliable criterion for distinguishing different classes of algae as there are several examples where morphologically alike algae come to be placed in different classes because they differ in their physiological and biochemical characters. After considerable debate it was thought proper that primary classification should be based on certain morphological and physiological features. The chief among those are; 1. Photosynthetic apparatus and pigments, 2. the nature of the food reserve, 3. the nature of the cell wall components, 4. number, morphology and orientation of flagella, 5. nature of eye spot, 6. nuclear compositions and 7. certain aspects of cell structure.

In general the details of vegetative structure and reproduction are not useful for the primary classification of algae but such details are important for classifying algae at the level of species, genus, family etc.

1. **Photosynthetic apparatus and pigments :** Each group of algae has its own particular combination of pigments and a characteristic colour which is not found in other groups. Infact various algal groups express striking variations in colour. Thus the nature of the pigments present in the algal cells forms a quick guide to the primary identification and classification of algae into different divisions.

PHOTOSYNTHETIC APPARATUS

Plastids are cytoplasmic intracellular organellels; bounded by a double plasma membrane and inside which photosynthetic lamellae are present. In the older literature a chloroplast that has a colour other than green is often called as **chromoplast**. In other words those which bear pigments chlorophyll **a** and **b** are usually known as **chloroplasts** while those without chlorophyll **b** but with chlorophyll **a** and chlorophyll **c, d or e** along with other pigments are usually known as **chromatophores**. Some plastids are colourless and are known as **leucoplasts.** A proplastid is a reduced plastid which usually develops into a chloroplast although in some heterotrophic algae it continues to remain a proplastid.

In Cyanophyceae chloroplasts or chromatophores bounded by double membranes are totally lacking. In Rhodophyceae, Chlorophyceae and Charophyceae the chloroplasts are bounded by the double membrane of the chloroplast envelope (Fig. 2.1). In the other eucaryotic algae the chloroplast envelope is surrounded by one or two membranes of chloroplast endoplasmic reticulum (C.E.R.) which bears ribosomes on the outer face. In the Euglenophyceae and Dinophyceae there is one membrane of C.E.R. In Cryptophyceae, Chrysophyceae, Bacillariophyceae, Xanthophyceae and Phaeophyceae there are two membranes of C.E.R., outer membrane of which commonly remains continuous with the outer membrane of nuclear envelope. The space present between C.E.R. and chloroplast envelope contains tubules, ribosomes and reserve product.

Series of flattened membranous **thylakoids** surrounded by matrix or **stroma** are present inside plastid, the basic photosynthetic apparatus. The thylakoids contain the chlorophylls and are sites of the photochemical reactions whereas carbondioxide fixation occurs in the stroma. The thylakoids can be free from one another or grouped to form thylakoid bands. In the Cyanophyceae and Rhodophyceae the thylakoids are usually free from one another with phycobilisomes on the surface of the thylakoids. The phycobilisomes appear as 35 nm granules when phycoerythrin dominates, or as discs when phycocyanin predominates. In Cryptophyceae the chloroplasts contain bands of two thylakoids and phycobiliproteins are dispersed within the thylakoids. In Euglenophyceae, Xanthophyceae, Chrysophyceae, Bacillariophyceae and Phaeophyceae the thylakoids are grouped in bands of three with a girdle or peripheral band running parallel to the chloroplast envelope. In Dinophyceae the thylakoids are also in bands of three but there is no girdle band. In the Chlorophyceae and Charophyceae the thylakoids occur in bands of two to six with thylakoids running from one band to the next. These groupings of algal thylakoids occur under normal growth conditions which vary during abnormal circumstances.

PHOTOSYNTHETIC PIGMENTS

The colour of algal thalli is due to relative abundance of one or more than one pigments. The earlier classification of these organisms into green, blue-green, brown and red algae was exclusively based on this character. However, the colour gets frequently changed with the changing environment and an accurate classification depends on chemical analyses of the photosynthetic pigments. There are three kinds of photosynthetic pigments in algae:

1. Chlorophylls
2. Carotenoids and
3. Biloproteins

According to Prescott (1969) there are 5 types of chlorophylls, 20 xanthophylls, 5 carotenes and 7 phycobilins in algae. The distribution of these is important in the classification of algae.

1. **Chlorophylls** : The photosynthetic algae have chlorophylls in their chloroplasts. Chlorophyll is composed of a porphyrin-ring system that is very similar to that of hemoglobin but has a magnesium atom instead of an iron. Due to different spectral properties five kinds of chlorophylls are known among algae. They are chlorophyll **a, b, c, d,** and **e**. The distribution of these chlorophylls amongst algal groups is shown in table 2.1.

Chlorophyll **a** is the primary photosynthetic pigment (the light receptor in photosystem I of the light reaction) in all photosynthetic algae and ranges from 0.3 to 3.0% of the dry weight. The molecular formula of chlorophyll **a** is $C_{55}H_{72}O_5N_4Mg$. It is insoluble in water and petroleum ether but soluble in alcohol, diethyl ether, benzene and acetone. The pigment has two absorption bands in vitro, one band in the red light at 663 nm and the other at 430 nm. While chlorophyll **a** is found in all the photosynthetic algae, the other algal

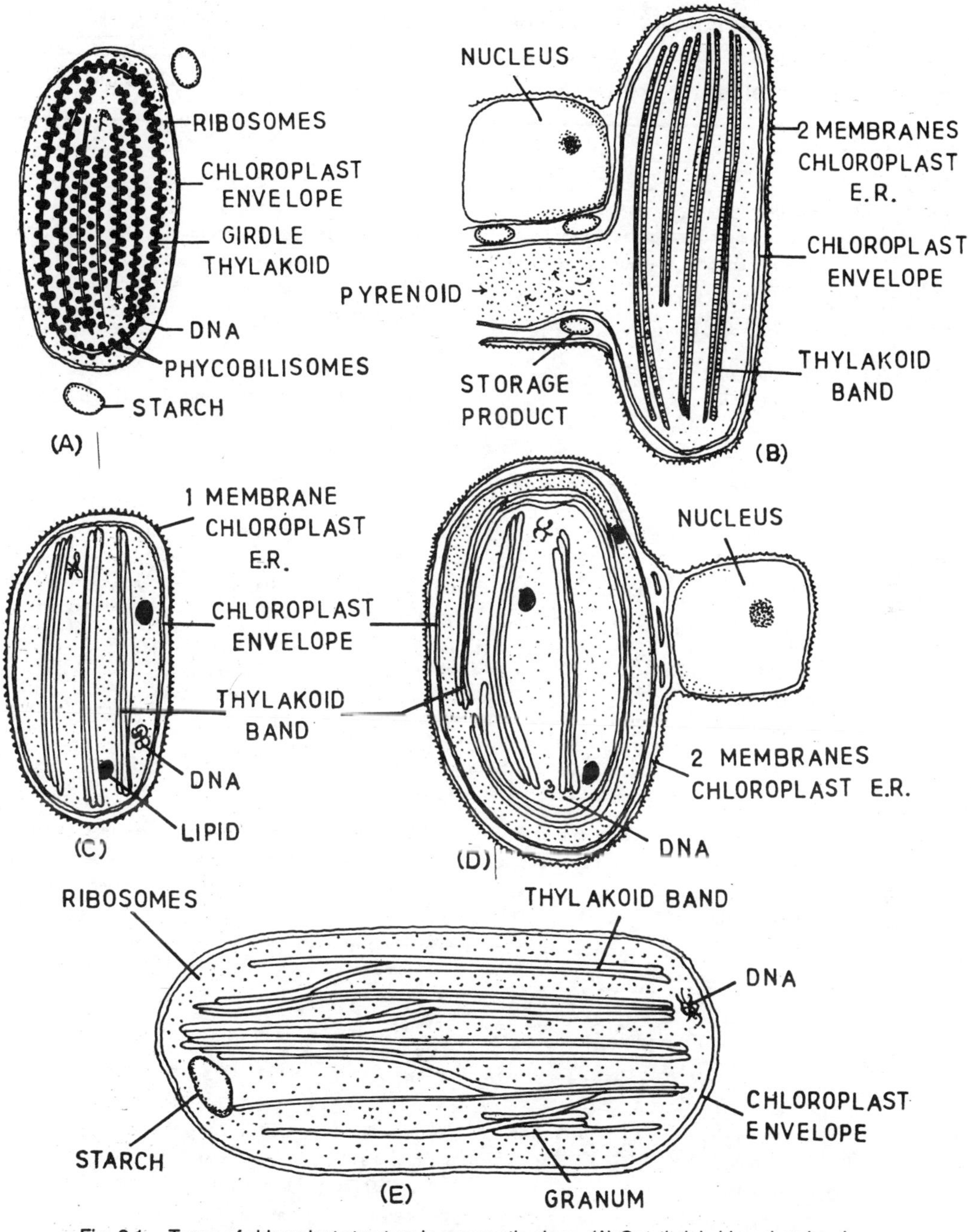

Fig. 2.1. Types of chloroplast structure in eucaryotic algae. (A) One thylakoid per band and no chloroplast endoplasmic reticulum (Rhodophyceae). (B) Two thylakoids per band and two membranes of chloroplast E.R. (Cryptophyceae). (C) Three thylakoids per band and one membrane of chloroplast E.R. (Euglenophyceae and Dinophyceae). (D) Three thylakoids per band and two membranes of chloroplast E.R. (Xanthophyceae, Phaeophyceae, Bacillariophyceae, Chrysophyceae, Prymnesiophyceae, Raphidophyceae, Eustigmatophyceae). (E) Two to six thylakoids per band and no chloroplast E.R. (Chlorophyceae, Charophyceae).

Table 2.1. Summary of the principal characteristics of the algal classes (South & Whittick, 1987)

Division/class	Principal pigments	Food reserves	Chloroplast features	Cell wall	Flagella
Prokaryota CYANOPHYCOTA Cyanophyceae	Chlorophyll a β-Carotene C-Phycoerythrin Allophycocyanin C-Phycocyanin	Phycocyanin granules (arginine and aspartic acid)	No chloroplast Thylakoids free in cytoplasm, unstacked Phycobilisomes	Four-layered peptidoglycan (murein) principal component	Absent
PROCHLOROPHYCOTA	Chlorophyll a Chlorophyll b β-carotene Zeaxanthin Cryptoxanthin		No Chloroplast Thylakoids free in cytoplasm, some stacking in pairs or more	As for Cyanophycota	Absent
Eukaryota RHODOPHYCOTA Rhodophyceae	Chlorophyll a Chlorophyll (d) γ-Phycocyanin Allophycocyanin C-Phycoerythrin α-and β-carotene	Floridean starch	Chloroplast envelope two-layered, chloroplast ER absent Thylakoids unstacked	Cellulose (xylan in some Bangio-phycidae) poly-saccharides; some calcified with $CaCO_3$	Absent
CHROMOPHYCOTA Chrysophyceae	Chlorophyll a Chlorophylls c_1 and c_2 Fucoxanthin	Chrysolaminarin (β-1 3-glucopyranoside)	Two additional membranes of chloroplast ER, the outer continuous with outer membrane of nuclear envelope Thylakoids in stacks of three	Naked; scles or lorica in some	Normally two, unequal; smooth posterior shorter; hairy anterior longer
Prymnesiophyceae (Haptophyceae)	Chlorophyll a Chlorophyll c_1, c_2 β-Carotene	Chrysolaminarin	Two additional membranes of chloroplast ER, the outer continuous with outer membrane of the nuclear envelope Thylakoids in stacks of three	Cell covering of organic scales in one to several layers	Usually two, unequal, smooth Haptonema

(Contd.)

Division/class	Principal pigments	Food reserves	Chloroplast features	Cell wall	Flagella
Xanthophyceae	Chlorophyll a (chlorophyll c in *Vaucheria)* β-Carotene Diatoxanthin Diadinoxanthin Heteroxanthin Vaucheriaxanthin ester	β-1, 3-linked glucan Fats and oils	Two additional membranes of chloroplast ER, the outer continuous with outer membrane of the nuclear envelope Thylakoids in stacks of three	Cellulose, glucose and uronic acids Many walls are bipartite	Usually two, unequal: the anterior longer, hairy; the posterior shorter, smooth
Eustigmatophyceae	Chlorophylls a, c β-Carotene Violaxanthin Vaucheriaxanthin	Oils	Two additional membranes of chloroplast ER, neither continuous with the nuclear envelope Thylakoids in stacks of three	Naked	Single anterior hairy flagellum, and second basal body
Bacillariophyceae	Chlorophyll a Chlorophyll c β-Carotene Fucoxanthin Diatoxanthin Diadinoxanthin	Chrysolaminarin Lipids	Two additional membranes of chloroplast ER, the outer continuous with outer membrane of nuclear envelope Thylakoids is stack of three	Silica frustule	Male gametes possess single hairy flagellum which lacks central microtubules
Dinophyceae	Chlorophyll a Chlorophyll c_2 Peridinin Neoperidinin β-Carotene	Starch	Two additional membranes of chloroplast ER, not continuous with nuclear envelope Thylakoids in stacks of three	Cellulosic theca Body scales (rare)	Usually two, transverse and longitudinal, heterokontic, highly distinctive: transverse, longer, helical. positioned in a groove or girdle; the shorter, longitudinal directed posteriorly
Phaeophyceae	Chlorophyll a Chlorophyll c_1, c_2 β-Carotene	Laminarin Mannitol	Two additional membranes of chloroplast ER,	Two-layered,the inner cellulose, the outer of	Usually two, unequal:the longer anterior,

(Contd.)

Division/class	Principal pigments	Food reserves	Chloroplast features	Cell wall	Flagella
	Fucoxanthin		the outer continuous with outer membrane of nuclear envelope Thyakoids in stacks of 2-6	alginic acid and fucoidan	hairy;the shorter posterior, smooth
Raphidophyceae	Chlorophyll *a* Chlorophyll *c* Diadinoxanthin Dinoxanthin Heteroxanthin ?Fucoxanthin		Two additional membranes of chloroplast ER Thylakoids in stacks of three	Naked	Usually two, unequal: the anterior longer, hairy; the posterior shorter, smooth
Cryptophyceae	Chlorophyll *a* Chlorophyll c_2 α-Carotene Diatoxanthin Phycoerythrin Phycocyanin	Starch-like	Two additional membranes of chloroplast ER Thylakoids in pairs	Periplast	Two apical or lateral, equal, hairy.
EUGLENOPHYCOTA					
Euglenophyceae	Chlorophyll *a* Chlorophyll *b* ß-Carotene Astaxanthin Antheraxanthin Diadinoxanthin Neoxanthin	Paramylon (β-1, 3-glucan)	One additional Chloroplast ER membrane, not continuous with nuclear membrane Thylakoids in stacks of three sometimes more	Proteinaceous Pellicle	Usually two, unequal, hairy, one often not emergent
CHLOROPHYCOTA					
Chlorophyceae	Chlorophyll *a* Chlorophyll *b* α, ß and γ-carotene Lutein Siphonoxanthin Siphonein	Starch (amylose, amylopectin)	Chloroplast ER absent Thylakoids in stacks of 2-6 or more Grana may be present	Naked (some) Cellulose, hydroxyproline, glycosides, xylans, mannans. Some calcified	Normally two (four) equal, smooth (may be tomentose) apically inserted
Charophyceae	Chlorophyll *a* Chlorophyll *b* α, ß and γ-carotene Various xanthophylls	Starch (α-1,4 glucan)	Chloroplast ER absent Thylakoids in stacks of 2-6 or more Grana present	Cellulose Many calcified	Male gamete only; two unequal, subterminally inserted; scaly
Prasinophyceae	Chlorophyll *a* Chlorophyll *b* ß-Carotene Siphonein Siphonoxanthin	Starch Mannitol	Chloroplast ER absent Thylakoids in stacks of 2-6	Naked, usually scaly	One or two, unequal, or four, more or less equal; scaly

chlorophylls have a more limited distribution and function as accessory photosynthetic pigments.

Chlorophyll **b**, the other chlorophyll of higher plants, is found in Chlorophyceae, Charophyceae and Euglenophyceae and in no other class of algae. The molecular formula of chlorophyll **b** is $C_{55} H_{70}O_6N_4$ Mg. It functions photosynthetically as a light harvesting pigment transferring absorbed light energy to chlorophyll **a**. The chlorophyll **a** to chlorophyll **b** ratio varies from 2:1 to 3:1. The solubility characteristics of chlorophyll **a** are similar to chlorophyll **b**, and in vitro chlorophyll **b** has two main absorption maxima in acetone or methanol, one at 645 nm and the other at 435 nm.

Chlorophyll **c** is the most widespread of the other chlorophylls, being present in Dinophyceae, Cryptophyceae, Chrysophyceae, Bacillariophyceae, Xanthophyceae and Phaeophyceae. Chlorophyll **c** has two spectrally different components: Chlorophyll $\mathbf{c}_1$ and $\mathbf{c}_2$. Chlorophyll $\mathbf{c}_2$ is always present but chlorophyll $\mathbf{c}_1$ is absent in Dinophyceae and Cryptophyceae. Chlorophyll **c** probably functions as an accessory pigment to photosystem II. The Chlorophyll **a** to Chlorophyll **c** ratio range from 1.2:2 to 5.5:1. The pigment is soluble in ether, acetone, methanol and ethyl acetate but is insoluble in water and petroleum ether. Extracted chlorophyll $\mathbf{c}_1$ has main absorption maxima at 634, 583 and 440 nm in methanol whereas chlorophyll $\mathbf{c}_2$ has maximum at 635,586 and 452 nm.

Chlorophyll **d** is a minor component and appears to be present in the extracts of many rhodophycean members. It does not occur in simpler red algae and has not been shown to exist in all higher red algae. It's photosynthetic function is unknown. It is soluble in ether, acetone, alcohol and benzene and slightly soluble in petroleum ether. It has three absorption bands at 695, 456 and 400 nm.

Chlorophyll **e** has been identified in only two genera of Xanthophyceae (*Tribonema* and the zoospores of *Vaucheria*).

2. **Carotenoids** : Carotenoids are group of yellow, orange, red and brown protective pigments which act as screen to light and pass on energy of light waves to chlorophyll. These occur usually inside the plastid but may be outside in certain cases. In general, naturally occurring carotenoids can be divided into two classes (1) carotenes and (2) xanthophylls.

Carotenes ($C_{40} H_{56}$) : Carotenes are linear unsaturated oxygen free hydrocarbon and xanthophylls are oxygen derivatives of these. There are six types of carotenes -carotene α, β, γ, ε, flavicine and lycopene. β carotene is present in most algae, although it is replaced by α-carotene in the Caulerpales of the Chlorophyceae, in the Cryptophyceae and to a lesser extent in the Rhodophyceae. β-carotene is replaced by two carotenes (lycopene and γ-carotene) in Charophyceae which are characteristic of photosynthetic bacteria. In Bacillariophyceae and in some members of Cryptophyceae, ε carotene is present.

Xanthophylls ($C_{40} H_{56} O_2$) : There are many differnt xanthophylls (about 20) in algae and since many are unique to particular algal groups, they are important in primary classification. The xanthophylls of Chlorophyceae most closely resemble with those of higher plants. Fucoxanthin is the principal xanthophyll in the golden-brown algae (Chrysophyceae, Bacillariophyceae and Phaeophyceae). Peridinin is found only in Dinophyceae, myxoxanthin and myxoxanthophyll in Cyanophyceae, taraxanthin in Rhodophyceae and antheraxanthin in Euglenophyceae. Like the chlorophylls, the carotenoids are soluble in alcohols, benzene and acetone but insoluble in water.

3. **Biloproteins** (or **biliproteins** or **phycobilins**) : Biliproteins are water soluble blue or red pigments located on (Cyanophyceae, Rhodophyceae) or inside (Cryptophyceae) thylakoids of algal chloroplasts. These are tetra-pyrrolic compounds joined to globilin proteins. During the extraction procedure the free pigment can not be separated from a protein moiety and the name of the pigments was therefore changed from phycobilins to biloproteins to indicate the existence of the pigment-protein complex. Biloproteins are present in only three algal classes, the Cyanophyceae, Rhodophyceae and Cryptophyceae. On the basis of absorption spectra, these pigments are of three types **phycocyanin, phycoerythrin** and

allophycocyanin. Each of these shows differences between the three groups. In general, those of the Cyanophyceae are of C-type, those of the Rhodophyceae are of R-type and those of Cryptophyceae are of a third type. Allophycocyanin (blue pigment) also occurs in Rhodophyceae.

The proportion of different pigments varies. The members of Chlorophyceae and Euglenophyceae appear dark green due to excess of chlorophylls over carotenoids. Yellow brown colour of Chrysophyceae, Bacillariophyceae, Phaeophyceae, Dinophyceae and yellow green colour of Xanthophyceae reflect an excess of carotenoids over chlorophylls. Characteristic blue-green colour of Cyanophyceae and red colour of Rhodophyceae are due to dominant phycocyanin and phycoerythrin pigments respectively. The proportion of one type of pigment to the other can vary with the change in environmental conditions.

2. **Nature of food reserve**: The primary product of photosynthesis are the same in all algae as the initial stages in carbon dioxide fixation are probably the same in all photosynthetic organisms. However, the insoluble products which accumulate over a longer period of time are more variable and they provide useful taxonomic criteria. Various polysaccharides are the most widespread and most helpful compound used in primary classification of algae. Polysaccharides in which glucose subunits are joined with α 1-4 linkages are **true starch, floridean starch** and **myxophycean starch.**

(i) **Starch** : True starch similar to that found in higher plants (composed of amylose and amylopectin) is only found in Chlorophyceae and Charophyceae. It occurs inside the chloroplast in the form of starch grains. In Cryptophyceae strach has unusually high content of amylose (45%) and is found as grains between the chloroplast envelope and the chloroplast endoplasmic reticulum. In the cytoplasm of Dinophyceae outside the chloroplast, starch also occurs but its structure is not known.

(ii) **Floridean starch** : This characteristic starch is found in Rhodophyceae and is similar to the amylopectin of higher plants. It occurs as bowl-shaped grains varying from 0.5 to 25 μm in size outside chloroplast.

(iii) **Myxophycean starch** : It is characteristic and found in Cyanophyceae. The structure is similar to glycogen. This reserve product occurs as granules (α-granules). The shape of these granules vary from rod-shaped (25 nm particles) to elongate bodies (31 to 65 nm).

Three polysaccharides with ß 1-3 linkages (Fig. 2.2) are now recognised as being important constituents of algae. These are **laminarin** (Phaeophyceae), **chrysolaminarin** or **leucosin** (Chrysophyceae and Bacillariophyceae) and **paramylum** or **paramylon** (Euglenophyceae). Laminarin occurs as an oil like liquid outside chloroplast in form of vesicle surrounding the pyrenoid. Chrysolaminarin also remains in

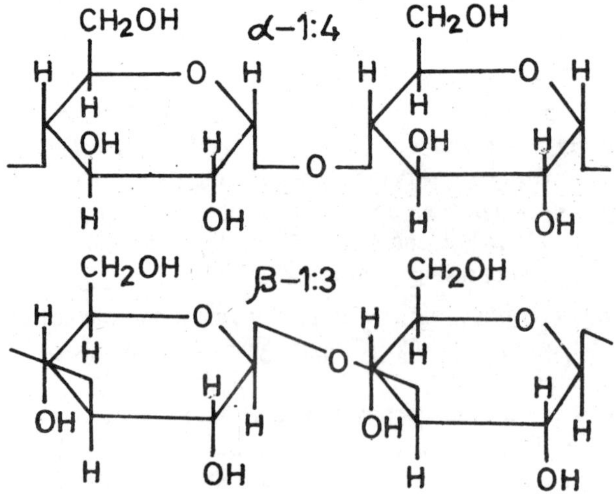

Fig. 2.2. α - 1:4 and ß - 1:3 linkages of algal polysaccharides.

vesicles outside chloroplast and has more glucose residues per molecule than laminarin. Paramylon occurs as water soluble single membrane bounded inclusions of various shapes and dimensions outside the chloroplast.

Other storage products of taxonomic importance are **floridoside** and **mannoglycerate** in the Rhodophyceae, preteinaceous **cyanophycin** in Cyanophyceae and **fructosans** in some members of Cladophorales (Chlorophyceae). **Mannitol,** a polyhydroxyalcohol is universally present in the Phaeophyceae. It is also present in some Chlorophyceae, Rhodophyceae and Dinophyceae.

Fats : In large number of algae there is accumulation of fats and the proportion of this to other storage products varies considerably. Cells of Chrysophyceae and Bacillariophyceae accumulate a high proportion of fat. Environmental conditions also affect the proportion of fat. The chemical nature of fat provides promising approach in algal classification. Degree of unsaturation can be correlated with certain algal divisions. Fats present in green algae (Chlorophyceae) resemble those of higher plants. Fats of Phaeophyceae are less unsaturated and those of some Rhodophyceae are more unsaturated.

Sterols : Although there is limited evidence but sterols have been found to be of help in the primary classification of algae. The members of Cyanophyceae (in common with bacteria) differ from all other organisms in lacking sterols. Sitosterol, the sterols most abundant in higher plants, is widespread in Chlorophyceae. Fucosterol is the most common sterol of the Phaeophyceae. The distribution of sterols has been shown in table 2.2.

Table 2.2. Distribution of Sterols in Algae

Sitosterol	-	Chlorophyceae, Xanthophyceae, Rhodophyceae
Fucosterol	-	Phaeophyceae, Chrysophyceae, Bacillariophyceae, Rhodophyceae, *Cladophora* (Chlorophyceae)
Sargasterol	-	Phaeophyceae
Cholesterol	-	Rhodophyceae
Ergosterol	-	Chrysophyceae, Euglenophyceae
Chondrillasterol	-	Bacillariophyceae, *Scenedesmus* (Chlorophyceae)

3. **Nature of cell wall components :** The chemical composition of cell wall varies from one class of algae to other. The cell wall is always formed outside the plasmalemma (as in higher plants). Its structure is highly variable and can be subdivided into two types : incomplete and complete. The former is characteristic for a variety of flagellates, the latter for all procaryotic algae and most members of Chlorophyceae, Xanthophyceae, Phaeophyceae and Rhodophyceae. In general, algal cell wall is made up of two components: (i) the inner water insoluble material known as cellulose and (ii) the outer pectic or mucilaginous substance soluble in boiling water. Although both inner and outer materials are polysaccharides, lipid and proteinaceous materials are also present. Cellulose is fibrillar component which forms the skeleton of the wall. It is present in walled species of all classes except Chrysophyceae, Bacillariophyceae where cell wall is silicified. X-ray diffraction patterns show the existence of different kinds of cellulose depending on the particular crystalline arrangement. Microfibrils in Chlorophyceae may be composed of only glucose molecules or of glucose and other related sugars. 30-200 A^0 wide cellulose microfibrils embedded in a smooth or slightly granular matrix are either laid down in two layers at right angles to each other or in three layers running in an obtuse angle to the other two. Deposition and orientation of microfibrils is the function of outermost layer of the cytoplasm. Cellulose takes two forms (i) Cellulose I, found in *Cladophora* which is characteristics to higher plants. (ii) Cellulose II, is found in *Ulva, Ulothrix.*

The variability in the cell wall composition in different algae is interesting and may have an evolutionary significance. Characteristic components of the cell wall are **polyuronic acid** and **alginic acid** in Phaeophyceae. In some larger members fucinic acid is found. In blue-green algae, a characteristic mucopeptide component is present which is absent in other classes of algae but present in bacteria. Thus, there are three categories of cell wall in algae (i) Cellulosic, (ii) Silicified & (iii) Mucopolymeric. Some additional compounds are also present in the cell wall like **hemicellulose** in Phaeophyceae, **chitin** in *Cladophora* and *Oedogonium*, silicon, iron and calcium etc. In place of true cell wall, covering of plasma membrane i.e., pellicle is present in some members (*Euglena*). In some members of Cyanophyceae and Rhodophyceae plasmodesmata is present between adjacent cells. In the Rhodophyceae the amorphous component of the wall is composed of galactans or polymers of galactose which include **agar, carrageenan, porphyran, furcelleran** and **funoran.**

4. **Number, morphology and orientation of flagella**: Flagella are present in all classes of algae except Cyanophyceae and Rhodophyceae (Fig. 2.3). Their number, nature and mode of orientation are important

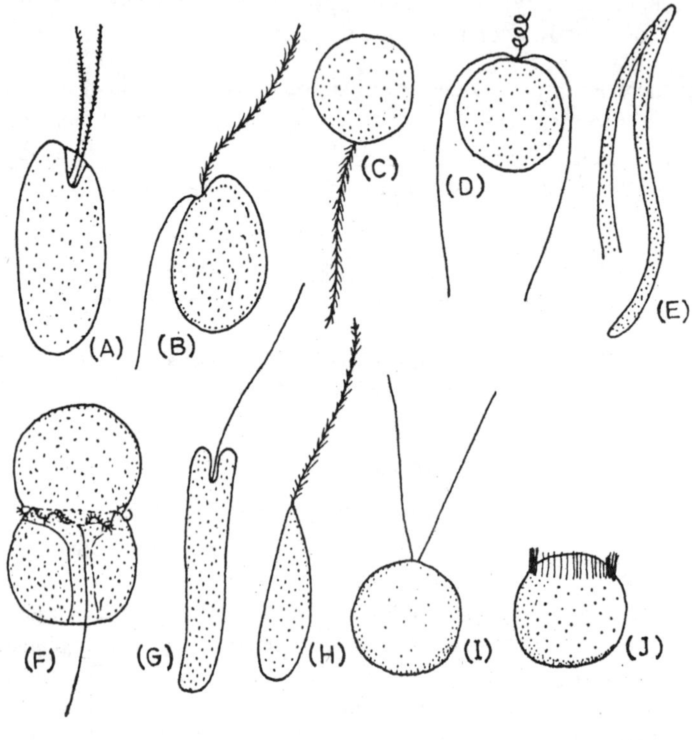

Fig. 2.3. Types of flagella and the common shape of eucaryotic motile algal cells. (A) Cryptophyceae. (B) Xanthophyceae, Phaeophyceae, Chrysophyceae, Raphidophyceae. (C) Bacillariophyceae. (D) Prymnesiophyceae. (E) Charophyceae. (F) Dinophyceae. (G) Euglenophyceae. (H) Eustigmatophyceae. (I, J) Chlorophyceae.

characters for primary classification of algae. Green algae were earlier classified into 4 major groups, i.e., Isokontae (cells possessing two equal flagella), Heterokontae (with unequal flagella), Akontae (without flagella) and stephanokontae (algae with motile stage possessing a whorl of flagella). Basically there are two types of flagella: (i) **acronematic** or **whiplash** type which are smooth and do not have hairs (mastigonemes) on its surface ii) **pantonematic** or **tinsel** type which have hairs on its surface. Flagellar hairs are of two types (1) fibrous solid hairs and (2) tubular hairs. The fibrous hairs are composed of solid fibrils, 5-10 nm wide and 1-3 µm long. The tubular or tripartite hairs are 20 µm wide, usually about 2 µm long and are composed of three regions: (i) a tapering basal region attached to the flagellar membrane, (ii) a microtubular shaft and (iii) a few terminal filaments. In addition to the hairs, electron microscopic investigations have revealed at least one other kind of flagella inwhich the flagellar surface is covered by minute hairs (different from pantonematic flagella) and scales (e.g., Prasinophyceae). A brief outline of the use of flagella in algal classification is given in table 2.3.

Table 2.3. Types of Flagella in Algae

Name of algal class		Flagellar conditions
Cyanophyceae	-	None
Chlorophyceae	-	2 or 4 anterior, equal, acronematic
Xanthophyceae	-	2 unequal, anterior, 1 acronematic, 1 pantonematic
Chrysophyceae	-	2,1 acronematic and 1 pantonematic
Bacillariophyceae	-	1 pantonematic, anterior
Dinophyceae	-	1 acronematic, 1 band shaped
Cryptophyceae	-	2 equal, lateral, pantonematic
Euglenophyceae	-	1 anterior, pantonematic
Phaeophyceae	-	2 unequal, lateral, 1 acronematic, 1 pantonematic
Rhodophyceae	-	None
Prasinophyceae	-	1, 2 or 4 flagella with fine hairs and scales on surface.

Flagella in algae and all other plants except bacteria are composed of nine peripheral doublet microtubules and two central single microtubules. This is commonly known as 9+2 pattern of fibrils of a flagellum. These are produced in a matrix and surrounded by flagellar membrane and the flagellar sheath (Fig. 2.4). On entering the cell body the two central microtubules end at a dense plate, whereas the nine peripheral doublets continue into the cell. There are also other structures between the microtubules in the basal region of the flagellum (basal body). There can be either microtubular roots or striated fibrillar roots attached to the basal body. After cell division the basal bodies move back to their position under the plasma membrane and extrude the 9+2 arrangement of microtubules to form new flagella.

Marine flagellates show an additional structure called haptonema which is similar to flagella. It consists of three concentric membrane enclosing a ring of 6 fibres.

5. **Nature of eye spot** : An eyespot (or stigma) is found in many motile algae or in the gametes and zoospores of non-motile algae for directing the flagellar movement. It mainly consists of osmiophilic, carotenoid-lipid globules often associated with a photoreceptive, rod-shaped or crystalline area near the flagellar axoneme. The photoreceptor is responsible for phototaxy. Eye spots are not found in the Cyanophyceae, Bacillariophyceae and Rhodophyceae while they are relatively uncommon in the Cryptophyceae and Dinophyceae. Majority of them retain throughout their life cycle while others may possess them for a brief period. Detailed studies have shown that eyespot granules develop as a result of fusion of smaller granules within the packet membrane in association with the endoplasmic reticulum.

Dodge (1969, 1973) classified eyespots into three main types and reserved a special category forthe Dinophyceae (Fig. 2.5).

1. The eyespot is part of a chloroplast but not obviously associated with flagella : Chlorophyceae, Prasinophyceae and some Chrysophyceae.
2. The eyespot is part of a chloroplast and abviously associated with flagella : Cryptophyceae, most Xanthophyceae and Phaeophyceae.

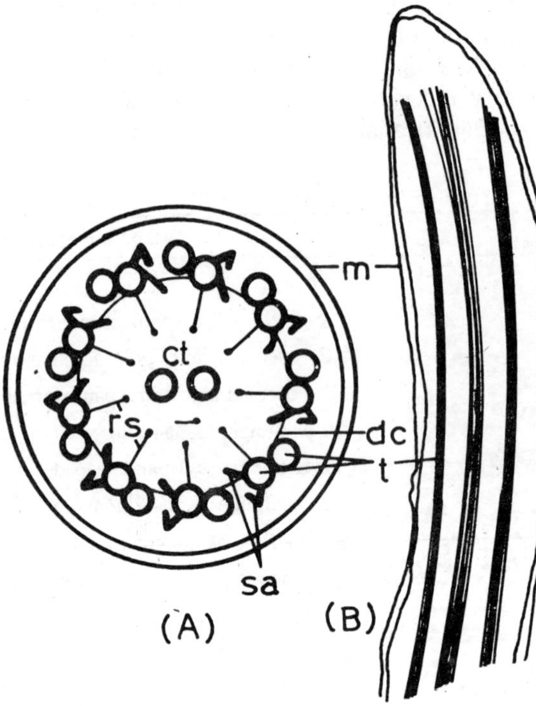

Fig. 2.4. Diagrammatic sketch of flagellar anatomy. (A) Cross section showing 9+2 pattern. (B) Longitudinal section (ct-central tubules, rs- radial spokes, sa- side arm, m- membrane, dc- introdublet connection, t- ring of doublet tubules). (After Hopkins, 1970).

3. The eyespot is independent of chloroplast but adjacent to flagella: Euglenophyceae, some Xanthophyceae and Eustigmatophyceae.

In the Dinophyceae, eyespots range from collections of lipid globules containing carotenoids and lacking membranes to more complex arrays of lipid-containing globules surrounded by membrane envelopes. Dodge (1984) described three types of dinoflagellate eyespot : independent and not membrane bound, independent and surrounded by three membranes,and situated at the periphery of a chloroplast. All are positioned behind the sulcus and there is a strand of microtubules between the eyespot and theca. He also suggested that structurally the eyespots link the dinoflagellates with the euglenoids and chrysophytes.

6. **Nuclear compositions**: Blue-green algae (procaryotic algae) do not have a well defined nucleus as are found in other classes of algae and higher plants. Their hereditary material is DNA embedded in an open nucleoplasm without having associated with histone. It does not possess nucleolus and is not demarcated by a nuclear membrane. The DNA fibrils are not organised into genes and chromosomes. Christensen's (1962) classification of the entire plant kingdom into 2 subkingdoms - Procaryota and Eucaryota is based on thses features. In eucaryotic algae, the DNA is found in the chromosomes which in addition to DNA contain nuclear proteins and small amounts of RNA in a complex known as chromatin. The chromosomes are confined within the nucleus by a double membrane envelope (the nuclear envelope) which is continuous with the cell ER.

Exceptions are the nuclei of Dinophyceae, Euglenophyceae and Cryptophyceae which possess unique features of the nucleus, nuclear division and cytokinesis. These types of nuclei have been termed

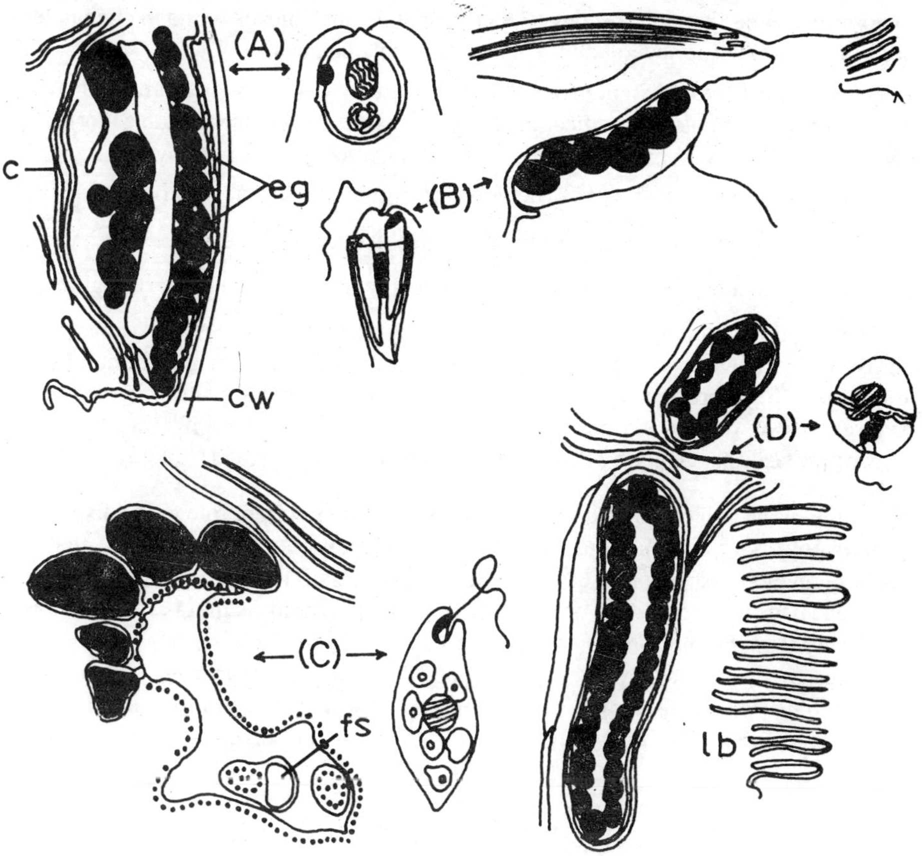

Fig. 2.5. Important types of eyespots found in algae. (A) Type A found in *Chlamydomonas*. (B) Type B found in *Dinobryon*. (C) Type C found in *Euglena* (D) Type D with lamellar body found in *Glenodinium* (c- chloroplast, eg - eyespot granules, cw- cell wall, fs- flagellar swelling, lb- lamellar body). (After Dodge, 1969).

mesocaryotic since they have chromosomes which remain condensed during all cell phases. Other features are the retention of the nuclear membrane during nuclear division (Dinophyceae and Euglenophyceae but not Cryptophyceae) and persistence of the nucleolus (or endosome) during division (Euglenophyceae).

Nucleus

Procaryotic	Mesocaryotic	Eucaryotic
Cyanophyceae	**Dinophyceae, Euglenophyceae & Cryptophyceae**	**Other classes of algae**

Majority of eucaryotic algal cells are uninucleate. Multinucleate cells are prevalent in vegetative cells of most Florideophycideae (Rhodophyceae), Xanthophyceae and some orders of Chlorophyceae. Algal nuclei and chromosomes are small except Dinophyceae and Euglenophyceae and as such it is difficult to study algal chromosomes using light microscopy. Generally single nucleolus is found in one nucleus but in order Conjugales of Chlorophyceae several nucleoli are found in a nucleus.

7. **Certain aspects of cell structure**: There are certain distinctive features in different classes of algae which also help in primarly classification of algae. Besides differences in nuclear organisation, Stanier and Van Niel (1962) have summarised the features of procaryotic cells as follows:

1. The absence of membrane bound structures like mitochondria and chloroplast.

2. Nuclear division does not take place by mitosis as in eucaryotic cells.

3. The presence of a cell wall which contains a specific mucopeptide as its strengthening component.

Superficial appearance of the cell is distinctive in many classes of algae. As in diatoms (Bacillariophyceae) the cell consists of two halves (valves) and valve surfaces have elaborate series of markings. In Euglenophyceae and some species of Cryptophyceae gullet is present. Presence of angular plates, characteristic longitudinal and transverse furrows on cell surface are the specific features of Dinophyceae. Electron microscopic observations of the scales present on cell surfaces in Chrysophyceae provide helping hand in classification.

Pyrenoids: Some of the cell organelles viz., pyrenoids play an important role in the distinction of different classes. A **pyrenoid** is a proteinaceous differentiated region within the chloroplast that takes the early products of photosynthesis and convert them to storage products. The majority of algae with the exception of Cyanophyceae have pyrenoids. The only other plant with pyrenoids is the *Anthoceros,* a bryophyte. The pyrenoid is denser than the surrounding stroma and may or may not be traversed by thylakoids (Fig. 2.6). Two basic types of pyrenoids have been identified: the embedded pyrenoids (Bacillariophyceae) and projecting pyrenoids (Phaeophyceae). A chromatophore may contain a single *(Chalmydomonas)* or many pyrenoids *(Oedogonium)*. They may develop *de novo* (Phaeophyceae) or by the division of existing pyrenoids.

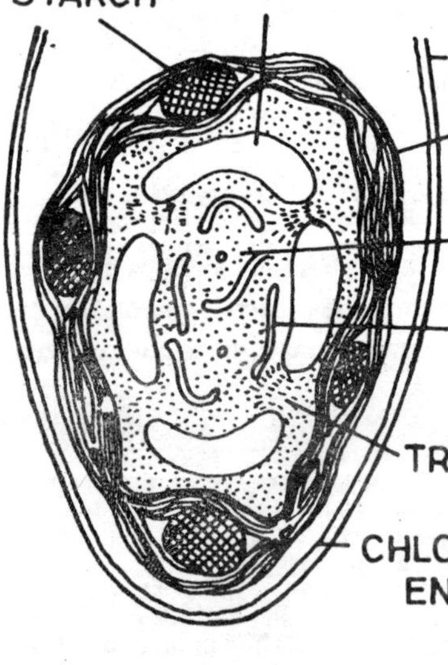

Fig. 2.6. Diagrammmatic sketch of electron micrograph of pyrenoid of *Chlamydomonas* sp.

In green algae, the pyrenoid is located as a dense mass in the chloroplast and a layer of starch plates encircles the proteinaceous core. In most other algae they do not seem to be associated with reserve food. Pyrenoids are invariably traversed by a variable number of chromatophore lamellae except in phaeophyceae.

Ultrastructural investigations do not reveal a limiting membrane between chloroplasts but differences in the density of the chloroplast matrix and the pyrenoid may exist (Sager and Palade, 1957). Matrix of the pyrenoid is made up of fibrils (60Å in diameter) representing a crystal like structure. The variation in different classes on the basis of pyrenoid is as follows:

Class		Type
Cyanophyceae	-	absent
Chlorophyceae	-	Pyrenoid with starch sheath
Chrysophyceae	-	Pyrenoid naked
Bacillariophyceae	-	Pyrenoid naked
Dinophyceae	-	Pyrenoid naked
Cryptophyceae	-	Pyrenoid naked
Rhodophyceae	-	Pyrenoid naked
Phaeophyceae	-	Pyrenoid naked and projecting
Euglenophyceae	-	Pyrenoid projecting with apical sheath of paramylum.

Apart from above factors, the process of reproduction, types of life histories and some specific conditions in the life cycle provide support mainly in secondary classification of algae.

Comparative Survey of Important Systems of Classification

From early times algae have been variously classified. In 1836 for the first time Harvey classified algae into three groups based on the variation in colour of the thallus. Since then a series of classification were proposed by many phycologists. Some of the important systems of classification of algae are those of Pascher (1931), Fritsch (1935), Smith (1950), Christensen (1962), Round (1963,73), Prescott (1969), Chapman & Chapman (1973), Dodge (1974), Trainor (1978), Lee (1980), Parker (1982), Silva (1982) etc. The important principles adopted in these classification are:

(1) Nature of pigments
(2) Phylogeny & interrelationship
(3) Structure of plant body
(4) Chemical nature of reserve food material
(5) Flagellar conditions
(6) Submicroscopic wall
(7) Chromosome number along with other cytological features

It will not be proper to give details of all these classifications here, however, outline of few most important systems are as follows:

Fritsch's System of Classification

F.E. Fritsch (1935) in his book "Structure and Reproduction of the Algae" Vol-I has classified algae into 11 classes.

These are:

(1) Chlorophyceae (Isokontae)
(2) Xanthophyceae (Heterokontae)
(3) Chrysophyceae

(4) Bacillariophyceae (Diatoms)
(5) Cryptophyceae
(6) Dinophyceae (Peridinieae)
(7) Chloromonadineae
(8) Euglenineae
(9) Phaeophyceae
(10) Rhodophyceae
(11) Myxophyceae (Cyanophyceae)

He considered the following criterias in his classification:

(i) Interrelationships of various forms of algae.
(ii) Structure of plant body and range of thallus structure
(iii) Nature of pigments
(iv) Nature of reserve food material
(v) Method of reproduction
(vi) Flagellar conditions
(vii) Occurrence and
(viii) Variation in life cycle

According to him all the groups show parallel evolution.

Demerits and merits of this system: Inclusion of some members of Siphonales and Charales in Chlorophyceae has not been accepted by many phycologists. Presence of characteristic xanthophylls like siphonein and siphonoxanthein, oil as food reserve and other conditions led phycologists to place Siphonales under Xanthophyceae. Similarly complex thallus structure, covering over sex organs and presence of protonemal stage in Charales raised the order upto rank of either class Charophyceae or division Charophyta. Fritsch's system, although not considered to be natural one but is based on phylogenetic relationship and is simple, efficient and widely accepted.

Chapman & Chapman's System of Classification

V.J. Chapman and D.J. Chapman (1973) in their book "The Algae" classified algae using a combination of biochemical characters together with morphological differences. According to them algae can be classified as follows:

	Division	Class
PROCARYOTA	CYANOPHYTA (blue-green 'algae')	Cyanophyceae
EUCARYOTA	1. RHODOPHYTA (red algae)	Rhodophyceae
	2. CHLOROPHYTA (green algae)	Chlorophyceae Prasinophyceae Charophyceae
	3. EUGLENOPHYTA (euglenoids)	Euglenophyceae

4. CHLOROMONADOPHYTA (chloromonads)	Chloromonadophyceae
5. XANTHOPHYTA (yellow-green algae)	Xanthophyceae
6. BACILLARIOPHYTA (diatoms)	Bacillariophyceae
7. CHRYSOPHYTA (golden-brown algae)	Chrysophyceae Haptophyceae
8. PHAEOPHYTA (brown algae)	Phaeophyceae
9. PYRROPHYTA (dinoflagellates)	Dinophyceae Desmophyceae
10. CRYPTOPHYTA (cryptomonads)	Cryptophyceae

Round's System of Classification

F.E. Round (1973) in second edition of his book "The Biology of the algae" classified

Algae

- PROCARYOTA
  - CYANOPHYTA
- EUCARYOTA
  - 12 PHYLA

(1) EUGLENOPHYTA
(2) CHLOROPHYTA
(3) CHAROPHYTA
(4) PRASINOPHYTA
(5) XANTHOPHYTA
(6) EUSTIGMATOPHYCEAE
(7) DINOPHYTA (= PYRROPHYTA)
(8) BACILLARIOPHYTA
(9) CHRYSOPHYTA
(10) PHAEOPHYTA
(11) RHODOPHYTA
(12) CRYPTOPHYTA

Parker's System of Classification

The scheme of S.P.Parker (1982) represents the collective view of a number of leading phycologists. It recognises a separation between procaryotic forms which include the bacteria, the Cyanophycota (or

Cyanobacteria, the 'blue-green algae) and the recently proposes Prochlorophycota (Lewin, 1977), and the eucaryotic forms which include algae and all other plants.

Kingdom	**Division**		**Class**
Plantae	Procaryota		Cyanophycota
			Prochlorophycota
	Eucaryota	Rhodophycota	Rhodophyceae
		Chromophycota	Chrysophyceae
			Prymnesiophyceae
			Xanthophyceae
			Eustigmatophyceae
			Bacillariophyceae
			Dinophyceae
			Phaeophyceae
			Raphidophyceae
			Cryptophyceae
		Euglenophycota	Euglenophyceae
		Chlorophycota	Chlorophyceae
			Charophyceae
			Prasinophyceae

Diagnostic features of algal Phyla : The characteristic features of common algal phyla are mentioned below:

(1) **CYANOPHYTA** (Blue-green algae)

1. Their cellular organisation is **procaryotic.**
2. Flagella are completely absent
3. When movement occurs, it is by a characteristic gliding motion.
4. Their photosynthetic pigments include characteristic bilo-proteins together with unique carotenoids like myxoxanthin and myxoxanthophyll.
5. Storage products include the proteinaceous material, **cyanophycin.**
6. Chromatophores, chlorophyll **b** and pyrenoids are absent. The thylakoids occur singly.
7. Sexual reproduction is absent.

(2) **CHLOROPHYTA** (Grass-green algae)

1. The photosynthetic pigments include chlorophyll **a** and **b.**
2. Starch accumulates as reserve product.
3. The cell wall invariably contains cellulose.

4. Flagella, when present, are normally 2 or 4 in number and of acronematic type.

(3) **XANTHOPHYTA** (Yellow-green algae)

1. There is an excess of carotenoids over chlorophylls. Chlorophyll **e** is found in two genera of this division only.
2. The two flagella are normally of unequal length, the longer being pantonematic and the shorter acronematic.
3. The food storage products are normally oil and fat, but starch never appears to accumulate.
4. A cell wall is frequently absent and when present it generally has a higher content of pectic material than the cell walls of Chlorophyta.

(4) **CHRYSOPHYTA** (Brown or orange algae)

1. The photosynthetic pigments include chlorophyll **a** and carotenoids such as fucoxanthin and dia-dinoxanthin.
2. The flagellation is variable. The cells may be either uniflagellate or biflagellate, and when biflagellate, the flagella may be either equal or unequal in length i.e., similar or dissimilar in nature.
3. The chief reserve products are oil and chrysolaminarin (leucosin). Starch never accumulates.
4. Cellulose wall is absent and the membrane shows a marked tendency to become silicified.
5. Production of endogenous cysts is common.

(5) **BACILLARIOPHYTA** (Yellow or golden- brown algae)

1. The cell walls are silicified and partly composed of pectic substances which show characteristic secondary structures. It consists of two halves.
2. The photosynthetic pigments include chlorophylls **a** and **c** together with xanthophyll, fucoxanthin.
3. Food storage products include fats and chrysolaminarin.
4. The motile stages have a single pantonematic flagellum.
5. Sexual reproduction is of special type in which auxospores are formed.

(6) **CRYPTOPHYTA** (Nearly brown algae)

1. Cells are flattened in dorsiventral plane and cell wall is periplastic.
2. Food reserve is true starch or oil or both .
3. Chromatophores show diverse pigmentation having some shade of brown. Major carotenes are α-carotene and diatoxanthin. Phycobiliproteins (Phycoerythrin and phycocyanin) are also present which differ from those of Cyanophyta and Rhodophyta.
4. Two apically or laterally inserted, somewhat unequal flagella normally arise from the base of a depression (gullet).
5. Ejectosomes (trichocysts), different from Dinophyta, are found.

(7) **DINOPHYTA** (Dark yellow or brown algae)

1. Photosynthetic pigments include chlorophyll **a** and **c** together with several carotenoids (dinoxanthin and peridinin) unique to Dinophyta.
2. The cells are biflagellate with one band shaped flagellum and another apparently of the acronematic type.
3. Nucleus is large and prominent with characteristic structures.

4. Most cells contain conspicuous membrane-bound pusules which resemble the contractile vacuoles but have no contractile activity.
5. Cell surface is covered with distinct angular plates.

(8) **EUGLENOPHYTA** (Pure-green algae)

1. The photosynthetic pigments resemble those of the Chlorophyceae since they contain chlorophylls **a** and **b**. Carotenoids such as ß-carotene, lutein, neoxanthin, astaxanthin and antheraxanthin are also present.
2. All euglenoid cells are naked.
3. They have one, two or three flagella (all of pantonematic type) arising from an invagination at the anterior end of the cell.
4. The reserve product is paramylum, a polysaccharide.

(9) **PHAEOPHYTA** (Brown algae)

1. The photosynthetic pigments include chlorophylls a and c together with xanthophylls such as fucoxanthin and diatoxanthin.
2. Food reserve products are laminarin and mannitol.
3. Cell wall constitutes characteristic alginic acid and fucinic acid.
4. The motile stages are mostly pear-shaped with two lateral flagella (one acronematic and other pantonematic).
5. In many forms characteristic fucosan vesicles are present.

(10) **RHODOPHYTA** (Red algae)

1. The photosynthetic pigments include the characteristic biloproteins: R-phycocyanin and R-phycoerythrin together with chlorophyll **d** and the carotenoid taraxanthin.
2. No flagellated stages are found.
3. Food storage products include floridean starch and the galactoside floridoside.
4. Polysulphate esters are characteristic components of the cell wall.
5. Highly specialised and characteristic process of sexual reproduction is found.
6. Except in few forms, cells show protoplasmic connections.

(11) **CHAROPHYTA** (Stone-worts)

1. Macroscopic plants with short nodal and long internodal cells.
2. Cell walls are composed of cellulose and are frequently heavily calcified.
3. Reproductive bodies are enclosed within a sterile envelope.
4. Growth is apical commencing from a protonemal stage and leads to nodal-internodal structure. Whorls of branches are formed from the node and the internodes are composed of a single enlarged cell that may be corticated.

(12) **PRASINOPHYTA**

1. Flagella and cell body are covered with one or more layers of scales of several types.
2. Principal storage products are starch and mannitol.
3. The photosynthetic pigments may include siphonoxanthin (known elsewhere only in the Chlorophyceae).
4. Members are mainly flagellate with four (occasionally 2 or 1) thick flagella aring from an apical pit surrounded by four lobes.

3

DIVISION - CYANOPHYTA

Class - Cyanophyceae (Myxophyceae, Schizophyceae, Blue-green algae)

GENERAL FEATURES

The blue-green algae are the only group of algae which are procaryotic. They have chlorophyll **a**, phycobiliproteins and glycogen as a storage product. Their cell walls are made up of amino sugars and amino acids.Because of their cellular organisation and biochemistry, Stanier et al. (1971) regard the blue-greens to be bacteria (**cyanobacteria**). Flagella are completely absent and the movement in the members is accomplished by gliding motion. Presence of **myxoxanthin** and **myxoxanthophyll** as unique carotenoids and proteinaceous **cyanophycin** as storage products are some of the distinctive features of Cyanophyceae. A number of blue-green algae have a capacity to change their colour in relation to the wave-length of the incident light. This change in colour is often described as **complementary chromatic adaptation** or **Gaidukov phenomenon**.

OCCURRENCE AND HABITAT

The blue-green algae are primarily freshwater except for the members of Chamaesiphonales. They are ubiquitous in waters showing wide range of salinity and temperature. These algae are also common on moist soil and rock surfaces. Blue-greens prefer a relatively neutral or slightly alkaline habitat. Members are both planktonic and benthic. In the marine environment planktonic forms are usually lacking (except *Trichodesmium*) and are commonly found in tidal pools. A fairly large number of members like *Scytonema, Cylindrospermum* etc. are able to fix atmospheric nitrogen and are common in rice fields of tropical and subtropical countries. Some strains of *Microcystis aeruginosa, Anabaena flos-aquae* are known to produce toxins (**phycotoxins**) and are thus responsible for acute poisoning of various animals (Carmichael and Gorham, 1981). Moikeha and Chu (1971) have isolated a toxic factor from *Lyngbya majuscula* that causes dermatitis.

Blue-greens along with some bacteria occur in different hotsprings where they can survive in water temperature up to 85°C. Many blue-green algae grow in association with other organisms showing endophytic, endozoic and symbiotic associations. *Gloeocapsa* and *Nostoc* are the phycobionts of lichens while *Nostoc* and *Anabaena* occur within the plant bodies of certain liverworts, water ferns and angiosperms.

CELL STRUCTURE

Cellular architecture of blue-green algae is simple. The protoplast which is delimited by cell wall, possess two regions, the peripheral **chromoplasm** and the central **centroplasm**.

CELL WALL AND SHEATH

Three types of investment are reported around blue-green algal protoplast-(1) sheath which is outside the cell wall, (2) multilayered cell wall and (3) plasmalemma. The chemistry of outer investment of blue-greens has not been investigated adequately. They are composed of pectic acids and mucopolysaccharides which after analysis have been found to be composed of glucose, hexuronic acids, D-xylose, ribose, galactose, rhamnose and arabinose (Wang and Tischer, 1973).

Mucilaginous sheath or **slime layer** is fibrillar and composed of polysaccharides. Non-cellulosic microfibrils are arranged in a reticular manner and embedded in a gelatinous matrix of homogenous appearance. The importance of sheath is in 1) colour variation, ii) gliding movement, iii) withstanding long period of dessication by water retaining capacity and iv) in perennation. Sheath is variable in thickness ranging from wide one to extremely delicate.

The cell wall of blue-green algae which lies between the plasmalemma and mucilaginous sheath is a multilayered complex structure (Fig. 3.1). It is composed of four layers-LI, LII, LIII and LIV with LI being next to plasmalemma. The LI and LIII layers are electron transparent (white in electron micrographs). The outer layer is a lipopolysaccharide similar to that of Gram-negative bacteria. This surrounds a muco-polymer layer of peptidoglycan which is dissolved by lysozyme and also digests the cell wall of Gram-negative bacteria. This mucopolymer has been shown to be composed of muramic acid, glucosamine, alanine, glutamic acid and α- and ε diaminopimelic acids.

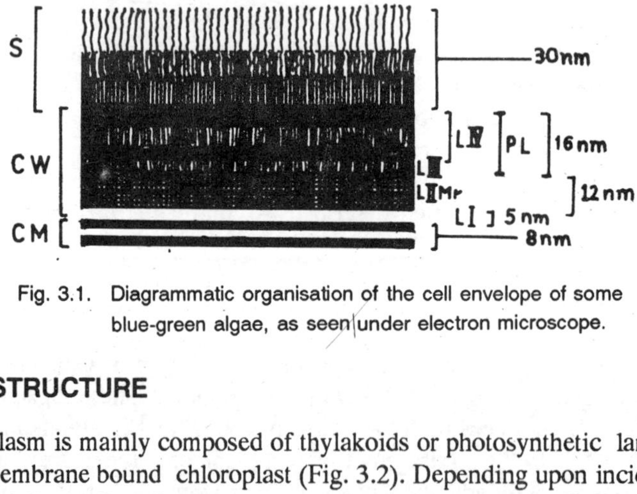

Fig. 3.1. Diagrammatic organisation of the cell envelope of some blue-green algae, as seen under electron microscope.

PROTOPLASMIC STRUCTURE

The peripheral protoplasm is mainly composed of thylakoids or photosynthetic lamellae which are not in form of group or membrane bound chloroplast (Fig. 3.2). Depending upon incident light intensity and age of the cell, these may be distributed throughout the protoplast. The thylakoids are the site of chlorophyll **a** and the accessory pigments are also found on their surfaces in the shape of small particles, the **phycobilisomes**. The accessory pigments are **C-phycocyanin, C-allophycocyanin** and **C-phyco-erythrin**. These accessory pigments apparently trasfer the light energy to chlorophyll **a.** It has also been suggested that the accessory pigments have a protective shading function that prevents oxidation in intense light. A variety of carotenoids are also present (Table 2.1).

In the central protoplasm the circular fibrils of DNA are present which lack basic protein (histones). In addition to the thylakoids and DNA, cyanophycean protoplast contains a variety of other structures including 70s ribosomes, gas vacuoles, polyglucan granules, cyanophycin granules, polyhedral bodies.

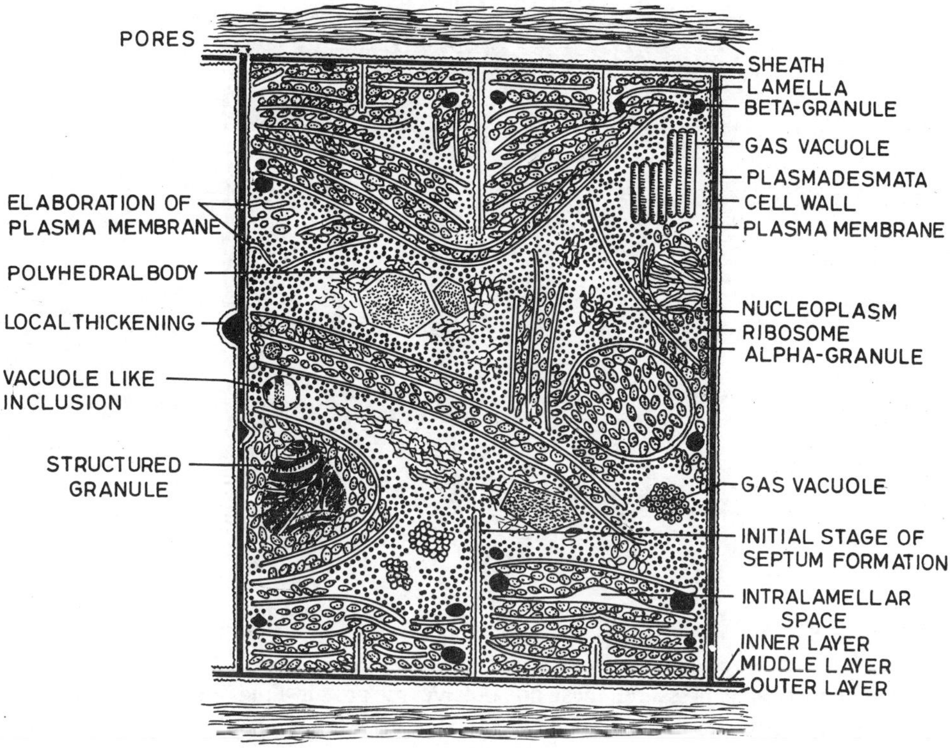

Fig. 3.2. Diagrammatic representation of a procaryotic cell (*Symploca muscorum*). (After Pankratz and Bowen, 1963).

lipid droplets etc. **70s ribosomes** are dispersed throughout the cell but are present in the highest density in the central region around the nucleoplasm. **Polyglucan granules** are polymers of glucose similar to animal glycogen and are also known as **α-granules** or **polyglucoside granules**. These are less dense, longer than robosomes and are commonly associated with thylakoids . The number and size of these granules depend upon nitrogen lebel, light intensity and age of the cell. These granules contain a carbohydrate composed of 14 to 16 glucose molecules like amylopectin (Frederick, 1951).

Cyanophycin granules are large dense bodies composed of copolymers of aspartic acid and arginine in 1:1 ratio and apparently serve as a nitrogen reserve (Simon, 1973). The cyanophycin granule polypeptide varies in amount with the algal growth cycle and is produced by a nonribosomal enzymatic system. It is also known as **structural granules. Polyhederal bodies** are polyhederal in shape and are usually associated with the DNA in the central region of the cell. They are composed of RNA. Another theory based on their function relates them to structures known as **carboxysomes** found in bacteria. Carboxysomes contain carbon dioxide-fixing enzyme ribulose 1, 5-bisphosphate carboxylase in the chemoautotrophic bacterium *Thiobacillus*. These carboxysomes are similar to polyhedral bodies in Cyanophyceae.

Polyphosphate bodies (Metachromatic or Volutin granules) are spherical and appear similar to lipid bodies of the eucarytotic cells under electron microscope. These contain stored phosphate and are absent in young growing cells but are prominent in older cells.

Gas vacuoles (pseudo vacuoles) are hollow cylindrical vesicles with conical ends. These are found in cytoplasm of all the Cyanophyceae except the Chamaesiphonales (Fig. 3.3). Gas vacuoles appear as reddish granules in many planktonic blue-green algae under light microscope. They disappear when cells

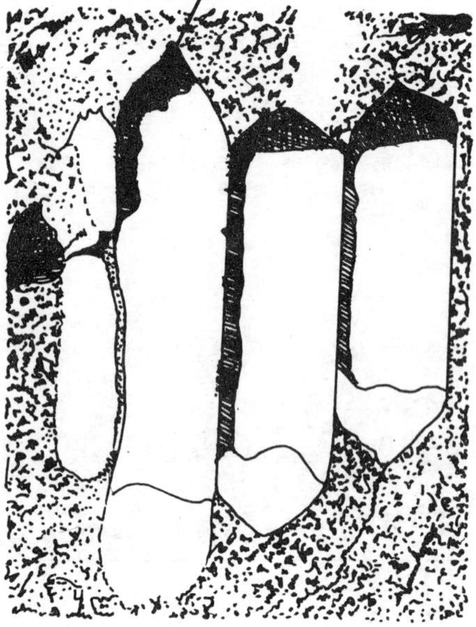

Fig. 3.3. Gas vacuoles showing cylindrical outline with conical ends.

are subjected to increasing pressure. These do not have true protein-lipid membranes but are composed of exclusively protein ribs or spirals arranged identically to the hoops on a barrel. The membrane of the gas vesicle is rigid with the gas inside having pressure of 1 atmosphere.

Gas vacuoles may function in shading the photosynthetic pigments (Porter and Jost, 1976) and in cellular buoyancy. The blue-green algae in which gas vaculoles are present are usually found on the surface of water and therefore they get high light intensity. Due to increased photosynthetic activites more sugar is produced leading to higher osmotic pressure. This results in collapse of gas vacuoles at the surface level of water which then sink down to the bottom and regulate the buoyancy of the planktonic forms. On basis of gas vocuoles, blue-green algae can be divided into two physiological-ecological groups (Walsby, 1973): (a) Those form which contain these vacuoles only at certain stages of their life cycle or only in certain types of cells. (*Gloeotrichia*, *Tolypothrix* and *Calothrix*) (b) The second group comprises planktonic members (*Anabaena, Microcystis* etc.) which derive positive bouyancy from their gas vacuoles and thus from blooms on the surface of water.

MOTILITY

The only type of motilty in the blue-green algae is by gliding movement. Many filamentous algae and some unicellular forms undergo movement when they come is contact with the substrate. There is neither a visible organ responsible for the movement nor any distinct change in the shape of the organism (Jarosch, 1962). Exact mechanism of gliding movement is not known but it may be due to a sort of propulsion caused by the secretion of slime and also by contractile waves on the surface of the cell (Halfen,1973; Fogg et al., 1973). Minute pores visible under electron microscope have been suggested to be the probable pathways for mucilage secretion.

HETEROCYSTS

Heterocysts are the cells which are easily distinguishable from the rest of the cells due to their thick walls, homogenous and pale-yellowish contents (Fig. 3.4). In most filamentous blue-greens (Nostocales and

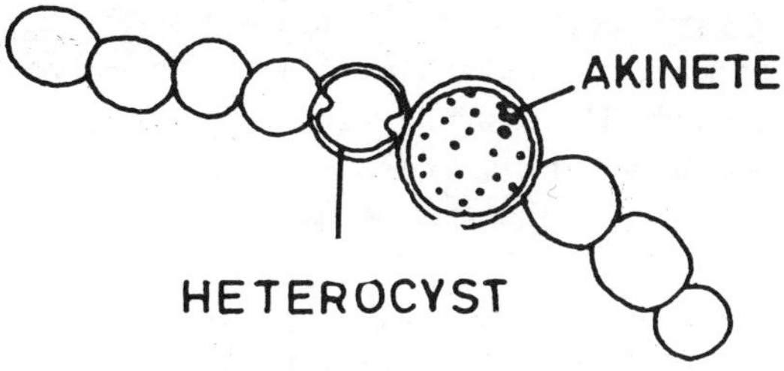

Fig. 3.4. Trichome of a vegetative cell with one heterocyst and one akinete.

Stigonematales) except in the members of Oscillatoriaceae these are found either in intercalary position or at the terminal end of trichome. Terminal heterocyst may be basal (*Rivularia*), pedicellate or lateral (*Nostochopsis*).Heterocysts normally occur singly, rarely in pair (*Anabaenopsis*) and sometimes may be arranged in small chains. In terminal heterocysts, one pore is present on one side at which it is attached to trichomes whereas intercalary heterocysts are marked by tapering of the two neighbouring vegetative cells and are distinguished by two pores, one at each end.

Heterocysts are formed from vegetative cells by the dissolution of storage granules, the deposition of a multilayered envelope outside the cell wall, the breakdown of the photosynthetic thylakoids and the formation of new membranous structures (Fig. 3.5). Mostly heterocysts are produced from newly divided

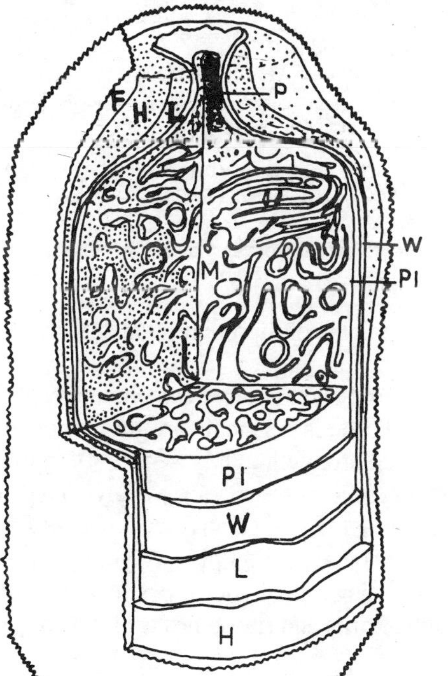

Fig. 3.5. Heterocyst in a three dimensional view (F-fibrous layer, H-homogenous layer, L-laminated layer, W-cell wall, Pl-plasmalemma, P-pore channel, M-membranes). (After Lang and Fay, 1971).

cells. Usually a single cell develops into heterocyst but rarely both the daughter cells may develop into two heterocysts. The frequency of heterocyst production is influenced by the concentration of molybdenum and combined nitrogen in the medium. The formation of heterocysts is inversely related to the amount of nitrogen in the algae. Heterocysts are unable to divide although there are some reports of germination. These appear to have a limited life and thus have a restricted period of physiological activity (Tischer, 1957).

Heterocyst structure can be divided into envelope, photosynthetic lamellae and other cytoplasmic inclusions. Wall is distinct into three layers the outermost fibrous, the middle homogenous and innermost laminated layers. The chemical composition of heterocyst envelope shows carbohydrate (62%), lipid (15%) and amino compounds (4%). Cellulose is absent. The contents of heterocysts are generally dense and uniform and the colour is yellowish due to disappearance of photosynthetic pigments leaving only the carotenes. The pores (1 or 2) present in the wall of heterocyst have prominent granules considered to be cyanophycin granules by some workers. Through these pores heterocysts are said to be in protoplasmic continuity with the neighbouring cells (Fig. 3.6).

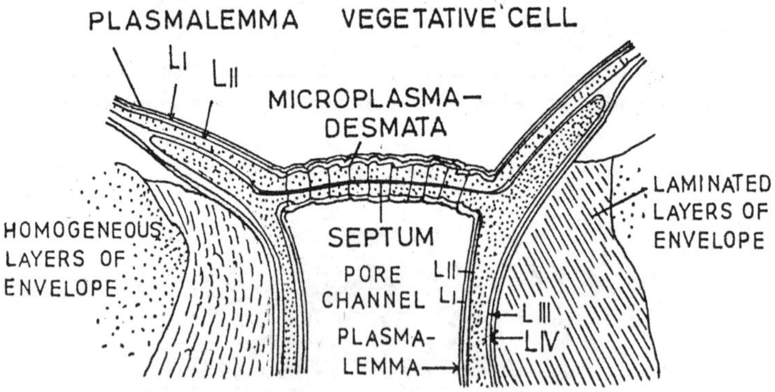

Fig. 3.6. Polar region of heterocyst showing microplasmadesmata and connection with nearby vegetative cell.

In heterocysts there are no phycobiliprotein pigments, no short -wavelength form of chlorophyll **a** (chl. 670), no polyphosphate but a higher concentration of P700 is found which is the form of chlorophyll required in photosystem I and therefore can photophosphorylate. They also carry out oxidative phosphorylation. As the phycobiliprotein pigments are the principal light absorbers and energy transfer depends on short-wavelength chlorophyll, the absence of these pigments indicates the lack of photosystem II in heterocysts. Heterocysts are unable to evolve oxygen and they can not fix carbon dioxide. These have a higher respiratory rate than vegetative cells. Heterocysts lack the normal lipid component present in the vegetative cells but contains two lipids (glycolipid & acyl lipid) not found in vegetative cells. Electron micrographs have shown that heterocysts lack structured granules and polyhedral bodies. Unlike vegetative cells, the DNA fibrils and ribosomes are uniformly distributed throughout the heterocyst protoplasm.

FUNCTIONS

The exact function of heterocysts is not clear although many theories have been proposed. Some of the important ones are: 1) they function as points of breakage of the trichome. This is based on the fact that breakage of the trichome into hormogonia frequently occurs at the heterocysts, ii) they stimulate the production of akinetes. This is based on this fact that in many species the akinetes are present just near heterocysts, iii) they stimulate growth and cell division. During reproduction, the nature of secretion changes and the production of spores is stimulated. This suggests that heterocysts are archaic reproductive bodies which have lost their functions. Because of these differing opinions, Fritsch (1951) called the heterocysts of Cyanophyta as a "botanical enigma". Fogg (1951) found that the combined nitrogen in the medium suppresses the production of heterocysts and it is related to nitrogen-metabolism of a blue-green alga.

NITROGEN FIXATION AND METABOLISM

The facts suggesting heterocysts as the sites of nitrogen fixation are: (1) that most of the species which have the capability of fixing nitrogen are heterocystous, (ii) that combined nitrogen suppresses both the formation of heterocysts as well as nitrogen fixation and (iii) that nitrogen fixation is a reductive process and the enzyme nitrogenase which is essential for nitrogen fixation and highly sensitive to oxygen is found in heterocysts only and not in the vegetative cells.

There are three main groups of nitrogen fixing blue-green algae: i) Heterocystous species which fix nitrogen aerobically and have been studied extensively. These species are the most important contributors of combined nitrogen in aerobic environments. ii) Nitrogen fixing unicellular aerobic forms, such as *Gloeocapsa*. iii) The filamentous forms (e.g., *Plectonema boryanum*, strain 594) which do not fix nitrogen in air but do so under microaerobic conditions. About the fixation of nitrogen by vegetative cells of heterocystous blue-green algae, there are three theories- i) Heterocysts have nothing special to do with nitrogen fixation and that the vegetative cells are responsible for fixing most or all of the nitrogen, ii) Vegetative cells may fix nitrogen under microaerobic conditions, as certain non-heterocystous filamentous algae do, but are unable to fix it in air due to the inactivation of nitrogenase by oxygen. iii) Vegetative cells of the heterocystous algae do not fix nitrogen either under aerobic or microaerobic conditions. Although heterocysts are the good locii for an efficient anaerobic functioning of nitrogenase because these have protective mechanism against oxygen inhition of the enzyme nitrogenase, vegetative cells also possess nitrogenase. Unicellular blue-green alga like *Gloeocapsa*, nonheterocystous filamentous forms like *Oscillatoria*, *Plectonema*, *Trichodesmium* etc. have shown to fix atmospheric nitrogen. The detailed mechanism of biological nitrogen fixation by blue-green algae has been given in chapter XIII (Algae in Biotechnology).

NUTRITION

Initially it was thought that blue-green algae were obligately photoautotrophic but in many forms heterotrophic growth on glucose, fructose or sucrose have been observed. Several blue-green algae like other algal groups require vitamin B_2 for growth. It is believed that due to lack of α-ketogutarate dehydrogenase and nicotinamide adenine nucleotide oxidase enzymes, blue-green algae behave as obligate autotrophs.

CELL DIVISION

Cells divide by amitotic division or fission. There is no mitotic cell division in Cyanophyceae. The cytoplasmic membrane shows centripetal growth and the inner layer of cell wall invaginates with the membrane. The outer layer does not take part in the division. Since the procaryotic cells do not have dictyosomes and endoplasmic reticulum it is not clear what organelles are responsible for the production of a septum. This is in contrast to the division found in higher organisms where a new plate is formed through dictyosome and endoplasmic reticulum and the new wall develops from this plate.

RANGE OF VEGETATIVE STRUCTURE

Thallus organisation in blue-green algae varies from unicellular, colonial to filamentous branched type (Fig. 3.7). Unicellular genera are not very common which include *Chroococcus, Anacystis, Synechococcus, Gloeocapsa* etc. Aggregation of numerous cells in form of colony can be seen in *Aphanocapsa, Aphanoth-*

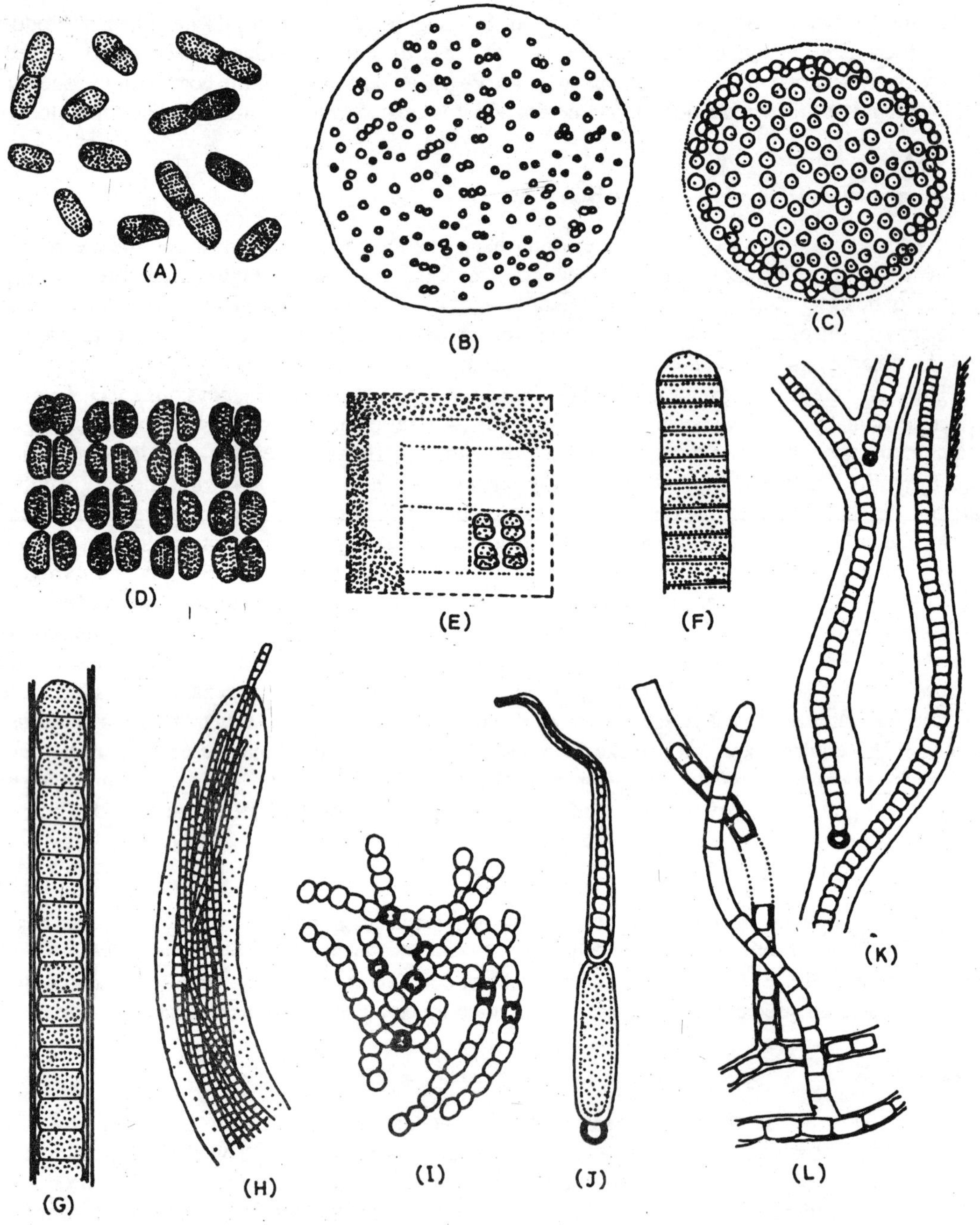

Fig. 3.7. Range of vegetative structure in blue-green algae. (A) Unicellular *Synechococcus*. (B) Colonial *Aphanocapsa*. (C) Colony with peripheral layers of cells in *Coelosphaerium*. (D) Flat colonial *Merismopedia*. (E) Cubical colony in *Eucapsis*. (F) Unbranched trichome of *Oscillatoria*. (G) Unbranched filamentous *Lyngbya*. (H) Group of trichomes in *Microcoleus*. (I) *Nostoc* showing differentiation of heterocysts. (J) *Gloeotrichia* showing differentiation of heterocyst and akinete. (K) False-branched *Tolypothrix*. (L) *Mastigocladus* showing true heterotrichous branching.

ece etc. Planktonic forms such as *Coelosphaerium* and *Gomphosphaera* show more elaborate colonies which consist of hollow sphere with peripheral layer of cells. Colony with diffluent mucilage sheath is found in *Microcystis*. In case of *Merismopedia* and *Eucapsis* the cells are arranged in definite shape giving the appearance of flat and cubical colony respectively.

Filamentous forms are more common in blue-green algae. A **trichome** is the basic structural unit consisting of a row of cells whereas the filament includes the trichome plus mucilaginous sheath. *Oscillatoria* and *Lyngbya* represent simplest unbranched forms without differentiation of cells into heterocysts and akinetes. In *Lyngbya* trichome is conspicuously enveloped within a mucilage sheath which is not distinct in *Oscillatoria*. In genera like *Microcoleus* more than one trichome may be embedded in the same mucilaginous sheath. Relatively more advanced filament with differentiation of cells into heterocyst, akinetes and spores are seen in *Nostoc, Anabaena, Cylindrospermum* etc. In these unbranched filamentous forms there is no differentiation into a base and an apex and growth is diffuse. In members of Rivulariaceae (*Rivularia, Calothrix,Gloeotrichia* etc.) trichomes are whiplike with broad bases which taper and usually terminates into a colourless hair. Growth is frequently trichothallic i.e., from meristem at the base of the hair.

Branched filaments may be either false or true. In true branching trichome itself branches and is characteristic of the Stigonematales whereas in false branching trichome is displaced to one side and passes into a branch of the mucilaginous sheath while the other trichomes of the filament continue in the original sheath (Fig. 3.7). False branching is characteristic of *Scytonema* and *Tolypothrix*. Most highly evolved thallus which is heterotrichous with true branching are found in *Hapalosiphon, Mastigocladus, Stigonema* etc.

REPRODUCTION

Cell division and subsequent separation of cellular progeny is the common mode of reproduction in the unicellular forms. Colonial and filamentous members reproduce by fragmentation. The fragmented portions get separated, float away and grow into new individuals. These short fragments of trichomes are called as **harmogonia** which are characteristic of all truly filamentous Cyanophyceae. In simple filamentous algae like *Oscillatoria* and *Cylindrospermum* the entire filament may break up whereas in others the hormogonia are produced at the tips of special branches. In some algae specialized biconcave **separation discs** or **necridia** are involved in breaking of the hormogone from the parent filament whereas in others the filament just fractures from any point.

Many of the filamentous blue-green algae produce unicellular reproductive agents. These are akinetes, endospores, exospores and heterocysts. An **akinete** or resting spore is a cell that is resistant to unfavourable environmental conditions and which can form new filaments after return of favourable conditions. An akinete develops from a vegetative cell that gets enlarged and filled with food reserves (cyanophycin granules) and augments its wall externally by an additional complex investment. Akinete envelope may be colored and ornamented. Akinetes of *Cylindrospermum* have 30 times as much DNA as do the vegetative cells. The development of akinetes in *Aphanizomenon* involved increase in cell size, gradual disappearance of gas vacuoles, increase in ribosomes, cytoplasmic density, cyanophycin and glycogen granules (Wildman et al., 1975). With the increase in cytoplasmic density the filament sinks to bottom sediment. During germination there is reversal of the above events.

Endospores are true reproductive cells formed in blue-green algae which are produced by internal division of the protoplast either forming a row of spores (*Chamaesiphon*) or an irregular mass (*Dermocarpa*, Fig. 3.8). **Exospores** are cut off one after another like fungal conidia from cellular apices of some genera like *Chamaesiphon*. **Nannocysts** are modified endospores which are formed by the division of cell contents without any visible enlargement of the vegetative cell as in *Gloeocapsa*.

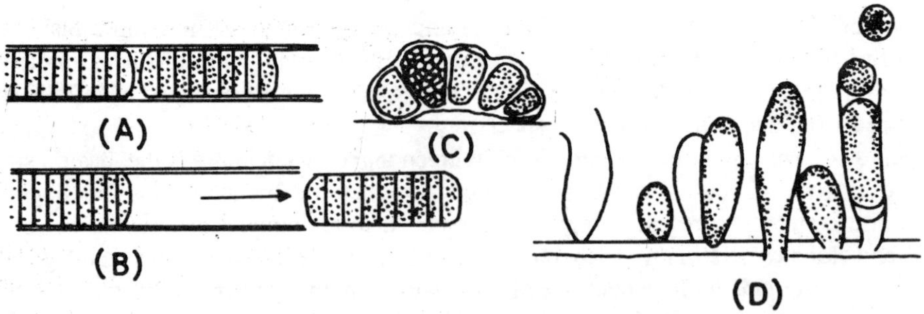

Fig. 3.8. Common modes of reproduction in blue-green algae. (A-B) Formation of a hormogonium in *Oscillatoria*. (C) Endospore formation in *Dermocarpa*. (D) Endospore formation in *Chamaesiphon*.

GENETIC RECOMBINATION OR PARASEXUALITY

Besides above modes of asexual reproduction, there are evidences of genetic recombination in Cyanophyceae, although plasmogamy has not been observed. It involves exchange of genetic materials among the two mating cells but this recombination differs from true sexuality due to lack of meiosis or syngamy and fusion of two gametes. The evidence is based mainly on studies of mutants of *Anacystis nidulans* (Kumar, 1962; Bazin, 1968; and Orkwizewski and Kaney, 1974). Singh and Sinha (1965) recorded genetic recombination in *Cylindrospermum majus* in culture.

CLASSIFICATION

Class Cyanophyceae has been divided into five orders (Fritsch, 1945) which is as follows:

(a) **Without hormogonia**

Order 1. **Chroococcales** : Unicellular or colonial, multiplication by cell division and by endospores *(Chroococcus, Gloeocapsa, Microcystis)*

Order 2. **Chamaesiphonales** : Epiphytes or lithophytes with thallus showing polarity, multiplication by endospores or exospores *(Chamaesiphon, Dermocarpa)*

Order 3. **Pleurocapsales** : Filaments without heterocysts, thallus divided into an erect and prostrate system (heterotrichy), multiplication by endospores (*Pleurocapsa, Hyella*)

(b) **With hormogonia**

Order 4. **Nostocales** : Filamentous thallus without true branching and with no heterotrichy, often showing false branching, heterocysts commonly present, multiplication by hormogonia, hormocysts and akinetes (*Oscillatoria, Anabaena, Nostoc, Rivularia, Scytonema*)

Order 5. **Stigonematales** : Filamentous thallus with true branching and heterotrichy, mostly with heterocysts and usually showing clear pit-connections between the adjacent cells, multiplication by hormogonia and hormocysts, rarely by akinetes (*Hapalosiphon, Stigonema, Mastigocladus*)

PHYLOGENY AND INTERRELATIONSHIP

Phylogenetic origin of Cyanophyceae is not known with certainty. The complete absence of flagellate stages lead to assumption that cyanophycean forms might have originated from such unicellular forms which were devoid of flagella. Many workers believe that abundance of terrestrial habit suggests a terrestrial rather than aquatic origin. Such conclusions, however, need further support.

Both blue-green algae and bacteria have their fossil records in Precambrian age (More than 3 billion years). One view is that the blue-green algae gave rise to bacteria by a complex evolutionary process with loss of biliproteins and change of chlorophyll **a**. Other presumption says that both the groups were evolved from some primitive unknown photosynthetic ancestors showing a sort of parallel evolution.

Relationship with Rhodophyceae

Some workers have sought an affinity with red algae due to these common features:

1. Presence of similar biloproteins
2. Similar 1:4 polyglucan linkage in floridean starch of Rhodophyceae and cyanophycean starch of Cyanophyceae
3. Absence of flagella and motile stages
4. Presence of pit connections in certain genera
5. Singly disposed thylakoids

Inspite of above similarities there are major differences between the two.

1. Cells of red algae are eucaryotic.
2. Sexual reproduction is present in Rhodophyceae.
3. Structures like heterocysts,akinetes, hormogonia etc. are not found in Rhodophyceae.
4. Presence of chlorophyll **d** is the unique feature of red algae.

Relationship with bacteria

Due to similarities between blue-green algae and bacteria both groups have been kept in Procaryota by many workers. The similarities are as follows:

1. Procaryotic organisation
2. In place of normal sexual reproduction, parasexuality is found in both groups.
3. Thylakoids are not membrane-bound both in blue-green algae and photosynthetic bacteria unlike eucaryotic conditions.
4. The cell wall possess characteristic mucopeptide.
5. They can fix atmospheric nitrogen.
6. Certain blue-green algae and bacteria thrive at temperature far higher than the cells of other organisms.
7. Some blue-green algae may be infected with viruses (**cyanophage**) which resemble bacteriophage in many respects.
8. Morphological similarities between the sulphur bacterium *Beggiatoa* and *Oscillatoria*.

Although there are strong resemblances between the two but some of the major differences are as follows:

1. Most of the blue-green algae are photosynthetic but bacteria are either saporphytic, parasitic or chemosynthetic.
2. Many bacteria possess flagella.
3. Heterocysts are not found in bacteria.
4. In blue-green algae main photosynthetic pigment is chlorophyll **a** while in bacteria it is bacteriochlorophyll.
5. Spores are formed endogenously in bacteria.

LIFE HISTORIES OF SELECTED MEMBERS

Order - Chroococcales

Family - Chroococcaceae

Genus - *Microcystis*

Microcystis is mainly found in freshwater ponds, lakes, rivers etc. as a planktonic form. It often causes **water blooms** in hardwater ponds and lakes. Some species have also been recorded from salt lakes, rice fields. *Microcystis aeruginosa* is the most commonly known bloom-forming algae. Some strains release toxins **(phycotoxins)** which have been found to be very harmful.

The colonies of this alga are always free-floating, spherical, ellipsoidal, irregularly elongate or clathrate masses of microscopic or macroscopic size (Fig. 3.9). Cells are spherical to slightly oval and lie close to one another with a common gelatinous matrix (Fig. 3.9C). They often have a blackish or reddish appearance because of the presence of numerous gas vesicles. Sometimes colonies are associated with daughter colonies. Presence of gas vaculoes (pseudovacuoles) has been found to be associated with certain ecological conditions.

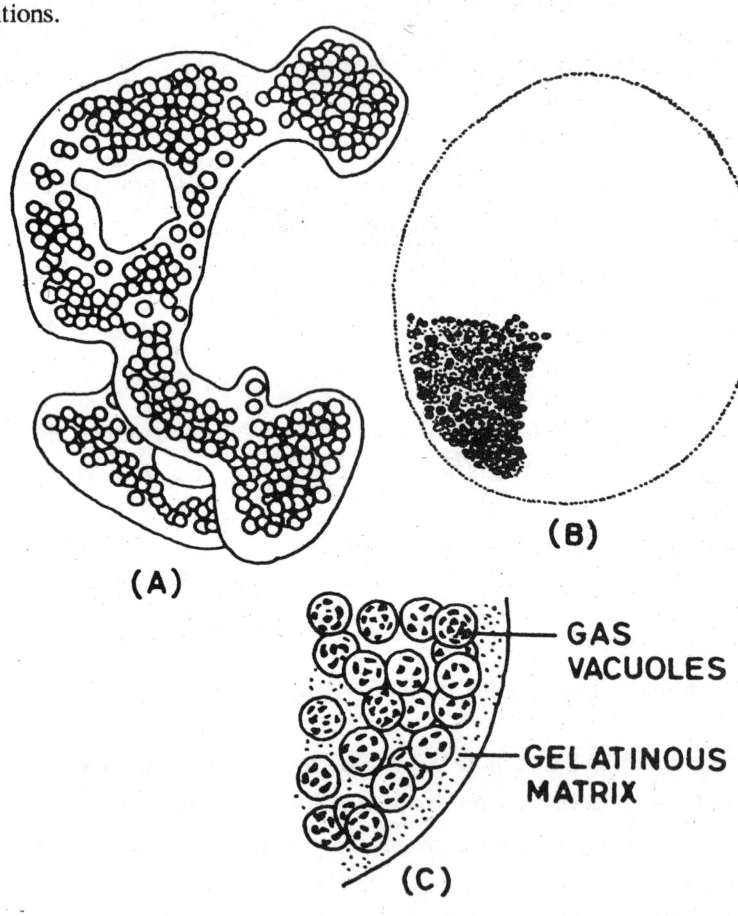

Fig. 3.9. *Microcystis aeruginosa.* (A) Clathrate colony. (B) Spherical colony. (C) A portion enlarged to show cells with gas vacuoles and common gelatinous matrix.

Cell division is multidirectional and the individual cells divide in all the three planes. In the elongated cells, the plane of division is transverse and portion of a colony separates apart to produce new colonies. In *M. flos-aquae*, reproduction also takes place by the formation of nannocytes.

Order - Nostocales

Family - Oscillatoriaceae

Genus - *Oscillatoria*

Oscillatoria is usually found on the moist surface of earth, on moist stones, muddy banks of ponds, streams and in pools of stagnant waters.

The trichomes are cylindrical and unbranched. They do not have sheath or have only a very delicate one. The trichomes often occur in floating masses or form a shiny mass on moist soil. Individual cells are often shorter than broad except for apical cell which may be rounded, attenuated and capped (Fig. 3.10 A-C). Apical cell may have walls like those of other cells or the free face may be thickened into a **calyptra**. According to the species, the protoplasts are homogenous, granulate or with numerous pseudovacuoles.

The trichomes of *Oscillatoria* exhibit gliding movement, rotation and an oscillatory motion. In narrow trichomes, the cells are cylindrical with length greater than the breadth. Electron microscopic studies of *Oscillatoria princeps* and *O.tenuis* have revealed that the central body is made up of a thin limiting membıane containing a homogenous matrix in which fine chromatin threads are arranged (Fig. 3.10 D).

Reproduction is accomplished by the fomation of hormogonia. Hormogonia are formed by the death of cell or group of cells anywhere in the trichome by accidental breaking. Biconcave discs of gelatinous material (**separation discs**) are formed between two vegetative cells in order to delimit the hormogonia (Fig. 3.10 E).

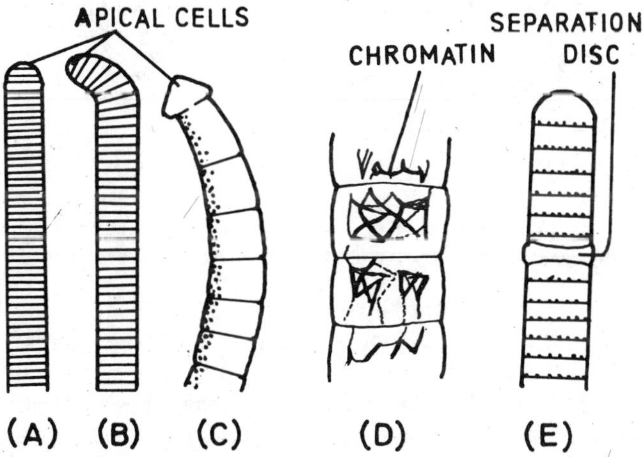

Fig.3.10. *Oscillatoria* sp. (A) Unbranched trichome. (B) Unbranched curved trichome. (C) Trichome with attenuated apical cell and cells having length less than width. (D) Cells containing chromatin. (E) Trichome with separation disc.

Lyngbya differs from *Oscillatoria* only in having a distinct sheath surrounding the trichome. Trichomes of *Phormidium* are always enclosed by a watery gelatinous sheath, thus it stands intermediate between *Oscillatoria* which lacks a sheath and *Lyngbya* which has a clearly defined one.

Order - Nostocales

Family - Nostocaceae

Genus - *Nostoc*

Nostoc is a common alga of both terrestrial and aquatic habitats. Terrestrial species grow on bare soil or intermingled with leafy plants. Aquatic species occur either free-floating in pools, lying on the bottom or attached to submerged vegetation. Some species prefer swiftly running water. Besides their symbiotic association with fungi in forming lichens they behave as space-parasites in the thalli of *Anthoceros*. Many species are known to grow in the paddy fields and in alkaline usar soils. They are also common on the rocky surface of small mountain streams.

The trichomes are always embedded within a gelatinous sheath and are contorted. The gelatinous aggregations of filaments have suggested such names as "star jelly", "witches butter" for species like *N. commune*. Innumerable threads are aggregated together within mucilage with almost firm boundry forming little balls -*Nostoc* **balls** (Fig. 3.11). Mature colonies have a smooth or verrucose surface, remain solid or hollow and are of macroscopic size. These may be few contimeters in diameter but they may attain a diameter of 50 cm is *N. amplissimum*. Heterocysts are generally solitary, intercalary and of much the same size and shape as vegetative cells. Hormogonia is formed very commonly in *Nostoc* due to breakage of trichome near the juncture of vegetative cell and heterocyst. Under favourable conditions the protoplast of a heterocyst may germinate to produce a new filament finally resulting into a colony (*N.ellisosporum*).

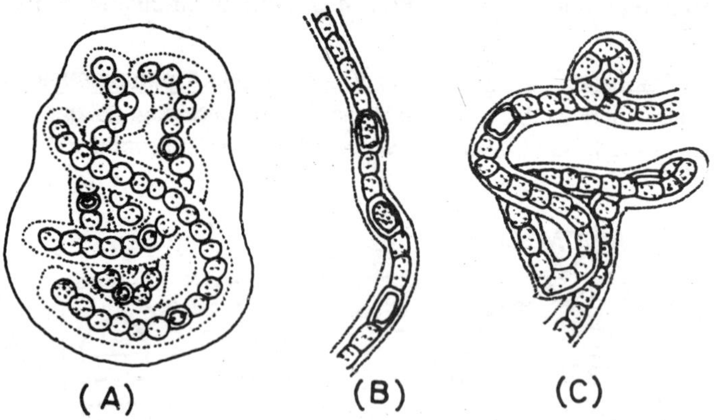

Fig.3.11. *Nostoc* sp. (A) *Nostoc* ball of aggregated filaments. (B) A portion of filament showing sheath, akinete and heterocyst. (C) A contorted filament with sheath and heterocyst.

Vegetative reproduction is accomplished by division of cells or by mechanical injury. Reproduction also takes place by **hormogonia** and spores. Short length trichomes (hormogonia) may be produced after breakage of the filaments which may come out through the colonial mucilage and ultimately form a new colony. **Akinetes (spores)** are formed after the maturity of colony. These are spherical, much larger and thicker than vegetative cells and are developed singly or in series between the heterocysts. After the onset of favourable season they germinate. Under certain conditions heterocysts also germinate to new filament and colonies.

Lazaroff (1973) summarized his own and other investigations of the life history of *Nostoc* as related to nutrition and light. There is an alternation between a heterocystous stage and a sporogenous one in the life cycle. When the alga is grown under complete darkness and is fed with organic sources of carbon by providing sucrose or glucose, the filaments break up into individual cells and few-celled hormogonia are formed. It is believed that the two hormogonia from two different filaments fuse in darkness and produce a heterocystous filamentous stage when exposed to light.

Order - Nostocales

Family - Nostocaceae

Genus - *Anabaena*

Most species of *Anabaena* are aquatic with planktonic nature which constitute important components of water blooms along with *Microcystis aeruginosa.* Many species show symbiotic association with variety of plants like *Azolla, Cycas* etc. Few species grow luxuriently in rice fields.

Filaments occur either singly or in floccose colonies. The trichomes are straight or irregularly contorted and of the same thickness throughout or slightly attenuated at their apices (Fig. 3.12). Trichomes occur singly within a sheath. Sheaths surrounding trichomes are hyaline and of watery nature. They may be broad or narrow and many planktonic species have sheaths several times broader than the vegetative cells. The cells are usually barrel-shaped and spherical. Protoplasts of vegetative cells are either homogenous, granulose or filled with numerous pseudovacuoles and variously colored. The shape of heterocysts are usually same as that of vegetative cells but are always larger than vegetative cells. Heterocysts are intercalary, generally solitary, sometimes in chain and several are usually present in any trichome. Akinetes may develop next to heterocysts, remote from them or in both positions.

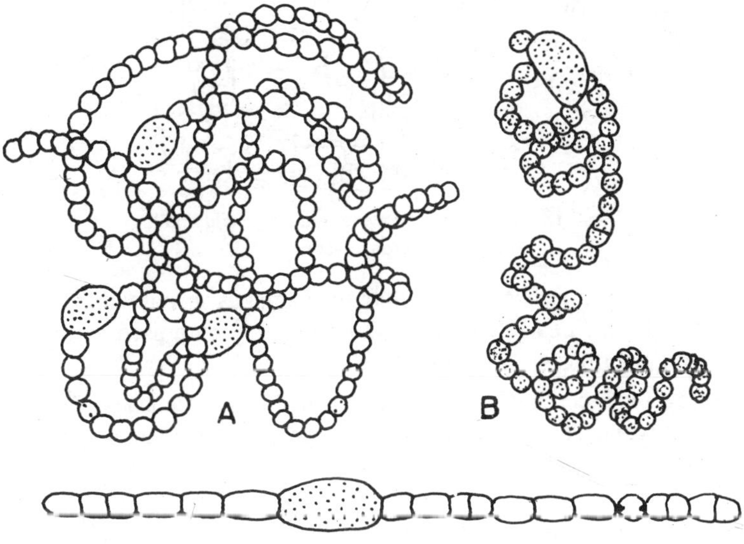

Fig.3.12. (A) *Anabaena circinalis* showing circinate trichome. (B) *A. spiroides* showing spiral trichome. (C) *A. levanderi* showing straight trichome.

Reproduction is achieved through hormogonia, akinetes and heterocysts like that of *Nostoc*. In many species of *Anabaena,* the number and position of akinetes is a specific character. They may be on one or both sides of a heterocyst. One of the factors for the production of akinetes is nitrogen deficiency. Akinetes are generally cylindrical or spherical in some species.

It is very hard to distinguish *Anabaena* from *Nostoc. Nostoc* can be identified by a firm gelatinous envelope in which the trichomes are always much contorted whereas *Anabaena* does not have contorted trichomes except a few planktonic forms. It has an extremely watery gelatinous sheath and never forms colonies of definite form.

Order - Stigonematales

Family - Stigonemataceae

Genus - *Stigonema*

Stigonema is generally characteristic of the cold acidic soils receiving high precipitation. It is found more frequently on moist rocks and soil than the aquatic habitat. The alga often grows in association with mosses and smaller plants.

Trichomes are truly branched that are partly or wholly multiseriate (Fig. 3.13). Branches may develop from all sides of an axis or may be borne unilaterally. The sheath surrounding the trichome is always of a firm texture with smooth or rough surface. Cells are usually spherical but in young axis or branches cells may be flattened due to mutual pressure. Heterocysts are present either in main axis or in the branches. They may be terminal or intercalary but lateral in position. If formed in multiseriate portion of a trichome they develop from superficial rather than deep-seated cells. Growth in *Stigonema* is largely apical and the branches are morophologically similar to the main axes. They are connected by coarse

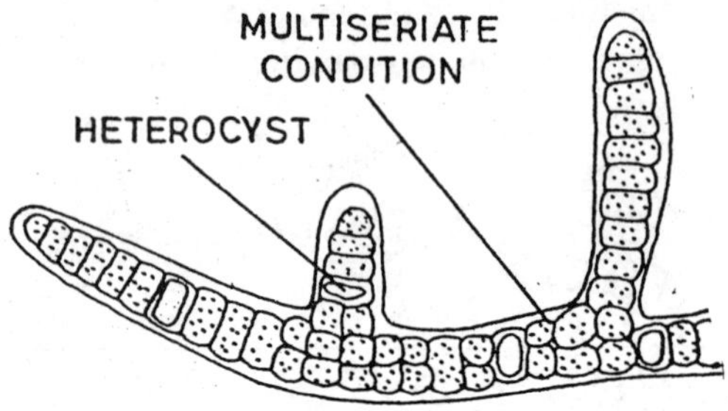

Fig.3.13. ***Stigonema*** **showing true branching and multiseriate organisation.**

protoplasmic strands. The younger trichomes have thin mucilaginous sheath while the mature parts show a prominent mucilage envelope. In early stage, the filaments may be uniseriate but later they become multiseriate due to longitudinal divisions of cells.

Formation of hormogonia is confined largely to the ends of young branches. They may originate with or without relationship to the position of heterocysts. The hormogonia develop into new filament independently.

4

DIVISION - CHLOROPHYTA

Class - Chlorophyceae (Green algae)

GENERAL FEATURES

Green algae constitute one of the most elaborate groups of algae in which the range of vegetative structure is much more wider than any other class of algae. Inspite of their great structural and reproductive diversity certain common features are :

(1) The photosynthetic pigments include chlorophylls **a** and **b**.

(2) Reserve product is starch.

(3) Cellulose constitutes the main component of the cell wall.

(4) Flagella are normally 2 or 4 in number and are acronematic.

(5) Chloroplast endoplasmic reticulum is not present around the chloroplasts.

(6) The cells possess eucaryotic type of cell organelles.

OCCURRENCE AND HABITAT

The Chlorophyceae include about 560 genera and 8600 species (Silva, 1982). About 90% of these are freshwater and the remaining are marine (Smith, 1955). Some orders are predominantly marine (Siphonales) whereas others are predominantly freshwater (Ulotrichales) or exclusively fresh water (Conjugales). The freshwater species are ubiquitously distributed with few species endemic in a certain area. They grow in waters exhibiting wide range of salinity. Some members are present in terrestrial and subaerial habitats. Species are both benthic and planktonic. Some members are found on ice, in symbiotic association with lichens and animals, as parasite, endophyte, epizoic and endozoic forms. Red snow on the mountains of U.S.A. is caused by *Sphaerella nivalis* due to the presence of haematochrome. A few members live as "space parasites" inside the tisse of higher plants.

CELL STRUCTURE

Cellular organisation in Chlorophyta as in all other algae except Cyanophyta is eukcaryotic (Fig. 4.1). Most of the unicellular and colonial members are spherical, oval or pear-shaped. Cells are almost cylindrical or barrel-shaped in filamentous members.

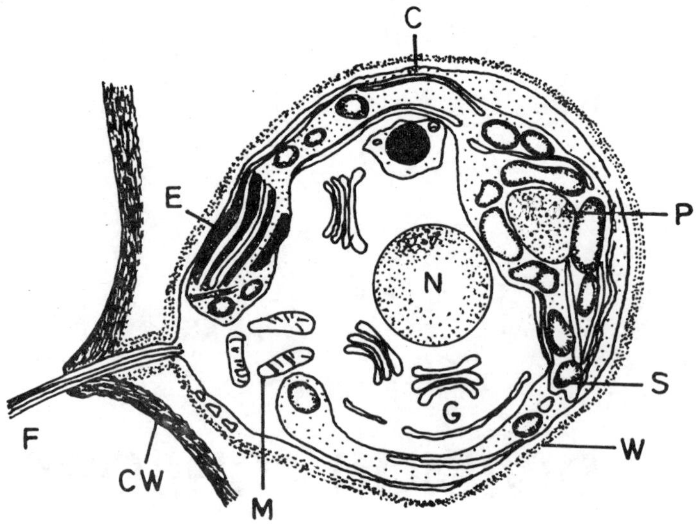

Fig. 4.1. Diagrammatic representation of a vegetative cell of *Volvox* sp. (F-flagellum, CW-colony wall, W-cell wall, S-starch, P-pyrenoid, C-chloroplast, E-eyespot, N-nucleus, M-mitochondria, G-Golgi body). (After Pickett - Heaps, 1970).

(i) **Cell wall :** The mature cell of the green algae possess a definite cell wall except in a few forms (e.g., *Polytomella).* Usually outer layer is pectic and inner cellulosic. In some members **lorica** is present (e.g., *Phacotus, Dysmorphococcus).* In these forms cell surface is separated from the outer firm envelope by a large space (Fig. 4.2A). In *Haematococcus* and the colonial *Stephanosphaera* the cell wall is separated from the protoplast by a mucilaginous space and is transversed by strands of cytoplasm (Fig. 4.2B).

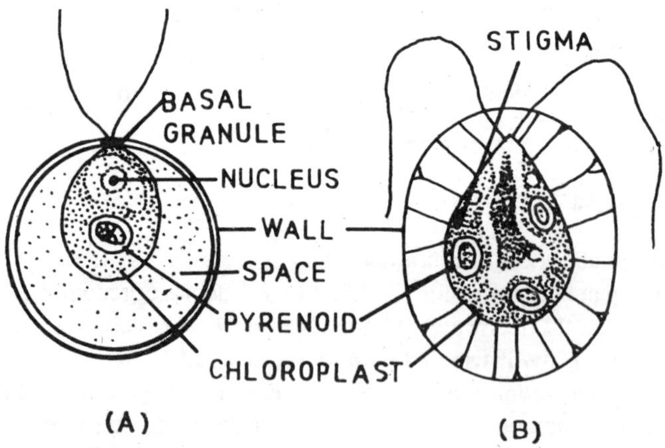

Fig. 4.2. Unicellular Volvocales. (A) *Phacotus lenticularis* showing large space between cell surface and outer firm envelope. (B) *Haematococcus pluvialis* showing mucilaginous space and cytoplasmic strands.

Cell walls generally have cellulose as the main structural polysaccharide except for the siphonaceous algae. Many of the green algae have a mucilage layer outside the cell wall. In some species cell wall is stratified or incrusted with calcium carbonate or it is cuticular.

(ii) **Chloroplast :** Besides basin-shaped chloroplast, reticulate peripheral chloroplasts are present in *Oedogonium, Haematococcus lacustris, Chlamydomonas reticulata;* axile stellate chloroplasts are found in *Zygnema, Chlamydomonas eradians;* parietal band-shaped in *Ulothrix;* spiral in *Spirogyra* and small discoid chloroplasts in *Chlamydomonas Alpina.* Some of the known colourless members are *Polytoma, Polytomella, Hyaliella etc.*

Only double-membrane chloroplast envelope without chloroplast endoplasmic reticulum is present outside chloroplast. The thylakoids are grouped into bands of two to six thylakoids. In some siphonaceeous members like *Caulerpa,* amyloplasts containing starch grains and few thylakoids are found with the chloroplasts. Chloroplast DNA fibrils are present in stroma region which are partially looped. Algae containing one or few chloroplasts have DNA in all the chloroplasts. In lower groups of green algae, usually a single chloroplast per cell is found but in Chlorococcales, Conjugales and Charales many discoid chloroplasts with or without pyrenoids are found. Characteristic spiral band of *Spirogyra* vary from one to many per cell. Only two stellate chloroplasts are present in *Zygnema.*

The pigments of green algal chloroplast are chlorophylls **a** and **b**, α-, β-, and γ-carotenes and several xanthophylls. In addition to these some vegetative cells lose their chlorophylls and develop secondary carotenoids.

(iii) **Pyrenoid :** In most green algae the plastids contain one or more specially differentiated regions called pyrenoids. These pyrenoids are the major sites of starch formantion in green algae. Griffiths (1970) has suggested that the pyrenoid is the region of temporary storage for early product of photosynthesis which after overproduction converts into starch. Presence of single pyrenoid is the common feature although these are entirely absent from some starch forming algae (e.g., *Microspora, Chloromonas).* Irregularly scattered numerous pyrenoids are found in *Chalmydomonas sphagnicola.* Pyrenoids may arise *de novo* or may be formed by the division of pre- existing ones. Electron microscopic observations reveal that the pyrenoids are traversed by thylakoids (Fig. 2.6).

(iv) **Flagella:** Motile stages of most of the green algae possess two acronematic anteriorly located flagella. Some members show quadriflagellate condition (e.g., *Carteria)* while a whorl of flagella are found in *Oedogonium.* In some organisms (e.g., *Chlamydomonas reinhardii)* hairlike delicate appendages arise from the flagellar surface while in others (e.g., *Pyramimonas)* the flagellar surface is covered by a layer or layers of minute scales. In some organisms the appendages of flagella are tubular and are called **mastigonemes.** Each flagellum arises from the anterior part of a cell of a basal granule called **blepharoplast.**

During normal forward movement, the flagella beat simultaneously in the same plane, Occasionally there is backward movement with undulating action.

(v) **Contractile vacuole :** Contractile vacuoles are found in vegetative cells of most Volvocales. In biflagellate forms, there are two contractile vacuoles at the base of the flagella. These are believed to be an organ of osmoregulation, removing excess of water and excretory wastes of the cells. When there are two contractile vaculoes, they contract alternately with a rapid contraction and slow distention. Contractile vacuoles are sometimes called as pulsating vacuoles because of their alternate filling and discharging action.

(vi) **Eye-spot (Stigma) :** Most of the flagellated cells that show phototatic movement have an eyespot. In green algae, the eye spot or stigma is always in the chloroplast near the flagellar bases in anterior portion (Fig. 4.1). It consists of one to many layers of lipid droplets usually in the stroma between the chloroplast envelope and the outer most band of thylakoids. An eyespot is orange-red in colour which consists of a curved coloured plate and a biconvex lens. The stigma is generally accepted to be the light receptor for the phototatic response of the cell. However stigma-less mutant strain of *Chlamydomonas* and many colourless motile forms lacking an eyespot also show phototatic movement.

(vii) **Mitochondria :** In green-algae, the number and size of mitochondria varies in different cells and species. They contain respiratory enzymes along with DNA, RNA and ribosomes. Similar to chloroplast, they are bounded by a double membrane. Inner membrane bears a number of finger like foldings.

(viii) **Golgi bodies :** Being composed of dictyosomes, golgi apparatus is responsible for the biochemical transformation and secretion of polysaccharide materials in cell wall.

(ix) **Endoplasmic reticulum :** It occurs in form of traversing tubules and vesicles in the cytoplasm and provides the surface for the attachment of the ribosomes and enzymes.

(x) **Ribosome :** In green algae two types of ribosomes are present. 80 s ribosomes as in other eukaryotic cells are present in the cytoplasm while in the chloroplasts 70 s type of ribosomes are encountered which is characteristic of a procaryotic cell. This gives an additional evidence that the photosynthetic procaryotes might have entered the eucaryotic cells as symbionts in the remote past.

(xi) **Vacuole :** The protoplast may possess a conspicuous vacuole of extremely variable size depending on species and habitat. The vacuole when present is traversed by cytoplasmic strands.

(xii) **Nucleus :** Majority of green-algae are uninucleate but multinucleate condition is not rare (Cladophorales, Siphonales, Charales and *Sphaeroplea).* Few members also show coenocytic conditions. Each nucleus has a definite nuclear membrane, one or more nucleoli and poorly developed chromatin network. Nuclear division is mitotic and its stages are quite similar to those of higher plants. Meiosis has also been recorded in many members. The morphology and number of chromosomes are variable and centromeres are median, submedian or terminal. In Conjugales and Siphonales nuclear apparatus is complex. Satellite chromosomes associated with nucleolar organising chromosomes and diffused centromeres have also been recorded in Conjulales.

Cell division

Along with nuclear division, cytoplasm also divides (cytokinesis) but cell plate (phragmoplast) is not formed. However, in some members (e.g., *Fritschiella)* similar cytokinesis and cell plate formation as seen in the land plants has been observed. Furrowing of cytoplasm, division of the protoplasm into two and formation of a cross wall take place in quick succession. Chloroplast and pyrenoid also divide during cytokinesis. Chloroplast divides longitudinally in unicellular forms having longitudinal symmetry (e.g., *Chlamydomonas).*

RANGE OF VEGETATIVE STRUCTURE

Members of Chlorophyta exhibit a very wide range of thallus structure which are as follows:

Motile forms: The most primitive thallus construction found in green algae is unicellular motile forms where unicellular, spherical, oval or pear-shaped motile plant body is the common feature (e.g., *Chlamydomonas).* In some members many such individuals assemble inside an envelope and form colonial habit (e.g., *Volvox).* Colonial form when behaves as a well-organized unit with a definite number of cells arranged in a specific manner, is called **coenobium.**

Nonmotile palmellate forms : In such type of thallus colonies are generally amorphous and often have definite shapes. It is a loose assemblage of cells which lie embedded in a common gelatinous matrix secreted by the cells forming a colony (e.g., *Tetraspoara).*

Non-motile coccoid forms : In this type plant body is small, isolated, unicellular and with non-motile spherical cell having sedentary habit (e.g., *Chlorella* and *Chlorococcum).* In some other forms (e.g., *Hydrodictyon)* such cells are united together to produce definite nonmotile coenobia.

Dendroid forms : In *Ecballocystis* and *Ecballocystopsis* mucilage is produced locally, generally at the base of cell which results in cell with a basal mucilage stalk.

Filamentous forms : Filamentous habit is an important advancement over the unicellular condition. In many higher forms, the plant body is filamentous which may be unbranched (e.g., *Ulothrix)* or branched (e.g., *Cladophora)*

Heterotrichous forms : This is an advanced type of construction where plant body is composed of creeping prostrate system and erect projecting system (e.g., *Stigeoclonium).* Sometimes prostrate system is more developed than the erect system or vice versa.

Parenchymatous thalloid forms : As a result of division of cells in more than one plane, parenchymatous thallus is produced. *Ulva* represents foliose parenchymatous condition while tubular parenchymatous habit is found in *Enteromorpha.*

Siphonaceous forms : This habit is believed to have originated by its enlargement accompanied with nuclear divisions but with no corresponding formation of septa (e.g., *Caulerpa, Protosiphon).* Such nonseptate multinucleate structure is termed as coenocyte.

REPRODUCTION

Like variation in thallus structure, the methods of reproduction in green algae show a wide range of diversity. These are:

Vegetative : Portions of the plant body get separated to form new individuals without any obvious change in the protoplasts. Common examples are the fragmentation into two or more pieces or accidental separation in pieces. The process of multiplication by ordinary cell division is also characteristic of some unicellular algae. In some members (e.g., *Pithophora)* akinetes are frequently formed.

Asexual : A sexual reproduction is commonly achieved by the formation of **zoospores.** These flagellated spores are produced either singly or in groups within a specialized cell (**sporangium**, Fig. 4.3A) or inside

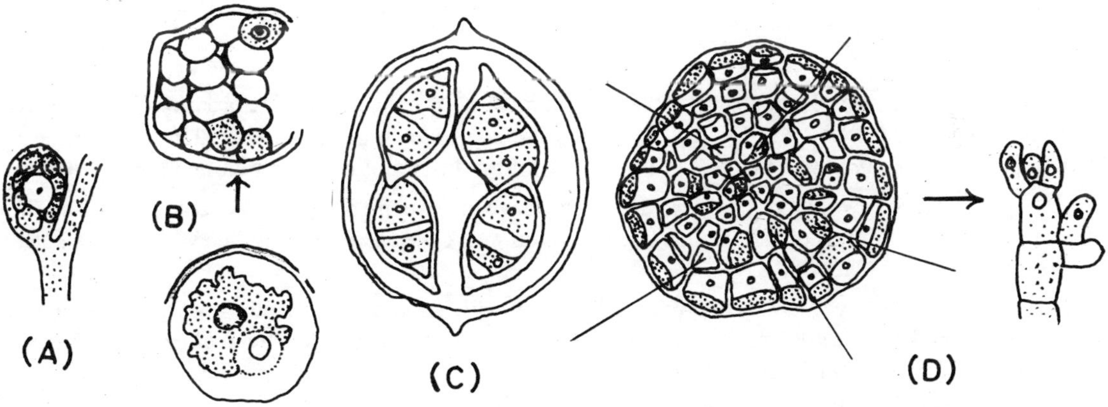

Fig. 4.3. Asexual reproduction in green algae. (A) Sporangium of *Derbesia.* (B) Aplanospore production in *Trebouxia.* (C) *Oocystis lacustris* before autospore formation. (D) *Coleochaete scutata* with specialized gametangia.

normal vegetative cell *(Ulothrix).* The number, shape, size and nature of these spores vary with species which actually depends upon the environmental conditions. After the onset of favourable conditions, the zoospores germinate into new plant. In some members non-motile **aplanospores** are produced *(Trebouxia,* Fig. 4.3B). With the addition of thick wall they are known as **hypnospores.** In the members of Chlorococcales **autospores** are commonly produced which are aplanospores having same shape as the parent cell *(Oocystis,* Fig. 4.3 C). **Coenobia** form daughter colonies frequently in the members of Volvocales.

Sexual reproduction : Sexual reproduction in the Chlorophyceae may be isogamous, anisogamous or oogamous. Usually gametes are specialized cells and not vegetative cells, although in unicellular Volvocales the latter may occur. In isogamous or anisogamous species, the gametes are usually not found in specialized cells, although in oogamous species gametes are normally formed in specialized gametangia e.g., *Coleochaete* (Fig. 4.3 D). Depending upon species, union of gametes may be either **aplanogamous** (non-flagellate) or **planogamous** (flagellate). Normally the two gametic nuclei fuse immediately after fusion but in many cases they do not fuse until before germination of zygote. Meiosis usually takes place before the germination of zygote as it represents diploidic or sporophytic phase in the life cycle. Sometimes the germination of zygotes take place soon after its formation but in many cases germination is delayed till ripening. In some cases unfused gametes may germinate parthenogenetically. Such phenomenon is recorded in form like *Ulothrix, Oedogonium* etc.

CLASSIFICATION

Classification of Chlorophyceae has been under a constant process of modification. It is because the members have a wide range of vegetative structure (much more than that of any other class). Although Fritsch's system (1935) is an old classification but is more commonly acceptable. According to this, class Chlorophyceae has been divided into nine (9) orders viz., Volvocales, Chlorococcales, Ulotrichales, Cladophorales, Chaetophorales, Oedogoniales, Conjugales, Siphonales and Charales. Smith (1955) preferred to classify division Chlorophyta into classes Chlorophyceae and Charophyceae. According to him class Chlorophyceae comprises twelve (12) orders - Volvocales, Tetrasporales, Ulotrichales, Ulvales, Schizogoniales, Cladophorales, Oedogoniales, Zygnemales (Conjugales), Chlorococcales, Siphonales, Siphonocladales and Dasycladales. In class Charophyceae only order Charales has been kept. He considered Charophyceae as most evolved group and elevated it to the rank of a class. With the accumulation of more informations based on ultrastructure, physiology and biochemistry, cytology etc. many authors have proposed new systems of classification. However, because of the simplicity and convenience Fritsch's system is being followed commonly.

INTERRELATIONSHIP

Because of their very simple thallus organization Volvocales form most primitive order of Chlorophyceae. It is widely believed that nonmotile unicellular forms like *Chlorella* might have evolved from simple, motile forms like *Chlamydomonas*. From this primitive stock four different lines of evolution may be traced.

(i) First line of evolution involves the progressive development of motile colonial forms from unicellular motile condition. This line of evolution is called as **volvocine** line which culminates **in** *Volvox.*

(ii) Second line of evolution is **chlorococcine** line in which nonmotile unicellular or colonial members develop from motile unicellular ancestor. This line is represented by *Chlorella, Pediastrum, Scenedesmus* and *Hydrodictyon.*

(iii) In **siphonalean** line of evolution, an extensive tubular or siphonaceous type of coenocytic thallus (e.g., *Caulerpa)* evolve from motile unicellular stalk.

(iv) In fourth line of evolution, multicellular **filamentous** conditions (unbranched & branched) and **thalloid** type of thallus originated from unicellular motile Chlamydomonad type of cell.

There are two common theories for the origin of the filamentous habit in Chlorophyceae. The first suggests an origin from unicellular types whereas the second suggests that they arose from palmelloid forms.

Besides these common tendencies of evolution, some orders represent the offshoot as Tetrasporales probably repersent an offshoot of the Volvocales where the vegetative cells become nonmotile. "Bryophytan" or "Charophycean" line of evolution led to the evolution of the higher plants and has the following characteristics: (1) asymmetric motile cells with somewhat laterally attached flagalla, (2) spindle of the nuclear apparatus does not disintegrate during cell division while the new cell wall is formed by coalescence of dictyosome (phragmoplast) and (3) the presence of enzyme glycolate oxidase.

Among other groups of algae, Chlorophyta show similarities with Euglenophyta with reference to pigments and cellular metabolism. Other groups of algae do not show real affinity with Chlorophyta. Some members of Chlorophyta (e.g., *Fritschiella* of Chaetophorales) due to many resemblances are supposed to be progenitors of higher land plants. According to some phycologists, the Chlorophyta are thought to be evolved from photosynthetic bacteria.

Along with the evolution in vegetative thallus, reproductive methods also evolved progressively. In primitive forms reproduction is asexual then the sexual reproduction developed; first isogamy, then anisogamy and finally oogamy.

In this chapter important features of vegetative and reproductive structures and life histories of selected genera will be discussed along with the basic characteristics of all 9 orders given by Fritsch.

ORDER 1 - VOLVOCALES

Members of this order are unicellular and coenobial which are normally motile by means of flagella. They move towards light showing positive phototaxis. In the life cycle of these members, temporary resting stage i.e., **palmella stage** may occur. On the basis of their habit and structure of thallus, Fritsch (1935) classified it into there suborders (i) Chloromonadineae (motile), (ii) Tetrasporineae (palmelloid) and (iii) Chlorodendrineae (dendroid).

Members of this order represent most primitive plant life which are widely distributed in freshwater and soil. Some members are found in brackish and marine waters. Certain Volvocales (e.g., *Carteria, Volvox*) are known to cause water blooms.

Most genera have spherical, ellipsoidal, pear-shaped or pyriform shape. All genera have the flagella at the anterior end. Unicellular members of the order have one, two, four or eight flagella, while the coenocytic members have two or four flagella. Flagella are whiplash type which show typical 9+2 pattern.

Except for the few members, the protoplasts of the cells are surrounded by cell walls that may be simple or complex in organization. The chlorophyll containing portion of the cytoplasm, the chloroplast is usually a massive cup-shaped structure occupying the greater portion of the cell. Many of these cup-shaped chloroplasts contain a single pyrenoid but some have more than one pyrenoid. *Polytoma* is an example of a colourless algae. A stigma or "red eyespot" is found in the anterior half of chloroplast of many members(Fig. 4.1). The cells are uninucleate and lack large central vacuoles. Except in marine species, most members contain one to many small contractile vacuoles that have an osmoregulatory role. These may be spherical at the flagellar bases or distributed over the surface of the protoplast.

The important features of the coenobial Volvocales is that the number of cells in the colony is determined at the beginning of their development and during growth the cell number does not increase. Colonial members show a series of increasing elaboration and specialization which are very apparent with reference to *Gonium, Pandorina, Stephanoon, Eudorina, Platydorina, Pleodorina* and *Volvox* (Fig. 4.4 A-G). The elaborations are in the shape and size of coenobium, morphological differentiation of the cells, specialization of reproductive cells and the method of sexual reproduction.

Both asexual and sexual reproduction occur in the members of Volvocales. In unicellular genera, asexual reproduction is performed as a result of mitosis and cellular division which may be repeated rapidly forming 2,4,8 or more daughter cells. Genera without a cell wall (e.g., *Dunaliella*) always have a longitudinal division while the genera with a cell wall may exhibit a longitudinal or transverse division of the protoplast. In wall-less species division may occur while the cells are motile but most walled organisms (e.g., *Chlamydomonas*) become transiently nonmotile during cell division which often takes place in darkness.

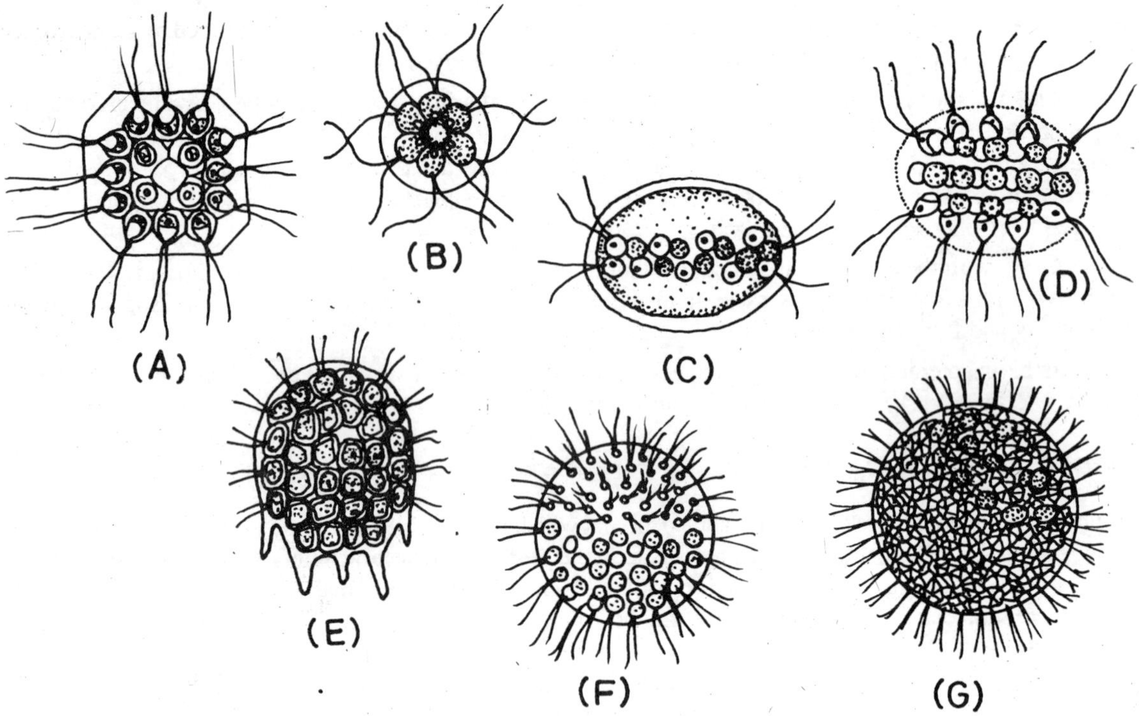

Fig. 4.4. Common colonial Volvocales. (A) *Gonium pectorale* with 12 peripheral cells and 4 central cells. (B) *Pandorina morum* with close cells. (C) *Stephanoon askenasyi* with cells arranged in equatorial plane. (D) *Eudorina elegans* with spherical coenobium and loosely arranged cells. (E) *Platydorina caudata* showing flattened horse shoe-shaped colony with conical projection towards posterior end. (F) *Pleodorina californica* showing anterior smaller vegetative cells and posterior larger reproductive cells. (G) *Volvox aureus* showing most highly evolved volvocalean genus with numerous cells.

Colonial members reproduce by repeated division of a cell to form a daughter colony. In primitive members (e.g., *Gonium, Pandorina, Eudorina)* all the vegetative cells function in the asexual reproduction whereas in *Pleodorina* only cells of the posterior half participate in reproduction. In *Volvox* only a few specialised cells, **gonidia** divide. After division the daughter protoplasts become arranged in a curved plate, the **plakea** before being arranged in the form of parent colony.

Sexual reproduction may be isogamous, anisogamous or oogamous depending on the species. Division of the zygotic nucleus is meiotic resulting in the formation of zoospores which produce haploid plants. Some unicellular and colonial members are discussed in the text.

Unicellular Volvocales

Family - Chlamydomonadaceae

Genus - *Chlamydomonas*

The *Chlamydomonas* cell represents most primitive member of the living green plants. It is widely distributed in variety of habitats mainly in freshwater systems or on moist soils. Few species are found in brackish waters *(C. halophila). C.nivalis* represents snow algae and is found in the polar regions imparting red colour. The red colour is due to red pigment, haematochrome.

Solitary free-swimming cells may be ellipsoidal, subcylindrical, spherical or pyriform (Fig.4.5 A). Two

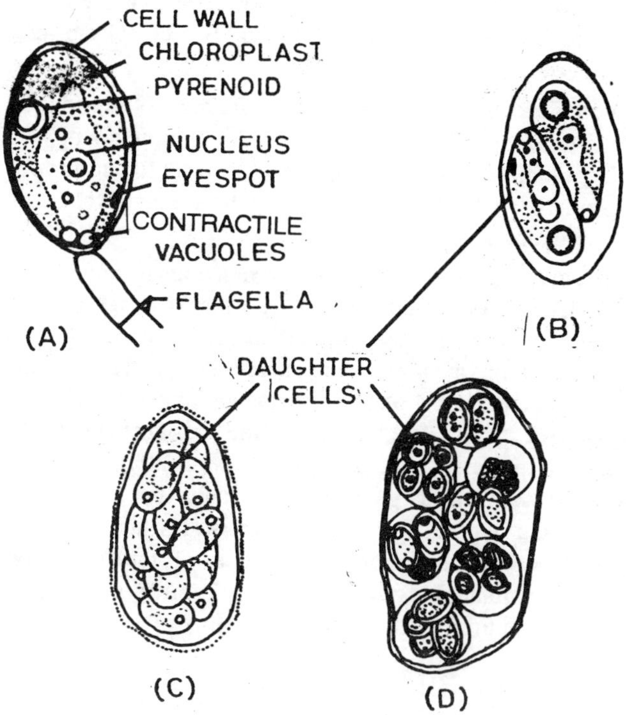

Fig. 4.5. *Chlamydomonas* sp. (A) A single cell showing the different parts. (B-C) Stages of cell division. (D) Palmelloid stage showing groups of cells within a common wall.

equal acronematic flagella are located in the anterior region of the cell which are responsible for motility. There is always a definite cellulose wall layer and some species have a distinct gelatinous layer external to the cellulose layer. In *C. reinhardtii* it has been shown to consist of seven layers (Roberts et al., 1972). Cells have delicate to massive cup-shaped chloroplast in most of the species. The chloroplast in some species are laminate, stellate or H-shaped in optical section. There is usually one pyrenoid within a chloroplast but some species regularly have more than one pyrenoid. Usually there are two **contractile vacuoles** near the base of the flagella but the number and position of the vacuoles are not constant. When there are two contractile vacuoles they are anterior otherwise they may be distributed over the surface of the protoplast. These vacuoles and a single nucleus lie in the colouries cytoplasm. The **stigma** or red **eye-spot** is embedded at the side of chloroplast and is composed of carotenoid subunits about 75 nm in dimension. It may be cicurlar, oval or sublinear in outline. Positive phototaxis is most intense in this genus during the exponential phase of growth. During movement the cells grow and ultimately divide. Two to eight or occasionally more daughter swarmers are produced within the cell (Fig 4.5 B, C). By the dissolution of the parent cell wall, these swarmers are liberated and form new plant. By the failure of daughter cells to escape, **palmella stage** is formed mainly in members growing on soil. As a result of which colony of many cells held together by a common mucilaginous investment (Fig. 4.5 D) is formed. Ultimately the cells escape and start active phase again.

Under electron microscopy, cell is bounded by cytoplasmic membrane which appears to be a typical unit membrane of two electron opaque layers separated by a comparatively lesser opaque zone. Cell wall is outside this and is stratified at the anterior region. Each flagellum shows 9+2 pattern. Cell organelles like chloroplast , mitochondria , nucleus and dictyosomes are bounded by a double membrane. 2-6 elonglate thylakoids form grana like bodies inside chloroplast. DNA elements are also present within the chloroplast and mitochondria. A pyrenoid with a granular core surrounded by tightly packed starch plates is observed.

An eyespot is composed of two or three parallel rows of fat droplets in a linear fashion. These are situated within the chloroplast and are not associated with flagella. The cytoplasm also includes endoplasmic reticulum and ribosomes.

In asexual reproduction, the cells of most species become nonmotile and their nuclei undergo one or more mitoses which are followed by cytokinesis. In this way 2,4 or more young cells are formed within the parental wall from which they ultimately escape by rupture of the wall. Released young cells are smaller than the parent which gradually enlarge to normal size and repeat the asexual cycle. *Chlamydomonas* may also form aplanospores, hypnospores and akinetes.

Sexual reproduction involves fusion of two haploid sexual units forming diploid cell or zygote. Some species are **homothallic** or **self-compatible**, others are **heterothallic** or **self-incompatible**. Gametic union may be isogamous, anisogamous or oogamous. Isogamous species usually have the cells capable of functioning as gametes when conditions are favourable. There is frequently an escape of the protoplasts from the enclosing walls before the two unite (Fig 4.6). In the anisogamous heterothallic species *(C. braunii)* a cell of a male clone divides to form 8 or 16 male daughter cells while one of the female clones forms 2 or 4 female daughter cells. In the oogamous species *(C. coccifera and C.oogamum),* a male cell divides to form 8,16 or 32 small cells, one of which fuses with a large female cell that has become immobile and lost its flagella. Immediately after plasmogamy the zygote secretes a primary wall which is supplemented by a secondary wall (Fig. 4.6 F) that is formed between the primary wall and the protoplast. As the zygote matures, large quantities of starch and lipids accumulate and obscure its protoplasmic organization. Secondary carotenoids also develop in some species so that the zygotes become orange-red.

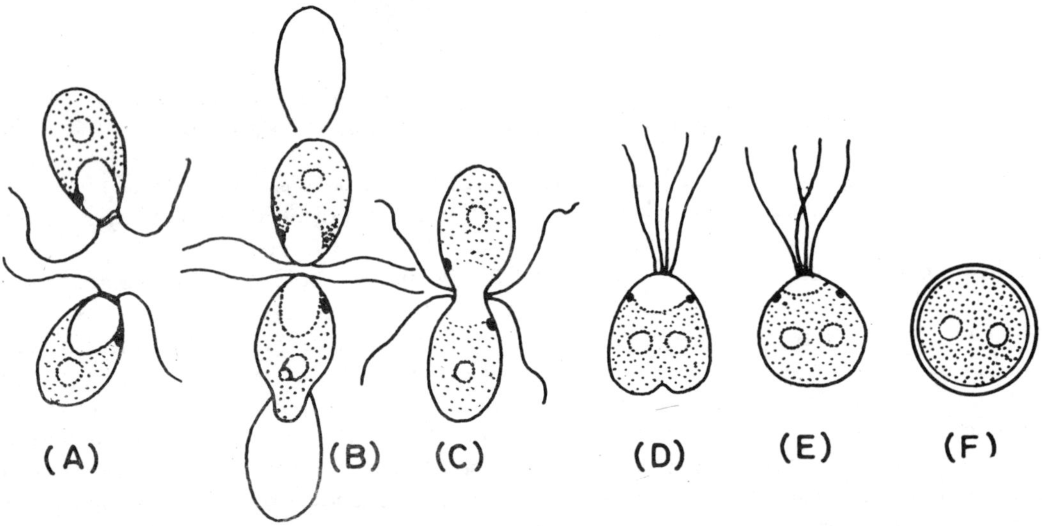

Fig. 4.6. Sexual reproduction in *Chlamydomonas* sp. (A) Two motile cells. (B-D) Gametic union. (E) Motile zygote. (F) Resting zygote.

Zygote germinates after a period of dormancy giving rise to 4 *(C. eugametos, C. moewusii)* or 8 *(C. reinhardtii)* zoospores by meiotic division.

Isogamy, anisogamy and oogamy occur among the species of single genus *Chlamydomonas*. Thus it illustrates all the stages in the evolution of sex in algae and can very well be arranged as isogamy —> anisogamy —> oogamy from the evolutionary point of view.

Colonial Volvocales

Important features of colonial Volvocales is that motile colonies have a fixed number of cells arranged in a constant manner. These colonies are known as **coenobia**. The series of increasing number of cells and differentiation of cells can be seen in the different genera of Volvocaceae. The genus *Gonium* is the simplest member with no distinction in the cells. All other genera have distinct anterior and posterior ends. In *Pandorina* and *Eudorina* the stigma tends to become less prominent in the posterior cells and it is absent from the posterior cells of *Pleodorina* and *Volvox*. In *Pleodorina* and *Volvox* the posterior cells are larger than the anterior.

Family - Volvocaceae

Genus - *Gonium*

Gonium is usually found intermingled with other algae of pools and ditches. The coenobium consists of 4,16 or 32 biflagellate cells arranged in a flat quadrangular plate (Fig. 4.4 A). The cells are connected to one another by gelatinous strands and are embedded in a common gelatinous matrix. 16-celled coenobium has 12 peripheral cells (three on each side) and 4 quadrately disposed central cells. Individual cells are ovoid to pyriform in shape with anteriorly placed two flagella. Two contractile vacuoles are present at the base of the flagella and the single eyespot lies in the anterior end of the cell. Cup-shaped chloroplast with one pyrenoid is present.

There is simultaneous formation of auto-colonies by all cells in a colony during asexual reproduction. In 16-celled coenobium, an accidental breakage leads to asexual reproduction immediately but 4-celled species do not reproduce immediately after the breakage. Each daughter colony has a colonial envelope of its own at the time when it is liberated by gelatinization of the envelople of the parent colony. Solitary cell which often gets detached from a colony may develop into akinete or into palmella stage.

During sexual reproduction the colonies produce individual, free swimming gametes that are isogamous. The method of gamete formation is almost identical to that of daughter colony formation but the gametes separate from one another when liberated from the parent colony. The quadriflagellate zygote takes rest for a short while and secretes a smooth thick wall. The zygote of *G. pectorale* germinates to form a 4- celled coenobium, each of whose cells gives rises to the typical 16-celled individuals in the next generation.

Genus - *Volvox*

Volvox is supposed to be most evolved genus of coenobial form. It is normally found in temporary or permanent freshwater pools and ponds. The colony is spherical and appears as small pinhead on the surface of water.It may be distinguished from all other members of the family by the large number of cells in the colony. The colonies of most of the species are visible to the naked eyes and are strongly phototactic.

The colony is composed of many small vegetative or somatic cells and relatively few larger aflagellate reproductive cells (**gonidia**). All are arranged at the periphery of the coenobium. Cell number in different species ranges from 500 and 50000. Colonies are spherical to ovoid and the biflagellate cells lie in single layer within gelatinous colonial matrix (Fig. 4.7). Each cell is surrounded by a gelatinous sheath of its own. Sheath may be confluent with or distinct from one another. In the latter condition they are angular due to mutual compression and usually hexagonal. Electron microscopic observations reveal that in all species except those with stellate cells *(V. globator)* there is the thickened layer of material of unknown composition lying immediately adjacent to the plasma membrane which is considered to be cell wall by some investigators. Most species have ovoid cells. In some species cells are joined to one another by conspicuous or delicate cytoplasmic strands, a connection which becomes established early in development of colonies.

The individual cell in many species reveals *Chlamydomonas* type organization. Broad intercellular connections give the cells a stellate appearance in others (Fig. 4.7). The whole coenobium moves and rotates by means of flagella of individual cells. Most of the cells in a coenobium are vegetative and do not have the capacity to form new individuals. Each vegetative cell is biflagellate with two contractile vacuoles near the base of the flagella or with 2 to 5 contractile vacuoles irregularly distributed in the anterior end of the cell.

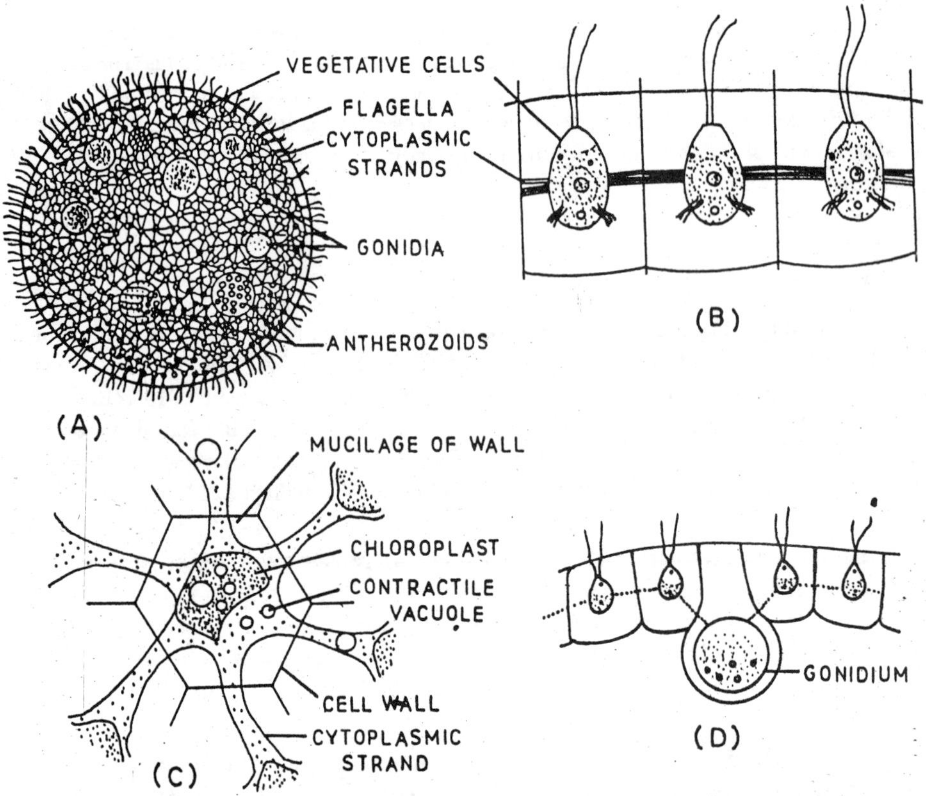

Fig. 4.7. *Volvox* sp. (A) A mature coenobium with vegetative cells, antherozoids and gonidia. (B) Individual vegetative cells of the coenobium with cytoplasmic strands. (C) A cell in a surface view. (D) A portion of coenobium showing the position of gonidium and vegetative cells.

Other features of each cell are cup-shaped to laminate chloroplast, single pyrenoid, centrally located single nucleus and anteriorly arranged eyespot. Eyespots present in the cells of anterior end are somewhat larger than those in cells of the posterior end and thus showing polarity.

For asexual reproduction few cells of *Volvox* coenobium differentiate as asexual reproductive cells **(gonidia)**by growing larger with the disappearance of flagella and gradual thickening. A number of daughter cells are produced by the division and redivision of gonidium which are held together as a new colony. Since the number of gonidia divide at the same time , many new colonies are formed simultaneously. By three successive logitudinal divisions, 8-celled cruciate plate of **plakea-stage** with some what concave surface towards the exterior is formed. Series of divisions then follow. After division ceases, the young colony turns itself inside out by inversion (Fig. 4.8) through a small pore or opening (**phialopore**). The individual cells then develop flagella. Immediately after the development the new coenobium may or may not escape from the parent coenobium.

Sexual reproduction is strictly oogamous. Depending on the species and strain, coenobia may contain either eggs or sperms (dioecious) or they may contain both (monoecious). Both kinds of gametes are present in small numbers and are developed especially from cells of the posterior half of the coenobia. The egg cells are usually dense and darker in appearance than gonidia. They enlarge considerably without undergoing any division and lose their flagella during development. Cells forming male gametes or antherozoids divide successively giving rise to numerous spindle- shaped, biflagellate antherozoids. Non-motile female gametes

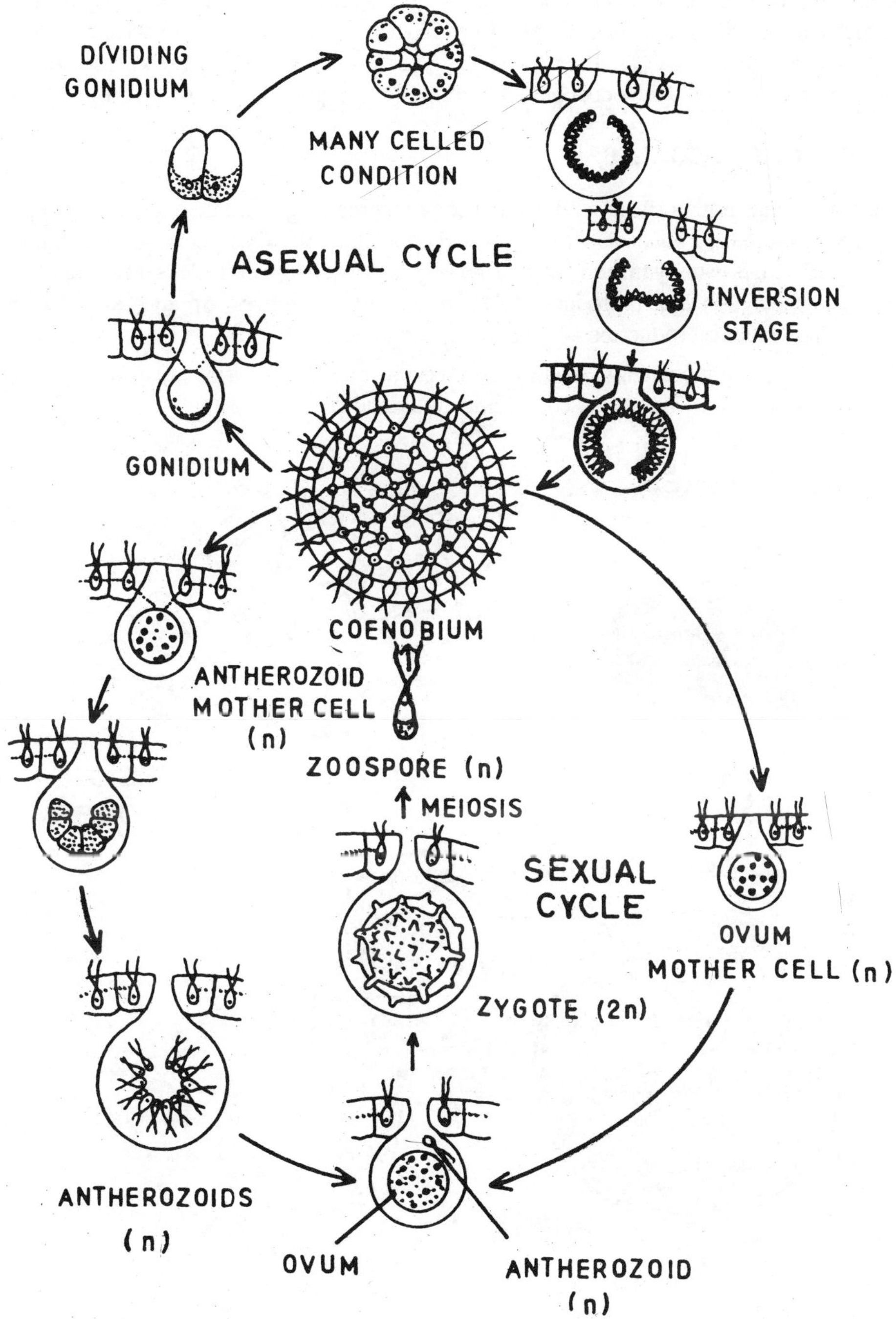

Fig. 4.8. Life cycle of *Volvox* sp.

are much larger than the male gametes and are surrounded by a gelatinous sheath. The groups of sperm are liberated from their parental colonies and presumably by chemotactic stimulation swim to the surface of female colonies (in dioecious species). Finally they dissociate into individual sperms and penetrate the female colonies and fertilize the eggs. The resulting zygotes very soon develop thick walls and become red. zygote divides by meiosis and a new coenobium is formed through the formation of single zoospore.

ORDER 2 - CHLOROCOCCALES

This order contains nonmotile unicellular and colonial green algae. Some of the colonial members are coenobic e.g., *Scenedesmus, Pediastrum* (Fig. 4.9 A,B). Members of this order have a tendency to become multinucleate. Cell wall is usually thin and there is no vegetative cell division. Cells lack flagella, contractile vacuoles and eye-spots. Asexual reproduction is achieved by **zoospores, aplanospores, autospores** or daughter coenobia. Sexual reproduction is by iso- or anisogametes.

These algae are mainly present in freshwater reservoirs and in terrestrial situations. Like Volvocales, the members of this order may be discussed as unicellular and colonial forms.

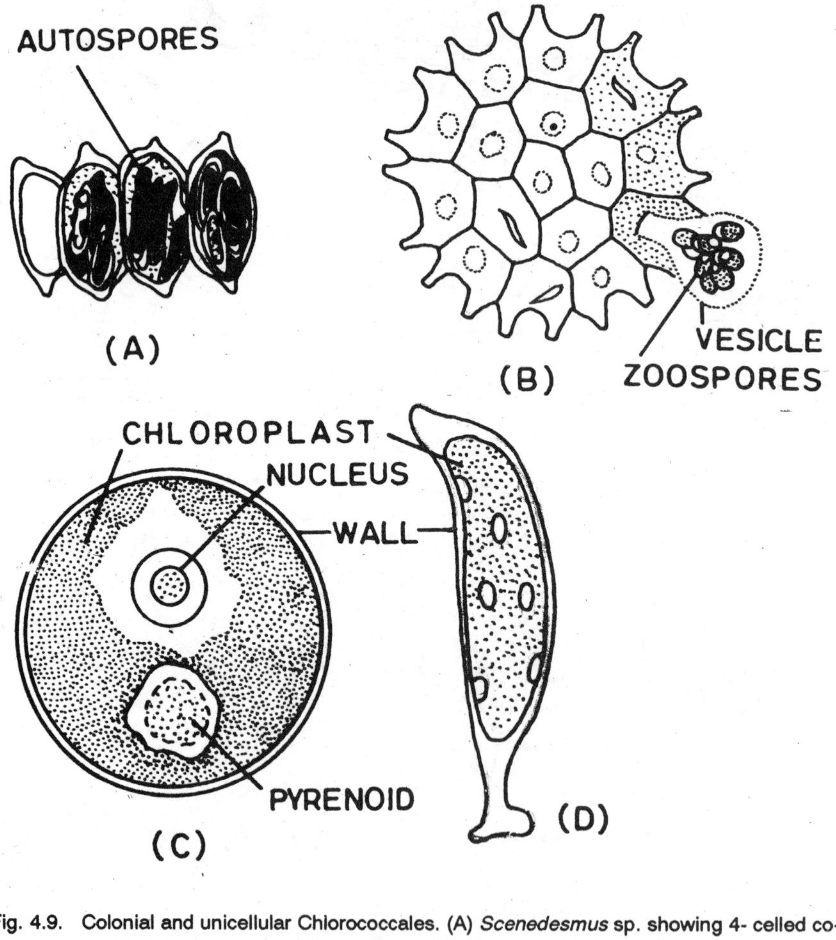

Fig. 4.9. Colonial and unicellular Chlorococcales. (A) *Scenedesmus* sp. showing 4- celled coenobium and formation of autospores. (B) *Pediastrum boryanum* with a vesicle containing zoospores. (C) *Chlorococcum multinucleatum* showing structural details. (D) *Characium* sp. with a basal attachment.

Unicellular Chlorococcales

Most of the unicellular Chlorococcales have spherical cells *(Chlorococcum)* with a well- defined cell wall, a single nucleus and a large parietal chloroplast having pyrenoid (Fig.4.9 C). This structure resembles most species of *Chlamydomonas* except that flagella, eyespots and contractile vacuoles are absent. Asexual reproduction is achieved by the formation of 8 or 16 naked biflagellate, ellipsoidal zoospores. During zoospore formation nucleus divides until 8-16 daughter nuclei are produced and successive cytoplasmic cleavage produces the zoospores. Thus, the cells become multinucleate for a while which is in contrast with the volvocales, where each nuclear division is followed by cytoplasmic cleavage. The tendency of cells to become multinucleate is rather more permanent in *Characium* (Fig. 4.9D).

Family - Chlorellaceae

Genus - *Chlorella*

The genus is ubiquitous and is found in various habitats. It is mainly present as planktonic form in fresh water bodies and on terrestrial habitats rich in organic matter. Some species are common on subaerial habitats. Symbiotic association of *Chlorella* species with lichens and animals *(Hydra)* are also found.

Cells are usually solitary or sometimes aggregated into groups being enveloped by mucilage. The cells are small, spherical or ellipsoidal. Each cell possess a single parietal and usually cup- shaped chloroplast. A pyrenoid may or many not be present in the chloroplast. Cells have thin cellulosic cell wall, centrally placed small nucleus and other eucaryotic organelles. Atkinson et al. (1972) reported **sporopollenin** (a polymer of carotenoid derivative) from the wall of *Chlorella* cell in the ultrastructural studies.

Chlorella is an important tool for physiological experiments. It has widely been used in the study of respiration and photosynthesis. It is also used for the purification of air in space capsule. An antibiotic, **chlorellin** is extracted from this alga. *Chlorella* is probaby the first alga to have been grown extensively in axenic culture.Cells have very high food value with about 50% protein, 20% fat and 20% carbohydrate and are rich in amino acids and vitamins. Because of this, scientists in different countries are trying to cultivate it on mass scale. It may act as food of future.

Reproduction is exclusively asexual which is achieved by the production of autospores. Zoospores are not produced. 4,8 or 16 autospores are produced by the division of mature cells (Fig. 4.10 A-C) and are liberated by rupture of the parent cell wall. These nonmotile spores develop into new individuals.

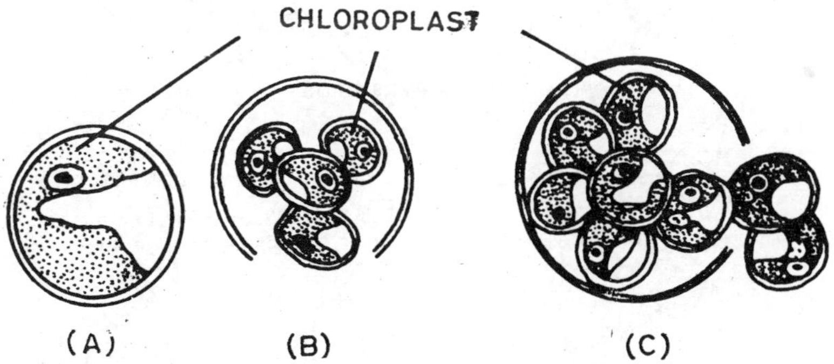

Fig.4.10. *Chlorella* sp. (A) Vegetative cell. (B-C) Formation of autospores.

Colonial Chlorococcales

Varying degree of morphological complexity can be found in colonial Chlorococcales but there is no progressive and complete series of elaboration as found in colonial Volvocales. In the members of Oocystaceae the shape and number of cells in the colony is indefinite. Colonial habit results by non-separation of autospores. In some genera like *Golenkinia* and *Micractinium* colonies are generally more fixed in shape and size. Besides autospore production , oogamous sexual reproduction is present in them. In *Scenedesmus, Pediastrum* and *Hydrodictyon* well defined coenobia are present *Scenedesmus* usually consists of a linear series of 4 or 8 cells (sometimes 2 to 32) joined together by the localised production of mucilage (Fig. 4.9 A).This genus is widely used in physiological studies. Outer cells of which frequently have spines or projections. It is an example of azoosporic Chlorococcales because motile stages are never produced.

Pediastrum is a common planktonic genus which consists of flat plate of cells varying from 2 to 128 depending upon the species. The outer cells have projections on the surface. A number of biflagellate zoospores are produced during asexual reproduction. These are liberated in a vesicle within which they loose their flagella and become arranged in the form characteristic of the parent colony. The biflagellate gametes are produced in the same way as zoospores but are liberated from the vesicle before fusing.

Family - Hydrodictyaceae

Genus - *Hydrodictyon*

Hydrodictyon, the water net is widely distributed in lakes and slow-flowing streams. Thalli float freely as planktons or sometimes remain as benthic form.

Coenobia have cylindrical to broadly ovoid cells united to form a meshwork. Cells are arranged end to end in polygonal configurations in which six members is a common number but cell number in a mesh may vary from 3 to 9. Each coenocytic cells after maturity contains large central vacuoles and the protoplasm which has a reticulate chloroplast with pyrenoids and many minute nuclei (Fig. 4.11).The angles of the cells are formed by the junction of three of the cells.Young cells contain a single nucleus, zonate entire chloroplast with one pyrenoid. Size of the nets is variable which may go up to 1m long. Total number of coenocytes composing a colony may vary from hundreds to several thousands depending on the species.

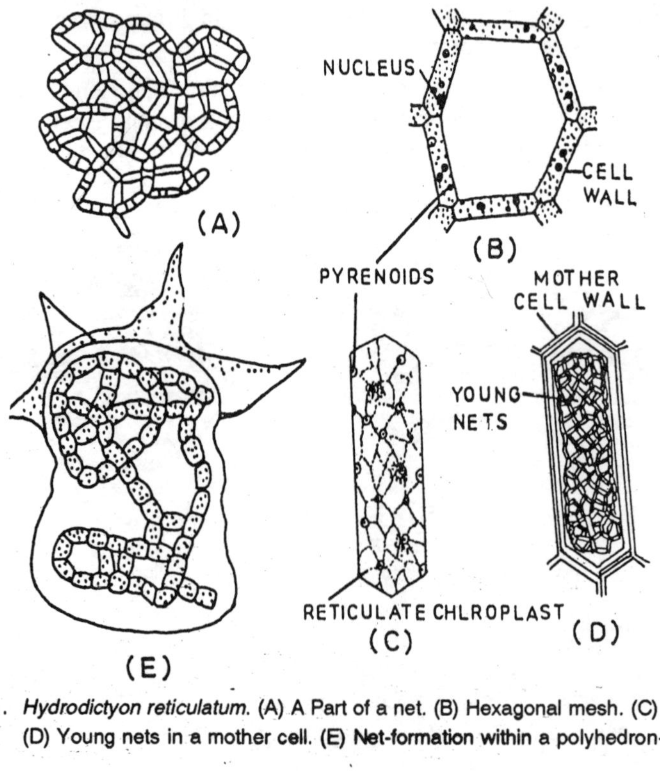

Fig.4.11. *Hydrodictyon reticulatum.* (A) A Part of a net. (B) Hexagonal mesh. (C) A single cell. (D) Young nets in a mother cell. (E) Net-formation within a polyhedron-stage.

Asexual reproduction is achieved by autocolony formation. Autocolonies are developed from biflagellate zoospores that are formed by progressive irregular cleavage of the protoplast of a coenocyte. The zoospores come to rest after swarming within the parent coenocyte and arrange themselves to form a new net. The young net is at first enclosed within the old mother cell wall but this soon gelatinizes and the new net is released out.

The genus reproduces sexually by isogametes and the plants are monoecious. Biflagellate gametes swim freely throughout the cell lumen before they are shed into the water through a pore in the parental cell wall. The resulting zygote, after a short period of rest undergoes reduction division forming 4 zoospores (Fig. 4.12). These zoospores after a period of motility get transformed into non-motile aplanospores, each of which develops in size and form polyhedral cell, known as **polyhedron**. When enlargement ceases, the contents of the polyhedron divide to form many zoospores which ultimately develop into new net.

ORDER 3 - ULOTRICHALES

In this order members are unbranched filamentous or parenchymatous in contrast to previous orders with unicellular and colonial forms. Members may or may not be differentiated into base and apex. Cells have distinct wall enclosing a single parietal chromatophore, pyrenoid and a single nucleus. Unlike Volvocales and Chlorococcales, members of this order have unlimited cell division in the vegetative phase. A septum is formed as an annular ingrowth from the logitudinal wall after a mitotic nuclear division. Cells may contain a large central vacuole (e.g., *U.zonata)* with the result that the cytoplasm is largely peripheral.

The order Ulotrichales is divided into 3 suborders on the basis of cell structure. In suborder **Ulotrichineae**, cells have parietal band-shaped chloroplast with single nucleus. This suborder has four families. The members of Ulotrichaceae have simple unbranched filament and single piece of cell wall (e.g., *Ulothrix).* Family Microsporaceae is represented by cell wall composed of overlapping H - pieces (Fig. 4.13; e.g.,*Microspopa).* In Cylindrocapsaceae, filaments are simple. Elliptical cells are with thick wall and cells in old filament divide obliquely forming multiseriate condition in some part. Thallus in Ulvaceae are large, macroscopic and parenchymatous. Second suborder **Prasiolineae** comprises single famaily Prasiolaceae with single genus *Prasiola.* The thallus is large, macroscopic and membranous and cells have stellate chloroplast. Suborder **Sphaeropleineae** has a single family Sphaeropleaceae with the genus *Sphaeroplea.* The cells of *Sphaeroplea* are much longer than width and contain several transverse cytoplasmic septa separated from each other by large vauoles, several nuclei and a band- shaped chloroplast or a number of discoid chloroplasts are present in each cytoplasmic septum.

Asexual reproduction in these algae may involve fragmentation of filaments and zoospore formation. The zoospores are biflagellate or quadriflagellate with contractile vacuoles (in fresh water habitats) and usually have stigmata. Akinetes are also produced by thickening of vegetative cells during unfavourable conditions.

Sexual reproduction is isogamous, anisogamous or oogamous type and has not been recorded in several genera. The uniting gametes are biflagellate and naked in isogamous and anisogamous conditions. Resulting zygotes assume spherical shape and secrete a wall. Germinating zygotes may produce either zoospores or aplanospores. 4 spores are usually formed at the time of germination but the number may be greater. The genera like Cylindrocapsa and Sphaeroplea exhibit oogamous sexual reproduction. In the former, antheridia produces biflagellate antherozoids and single ovum is formed inside large thick walled oogonium. Fetilization is achieved through a pore in the lateral wall. In *Sphaeroplea,* progressive cleavage of the cytoplasm produces a number of uninucleate protoplasts and each metamorphoses into a naked, spindle-shaped, biflagellate antherozoids. The number of eggs is variable and the antherozoids enter the oogonial cell through a number of pores. Many thick walled zygotes are formed inside one cell which normally undergo a resting period before germination.

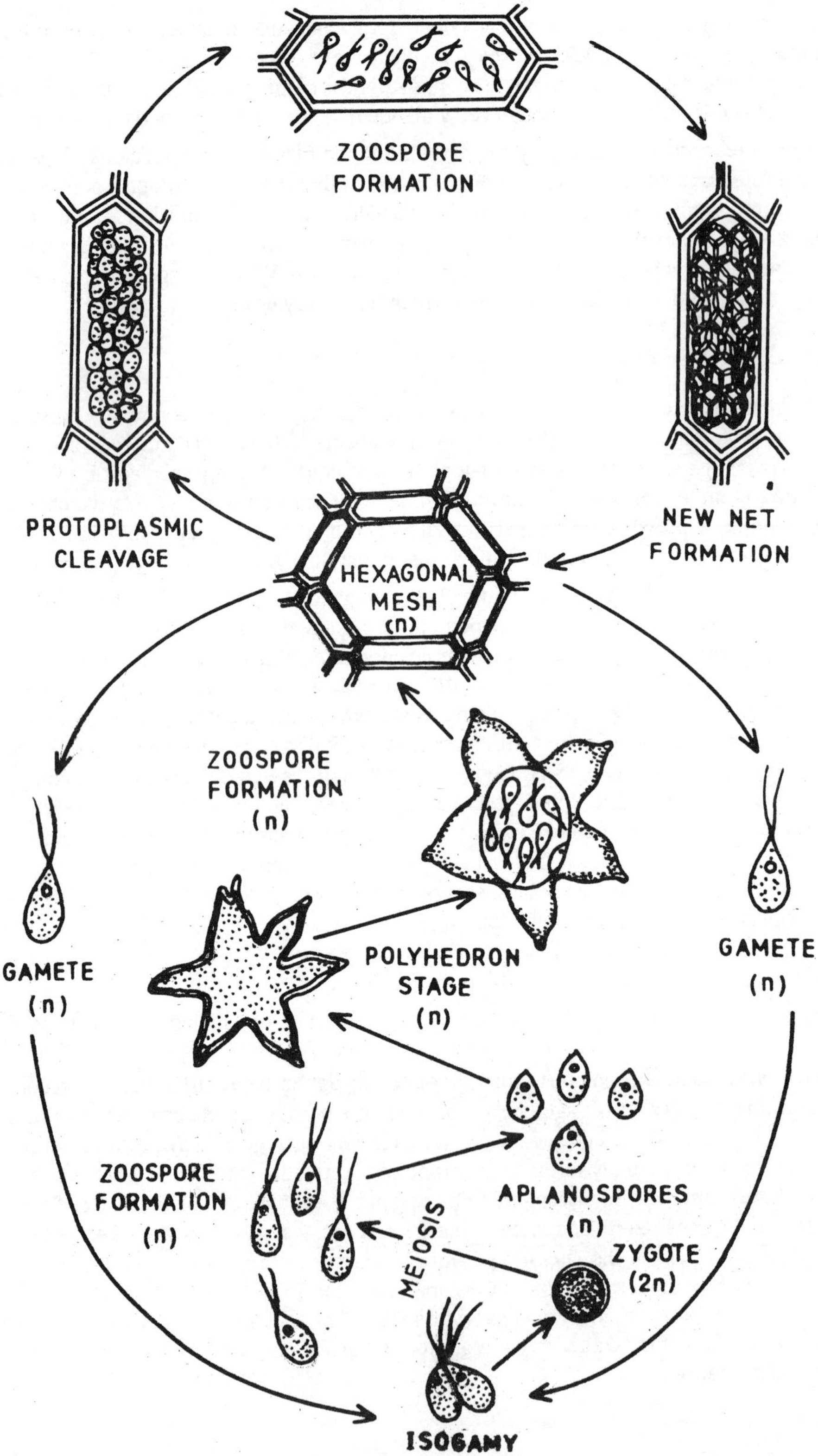

Fig.4.12. Life cycle of *Hydrodictyon reticulatum*.

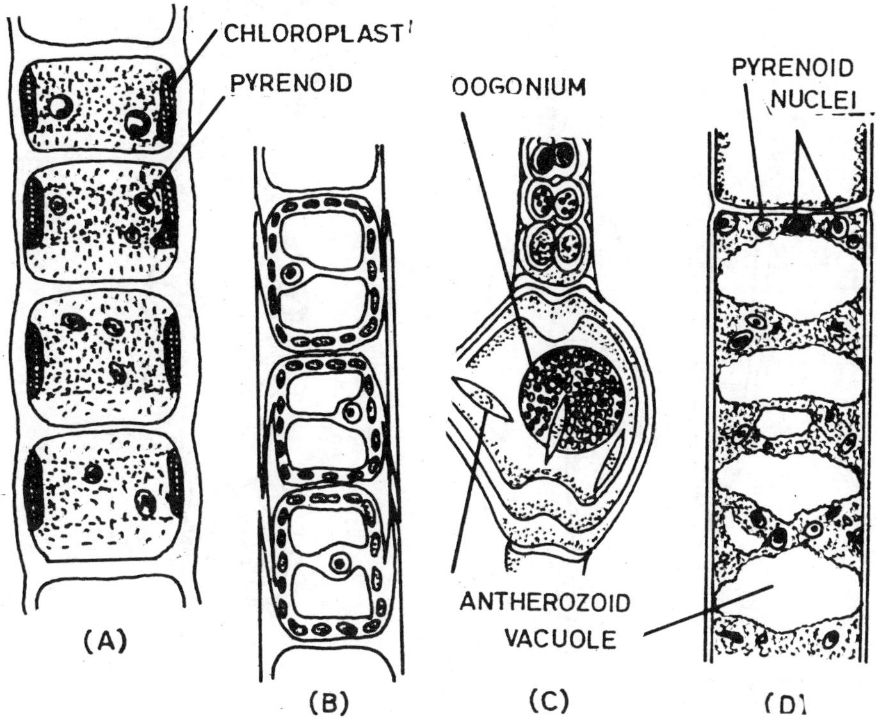

Fig.4.13 Members of Ulotrichales. (A) *Ulothrix zonata* with band-shaped chloroplast. (B) *Microspora quadrata* with overlapping H-pieces. (C) *Cylindrocapsa involuta* with elliptical cells and multiseriate condition. (D) *Sphaeroplea annulina* with vacuoles and annular chloroplasts.

Family - Ulvaceae

Genus - *Enteromorpha*

The members of family Ulvaceae are tubular or membranous. *Enteromorpha* is widely distributed in the intertidal zone of marine habitats where it grows on rocks, shells and other algae. Some species are found in freshwater also.The general habits of *Enteromorpha* vary so greatly in different environments as well as with age that the shape of the thallus is of little help in making specific distinctions.

Mature thalli are always in form of hollow tubes with one celled thick wall (Fig. 4.14). The tubular thalli may be constricted at various loci and are attached to the substrate by rhizoids that may or may not be multinucleate. Tubes of species growing in salt water are usually simple but those growing in fresh water are usually branched and with many lateral proliferations. In ontogeny the agents of reproduction (zoospores, parthenogentic gametes or zygotes) attach to the substrate and by cell division in one direction form uniseriate filament initially which subsequently becomes pluriseriate and later tubular. The cell walls are mainly composed of polysaccharides and some lipids. The two layers of cells separate along their contiguous walls to give rise to the tubular type of organization. The cells are angular by mutual compression and irregularly disposed or with a tendency to lie in vertical series. Each cell is uninucleate and with single parietal chloroplast. Usually there is one pyrenoid but in some species it may be more than one. Excised segments of *Enteromorpha* plant have polarity. From the proximal cut surface rhizoids are germinated and the distal (upper) cut surface gives rise to phtosynthetic papillae.

By an abscission of proliferous shoot, vegetative multiplication is accomplished. Asexual reproduction

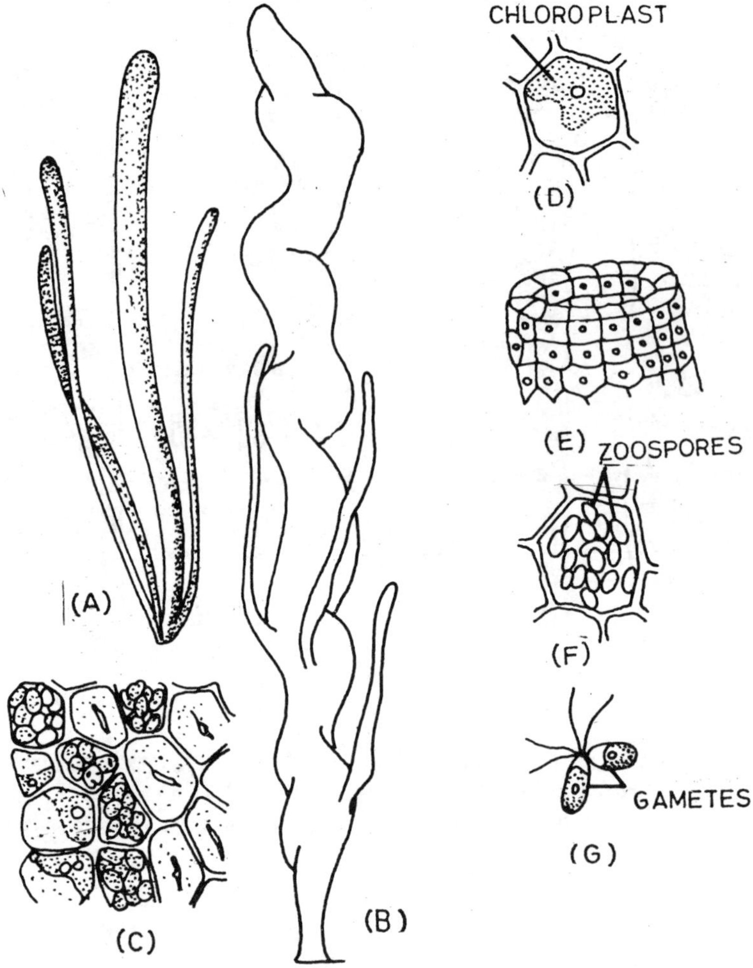

Fig.4.14. *Enteromorpha* sp. (A) A cluster of young plants of *Enteromorpha intestinalis.* (B) Habit of plant. (C) Fertile portion of gametophyte. (D) A cell with chloroplast. (E) Diagram of section through thallus. (F) Cell with zoospores. (G) Gametic union .

is achieved by quadriflagellate zoospores and they may be formed from any cell of the thallus except the lowermost ones. A cell may produce 4,8,16 or 32 zoospores and these escape through a pore or through a slit like opening in the cell wall. After settling down on some object the zoospores germinate.

Biflagellate gametes take part in sexual reproduction and species are heterothallic.Gametic union is isogamous in most of the species but some are anisogamous. In some species parthenogenetic germination of gametes have been recorded. Depending on the species the life cycle may be-(1) diplobiontic and isomorphic with quadriflagellate zoospores and biflagellate gametes, meiosis occurring during zoosporogenesis; or (2) only asexual by means of biflagellate or quadriflagellate zoospores.

ORDER 4 - CLADOPHORALES

Members belonging to this order have multinucleate cylindrical cells united end to end in simple or branched filaments. Each cell contains a reticulate chloroplast with numerous pyrenoids or disk like fragmented chloroplasts.

Many members of this order are freshwater. Some genera are strictly marine while large genera like *Cladophora* have both fresh water and marine representatives. The filaments are mostly attached by rhizoids.

Growth of the thallus may be terminal *(Cladophora, Pithophora)* or intercalary *(Rhizoclonium, Chaetomorpha)*. Cell division is independent of nuclear division. Cells of the filament have thick stratified walls mainly composed of outermost chitin, middle pectin and innermost cellulose. Due to absence of outermost pectic sheath, filaments provide suitable substratum for diatoms. Pyrenoids of Cladophorales are usually surrounded by a conspicuous sheath of starch plates. The nuclei lie internal to the chloroplast. The average number of nuclei in a cell depends more upon the volume of the cell than upon the genus conerned. The multinucleate condition arises ontogenetically in that mitosis and cytokinesis are not closely correlated in time as in uninucleate cells. Thus the agents of reproduction (zoospores, parthenogenetic gametes or zygotes) attach to the substrate and their single nucleus undergoes a series of mitoses without immediately following cytokinesis to give rise to multinucleate cells.

The asexual reproductive cells are biflagellate or quadriflagellate zoospores. Zoospore formation may take place in all cells of a filament but for the holdfast *(Chaetomorpha, Urospora)* or it may be restricted to cells toward the extremities of branches *(Cladophora)*.

Zoospores are produced by progressive cleavage of the protoplast. A small lens-shaped area is formed near the apex *(Cladophora)* or in the equatorial region *(Chaetomorpha)* along with the cleavage of cytoplasm. The gelatinization and gradual dissolution of this area lead to a small circular pore through which the zoospores escape. Presence and formation of akinetes is a regular feature in *Pithophora* and in *Cladophora* it is occasional.

Sexual reproduction is exhibited by biflagellate isogametes. Some species of *Cladophora, Chaetomorpha* and *Spongomorpha* are heterothallic and the gametes unite only when these are produced from different thalli.Some species show parthenogenetic germination of gametes that fail to unite. Zygote may germinate directly into a new plant without a period of dormancy.The nuclear division immediately preceding zoospore formation is meiotic except in *Cladophora glomerata.*

Generic separation has traditionally been made on vegetative characters such as the presence or absence of branches, the kind of branches and the form of the basal cell and its attachment rhizoids. The gradual series of progression can be seen in the series *Cladophora, Pithophora, Rhizoclonium* and *Chaetomorpha* (Fig. 4.15 A-D). The filament of *Cladophora* is relatively much branched than *Pithophora*. The filaments of *Rhizoclonium* are either unbranched or with one or few-celled rhizoidal branches. *Chaetomorpha* is unbranched. Highly stratified wall is found in *Chaetomorpha* with barrel-shaped cells. In *Urospora* (Fig.4.15 E), filaments grow attached to solid substrates by means of basal rhizoidal cells.

Family - Cladophoraceae

Genus - *Cladophora*

Most of the species are quite common in freshwater ponds, lakes and rivers. Some species are found in marine habitat. Members grow generally attached to some substratum such as rocks, living or dead aquatic plants, shells of snails etc. Some species are free-floating.

Filaments are typically branched and usually grow attached by fairly long rhizoidal branches. The branching is usually lateral but often appears to be dichotomous because of the pushing aside (evection) of the original axis of the branch. Sometimes two branches arise opposed to each other and provide trichotomous appearance due to evection. Branches originate from the upper end of a cell just beneath a septum. Cells are elongated (many times longer than the width), thick and are sometimes encrusted with carbonate of lime. Cytoplasmic strands traverse the large central vacuole of a mature cell (Fig.4.16). Each cell has reticulate chloroplast with many pyrenoids or discoid chloroplast are present. Numerous nuclei are embedded within the chloroplast. Wik-sjostedt (1970) found that the basic chromosome number was six and polyploidy was observed in many species. The cells have thick stratified walls consisting of an inner cellulose zone, a median pectic zone and an outer chitinous zone.

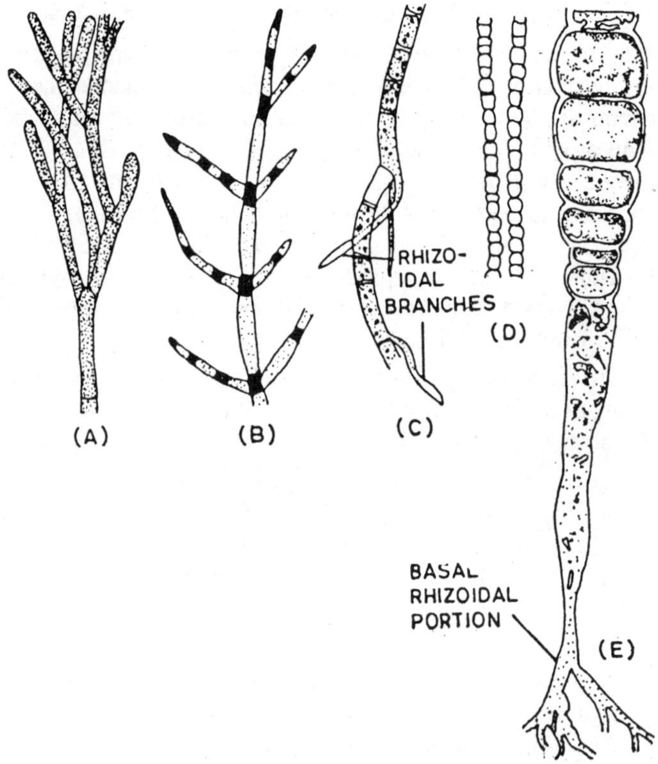

Fig.4.15. Members of Cladophorales. (A) *Cladophora* sp. (B) *Pithophora* sp. (C) *Rhizoclonium* sp. with rhizoidal branches. (D) *Chaetomorpha* sp. (E) Basal part of *Urospora* sp.

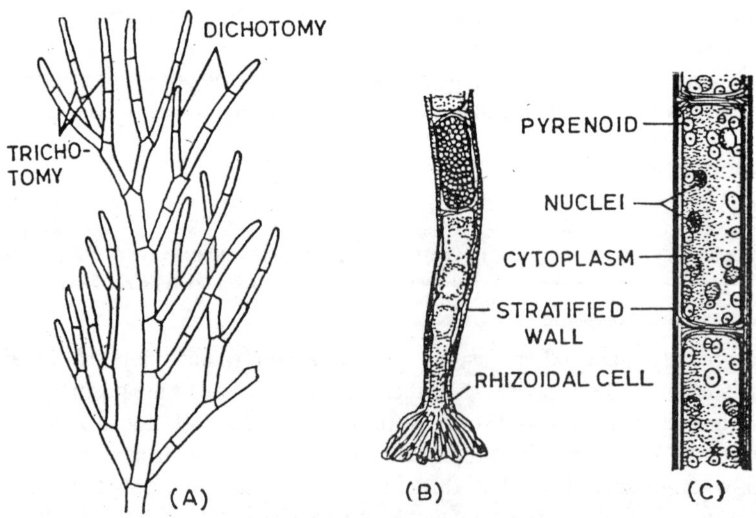

Fig.4.16. *Cladophora* sp. (A) Portion of filament showing dichotomous and trichotomous branching. (B) Lower portion of filament showing rhizoidal cell. (C) Structural details of a cell.

During unfavourable conditions rhizoidal branches of perennial species of *Cladophora* undergo repeated divisions forming short branched filaments, the cells of which are filled with food materials. These cells survive after the main parts of the filaments are dead and grow into new plant after the onset of favourable season. Akinete formation is also seen in some of the fresh water species.

Numerous small, pear-shaped quadriflagellate zoospores are formed during asexual reproduction. This formation is generally restricted to narrow terminal branches which starts from the apex to the basal part of the thallus. Before zoospore formation (zoosporogenesis), the nuclei undergo division and zoospores are formed which directly germinate into new filaments.

Sexual reproduction is isogamous and biflagellate gametes are formed like zoospores. Gametes from two different filaments fuse(heterothallic). Diploid sporophytic plants are formed from zygote without undergoing a period of rest (Fig. 4.17). Haploid zoospores are produced from these diploid plants after meiotic division which ultimately form haploid gametophytic plants having similar morphological features with diploid plant. Hence in the life cycle haploid and diploid plants alternate with each other which are morphologically similar, the difference being in cytological features. Such type of alternation is **homologous** or **isomorphic** alternation of generations. In *C. glomerata,* gametes and zoospores both are formed on diploid filaments and reduction division takes place only during gamete formation. Thus there is no haploidic generation in this species. Such type of life cycle is known as **diplontic** type.

ORDER 5 - CHAETOPHORALES

Members of this order are characterised by the presence of **heterotrichous** filament in which distinct prostrate system gives rise to upright or erect threads constituting projecting system. In some species prostrate system is dominant over projecting system while in others reverse condition is found. The cells are ulotrichalean in organization with cylindrical or curved parietal plastids usually with pyrenoids. In most of the members hair-like projections (**setae**) are found which are usually unicellular and colourless. Sometimes multicellular hair like branches or sheathed cytoplasmic protuberances may be found. Morphological plasticity of the thallus of Chaetophorales is considered to be of great biological significance.

The members may reproduce by fragmentation, biflagellate or quadriflagellate zoospores and by aplanospores. In some members (e.g., *Stigeoclonium)* vegetative cells may thicken their walls as akinetes which survive long periods of dessication. Sexual reproduction varies from isogamy to oogamy.

Members are mainly found in freshwaters. A few members are terrestrial (e.g., *Fritschiella),* epiphytic (e.g., *Aphanochaete)* and parasitic (e.g., *Cephaleuros).*

According to Fritsch's system, this order has been divided into 5 families. In Chaetophoraceae, members are clearly heterotrichous, its genera varying in degree of development of the prostrate and erect systems. Cells are with thin to gelatinous walls, usually uninucleate having a single dissected plate-like or ring-like chromatophore with pyrenoids. Some of the important members of this family are *Chaetophora, Stigeoclonium, Draparnaldia, Draparnaldiopsis, Fritschiella, Aphanochaete* etc. (Fig. 4.18). Members represent two lines of variants (i)erect system dominant and (ii) prostrate system dominant from a balanced type of heterotrichy*(Stigeoclonium).* In *Chaetophora*, feebly developed prostrate system is often represented by loosely connected rounded cells (Fig.4.18 B). Apices of projecting systems are terminated by long hyaline hair like structures known as setae(Fig.4.18 A). The aerial branches are held together with a compact mucilage forming a tough envelope. Each cell contains a single band-shaped chloroplast with several pyrenoids (Fig.4.18 C). In *Draparnaldia,* elaborate aerial system is developed against a very insignificant prostrate system (Fig.4.18 D). The cells of the main branches are much larger than those of the finer branches. The lateral branches may be borne on any cells of the main axis except the lower region from which numerous multicellular rhizoidal branches arise performing the function of anchorage. This is also shown by *Draparnaldiopsis* which has the additional characteristic of differentiation of cells of main axis into short and long cells. The laterals of limited growth here develop only from the short cells, thus showing a differentiation

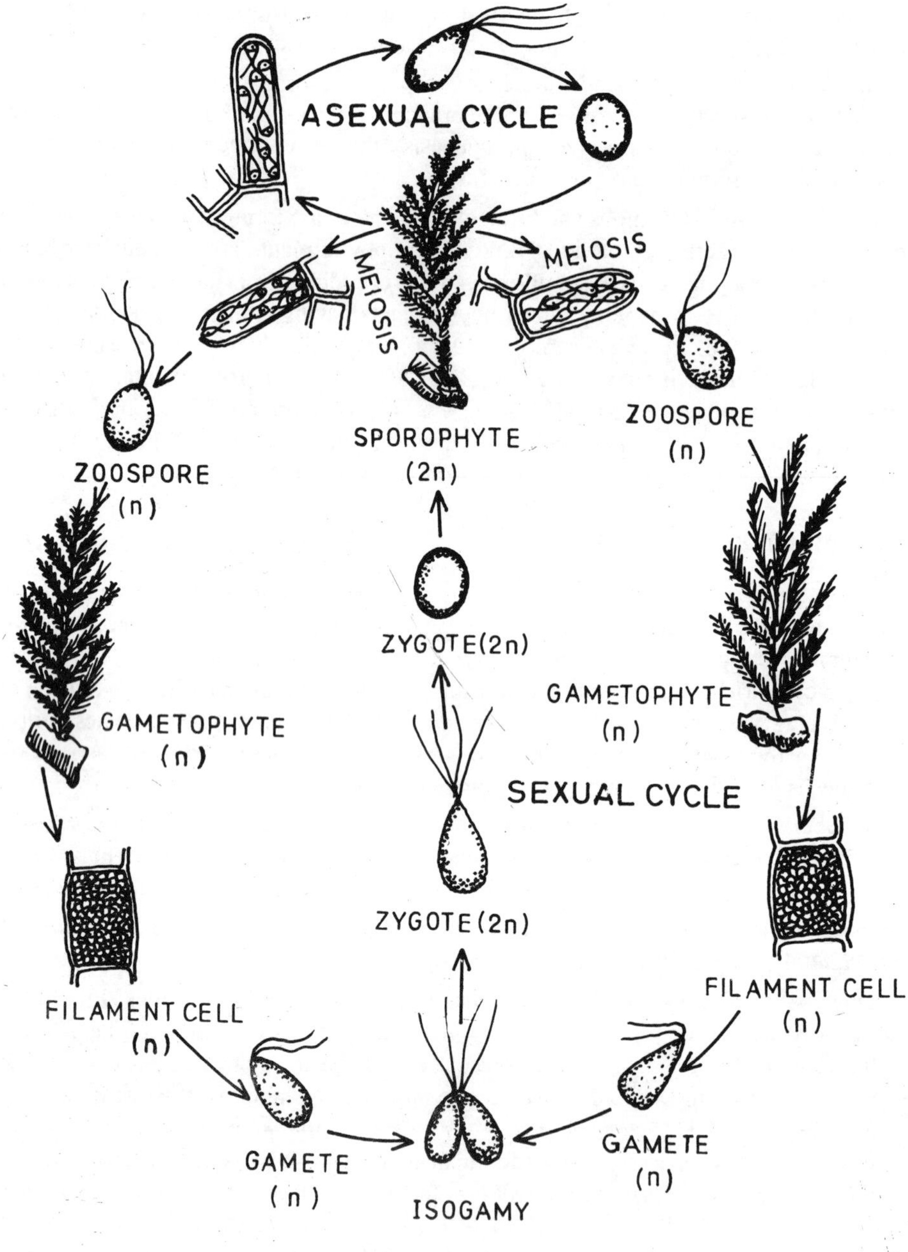

Fig.4.17. Life cycle of *Cladophora* sp.

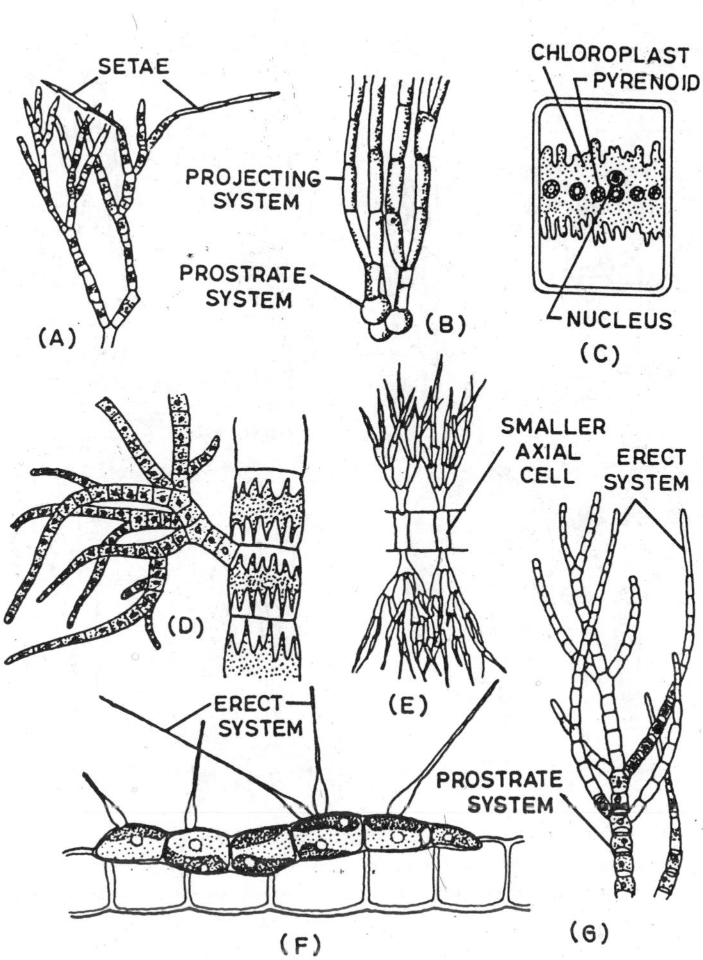

Fig.4.18. Members of Chaetophorales. (A) Projecting system of *Chaetophora* sp. (B) *Chaetophora* sp. showing prostrate system of rounded cells and branched aerial filaments. (C) A single cell of *Chaetophora* showing structural details. (D) *Draparnaldia* sp. showing elaborate erect system. (E) *Draparnaldiopsis* sp. showing differentiation of axial cell into smaller nodal and longer internodal cells. (F) *Aphanochaete* sp. showing reduced erect system. (G) *Stigeoclonium* sp. showing both well developed prostrate and erect system.

into nodes and internodes (Fig.4.18 E). In some species of *Draparnaldiopsis* the branches have the tendency to coil around the main axis giving somewhat corticated appearance.

In many other members one or other system is reduced considerably so that it becomes difficult to identify the heterotrichous nature of the algae. In *Aphanochaete* (Fig.4.18 F) the erect system is represented in form of hairs only whereas it is completely absent from *Endoderma*.

The members of Trentepohliaceae grow as epiphytes and parasites. The widely distributed genus *Trentepohlia* occurs attached to rocks, tree-trunks and even as a common phycobiont of lichens. The branches of aerial system are opposite, alternate or unilateral (Fig.4.19 A-C). The free ends of the apical cells are often covered by **pectose caps**. The cells have numerous band-shaped or discoid chloroplast without pyrenoids. The characteristic orange-yellow colour of the cells is due to the presence of haematochrome. Biflagellate gametes are produced in terminal or intercalary sessile gametangia (Fig.4.19 D). Zoosporangia are borne singly or in clusters and are stalked(Fig.4.19 E) which are disseminated by wind, rain or insects.The development of sessile and stalked sporangia in the life cycle of the alga is suggetive of the possibility of alternation of generation.

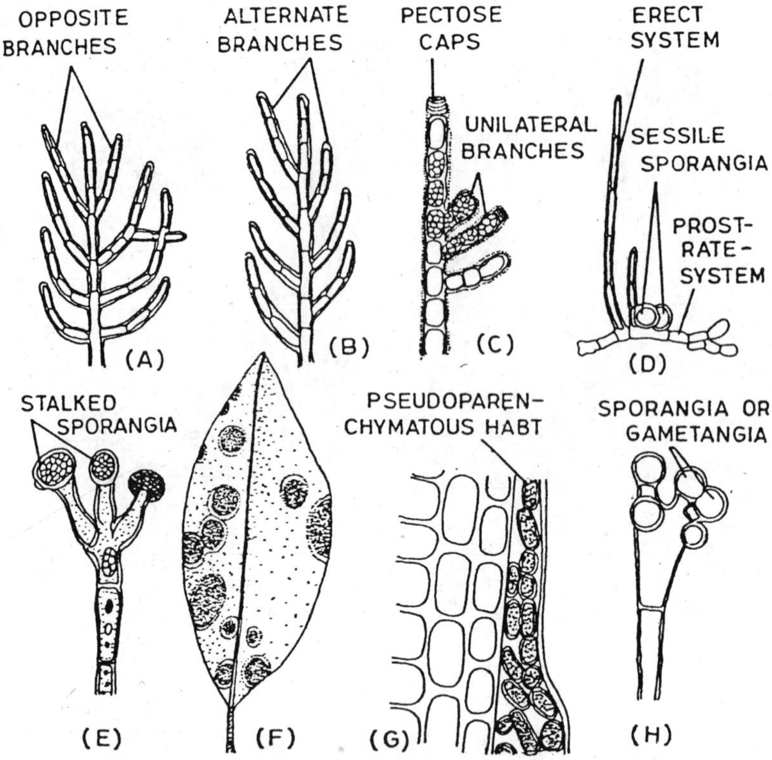

Fig.4.19 (A - E) *Trentepohlia* sp. (A - C) Different types of branches in the aerial system. (D) Presence of sessile sporangia on prostrate system. (E) Stalked sporangia on the terminal end of erect system. (F - H) *Cephaleuros virescens*. (F) Leaf of *Magnolia* with patches of *Cephaleuros*. (G) Vertical section of thallus showing pseudoparenchymatous habit. (H) Fertile filament.

Cephaleuros grows parasitically on leaves, young stems and fruits of tropical and subtropical plants including *Magnolia, Thea, Citrus, Rhododendron* etc. *Cephaleuros virescens* causes the red rust of tea.The infected areas appear as prominent, mostly circular, discolored spots. The thallus of the alga is pseudoparenchymatous (Fig. 4.19 F,G). Clusters of sporangia or gametangia are borne at the tips of sporangiate laterals (Fig.4.19 H) and as in *Trentepohlia* the abscised zoosporangia are dispersed by air currents and give rise to quadri-flagellate zoospores. Biflagellate gametes are also produced within gametangia resulting from enlargement of certain cells in the pseudoparenchymatous portion of a thallus.

The family Coleochaetaceae is represented by a single genus *Coleochaete* which is epiphytic on aquatic plants like *Hydrilla* and *Potamogeton*. Different species represent different degree of heterotrichy. Some species *(C. pulvinata)* has definite prostrate and projecting system while *C.scutata* shows discoid type of thallus showing elaboration of prostrate system only(Fig.4.20 A , B). Vegetative cells have a laminate chloroplast and sheathed bristles or hairs.Asexual reproduction is achieved by biflagellate zoospores while sexual reproduction is oogamous. Female gamete is formed inside oogonium with a neck, **trichogyne** (Fig.4.20 C). Zygote along with reddish-brown pseudoparenchymatous investment is termed as **spermocarp.**

The genus *Chaetosphaeridium* is an example of Chaetosphaeridiaceae. This epiphytic alga has spherical to ovoid cells, each of which bears a single long seta at its distal end (Fig.4.20 D). The family Pleurococcaceae is represented by genus *Pleurococcus*. The cells are globose in shape and occasionally branched. Cells may be unicellular or as many as 4 cells may be joined in a group (Fig.4.20 E,F). Species are

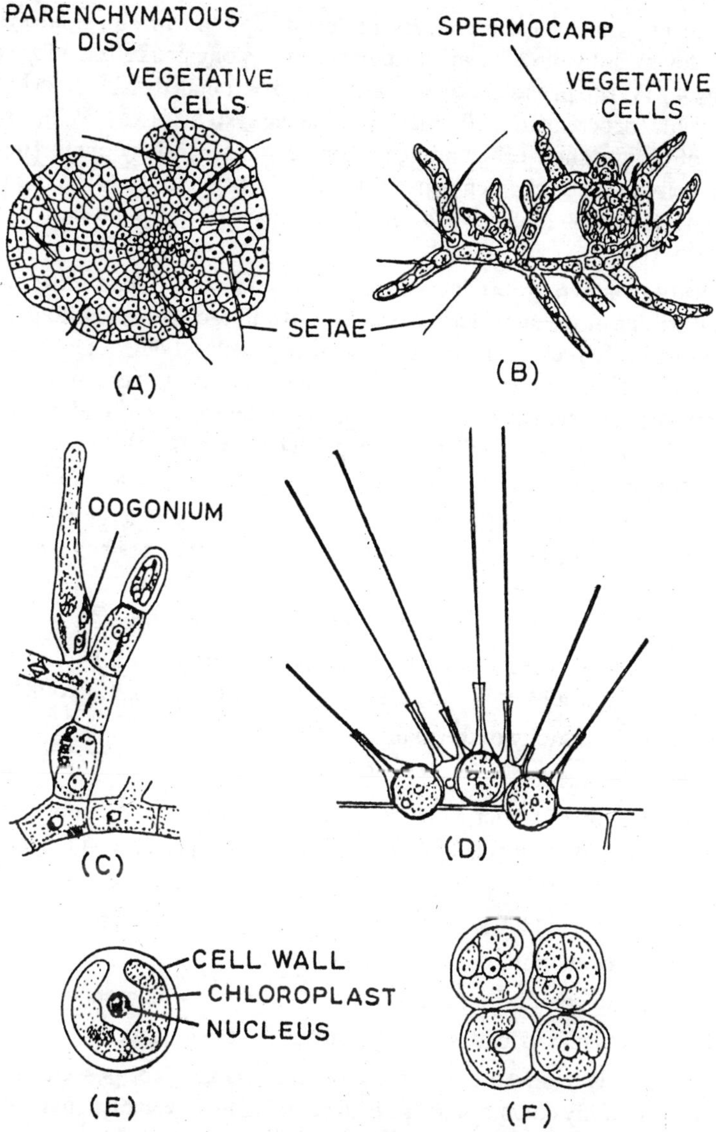

Fig.4.20 (A - C) *Coleochaete* sp. (A) Elaboration of prostrate system into pseudoparenchymatous discoid thallus. (B) A portion of thallus showing setae and spermocarp. (C) Branch with oogonium. (D) *Chaetosphaeridium* sp. showing spherical cell with long seta. (E - F) *Pleurococcus* sp. (E) A single cell with structural details. (F) Group of 4 cells after division.

present in all kinds of damp situations like stones, walls, tree-trunks etc. Vegetative division in three planes is the sole method of reproduction.

Family - Chaetophoraceae

Genus - *Fritschiella*

This alga grows on moist alkaline soil and occasionally on bark of trees. They show great tendency towards terrestrial habit. It has well developed heterotrichous plant body differentiated into rhizoidal system, prostrate system, primary projecting system and secondary projecting system. In **rhizoidal system** ,rhizoid like filaments are septate, downwardly directed and are composed of elongated cells having scanty contents and joined end to end (Fig. 4.21). The **prostrate system** is made up of rounded or irregularly swollen clusters of cells which are responsible for perennation. **Primary projecting system** arises from the prostrate system and consists of a number of upright short celled branched threads. In **secondary projecting system,** elongate branches have longer cells with bright green dense contents. This system is always found above the soil as a tuft. Cells of the plant body are uninucleate and cell contents have curved plate-like chloroplast containing 2 to 8 pyrenoids in each.

All the three types of swarmers are produced in prostrate system only. Quadriflagellate macrozoospores are formed singly, quadri-or biflagellate microzoospores 2 to 4 in each cell and biflagellate isogametes are produced in large numbers in each cell. Some sporophytic plant produces both the macro- and microzoospores whereas gametes are formed from different gametophytic plants.New plants, germinated from the germination of zoospores, are morphologically similar, though cytologically different. Plants produced from quadriflagellate zoospores are diploid i.e., sporophytic plant and those produced by biflagellate zoospores are haploid i.e., gametophytic plant, since reduction division takes place during the formation of biflagellate zoospores.

Isogametes are formed from two different haploid plants. Without undergoing a period of rest, zygote germinates into diploidic plant which is similar in morphology to haploidic plants. Thus this alga represents **isomorphic alternation of generation.**

Points in favour of *Fritschiella* as progenitor of higher land plants:

Order Chaetophorales is supposed to be the order and possibly from *Fritschiella* type of alga,evolution of higher land plants may be traced. Certain specific points in favour of such hypothesis are:

(1) Greater diversity and plasticity of the thallus

(2) Tendency of terrestrial habit

(3) Well developed division of labour-rhizoidal system for anchorage & support, prostrate system for production of reproductive structures and perennation, primary and secondary projecting systems for photosynthesis

(4) Resemblance of thallus with the gametophyte of some species of *Equisetum*

(5) Establishment of cortication in the erect system

(6) Close approach of the reproduction and method of life cycle with archegoniatae.

(7) High concentration of cell sap in prostrate system and its resemblance with rhizome of higher plants.

From these features it is believed that some chaetophoralean plant possibly *Fritschiella* was the ancestral plant which successfully colonized on the land resulting in the establishment of archegoniate plant. Besides *Fritschiella,* many features in *Trentepohlia* and *Coleochaete* are similar to the higher land plants. Mehra(1969) explains that some chaetophoralean member, the pre-land plant ancestors give rise to proto-archegoniate which again give rise to all the land plants (Fig. 4.22).

ORDER 6 - OEDOGONIALES

This order contains single family Oedogoniaceae comprising only three genera i.e., *Oedogonium, Oedocladium* and *Bulbochaete. Oedogonium* is unbranched whereas *Oedocladium* and *Bulbochaete* are branched. These filamentous freshwater, uninucleate algae are characterized by a unique type of cell division and the production of motile reproductive cells with a whorl of flagella at one pole (**stephanokonts**).

These algae are usually present in permanent bodies of water such as ponds or lakes. While growing in moving water, they are seldom in the fruiting condition. Usually fruiting takes place in the summer-time. The plants may be epiphytic on aquatic plants and other algae or they may be free-floating.

Chloroplasts are reticulate, extending from one end of the cell to the other. Many pyrenoids are at the

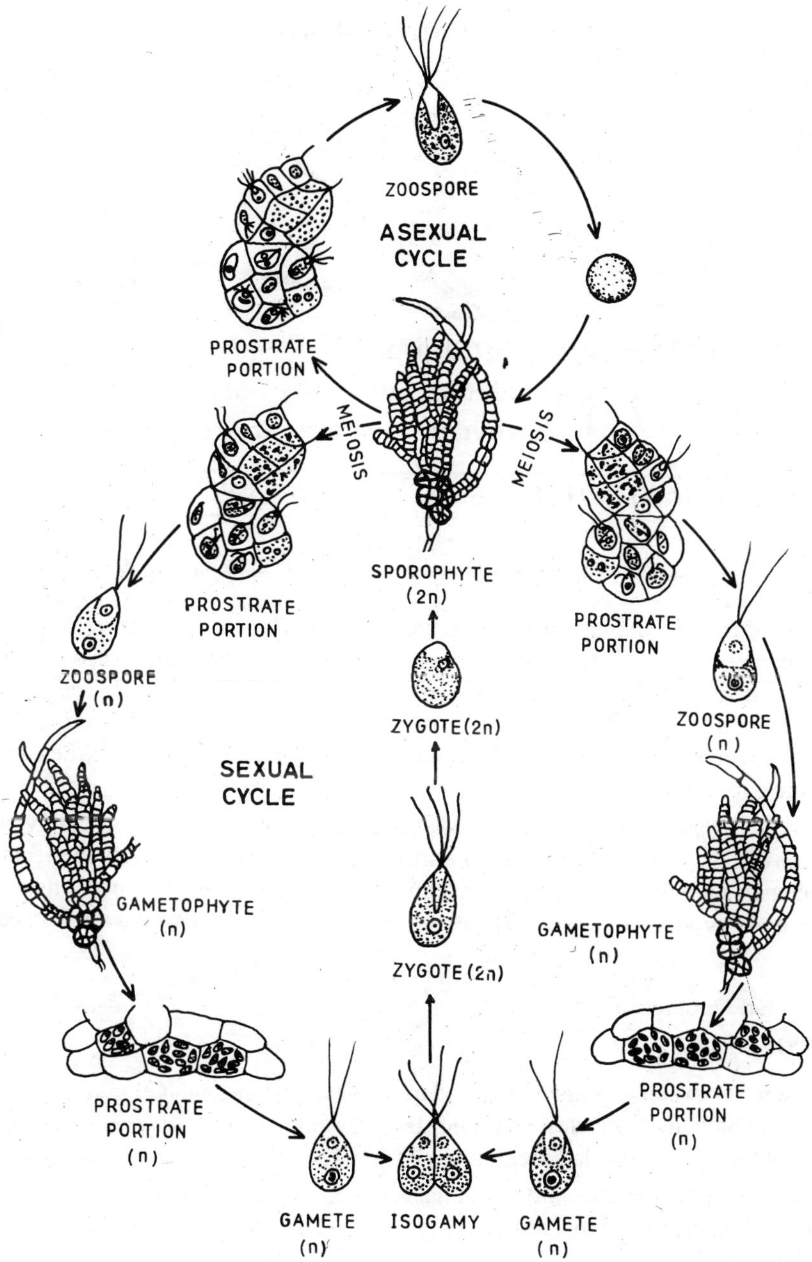

Fig.4.21. Life cycle of *Fritschiella tuberosa*.

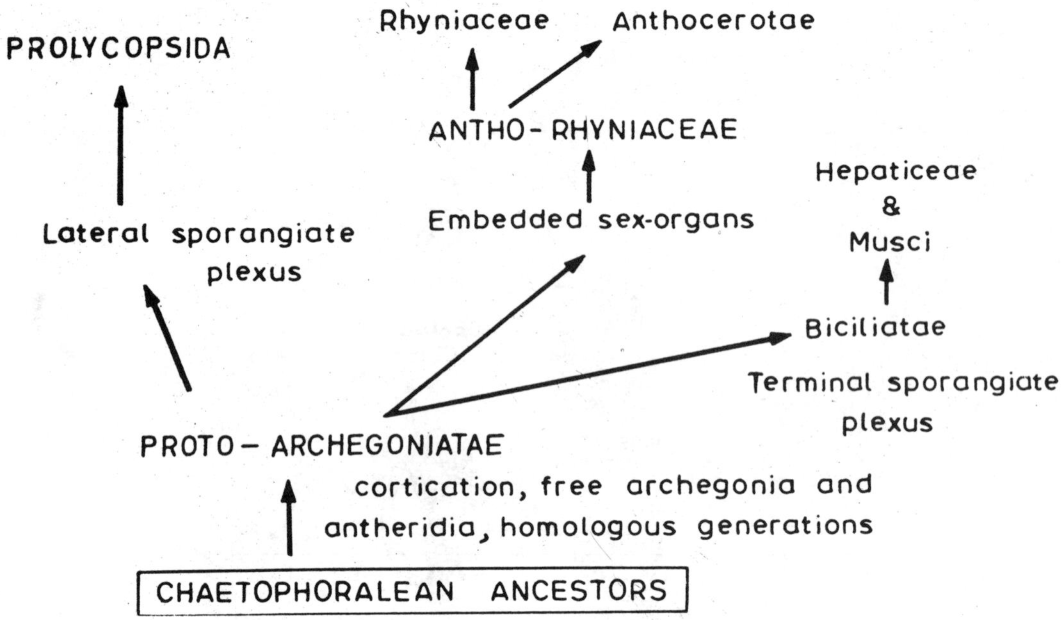

Fig.4.22. Phylogenetic tree showing the origin of land plants (Mehra, 1969).

intersections of the reticulum. Cells are uninucleate and they undergo several successive divisions, each leaving a characteristic hemicellulosic apical cap. Because of the repeated divisions of the protoplast, a series of striation at the end of the cell is formed. Plasmodesmata is present between cells and in *Bulbochaete* this structure is similar to those of higher plants (Fraser and Gunning, 1969). Numerous flagella are attached in the form of 'crown' at the anterior end of the cell in all motile reproductive bodies i.e., zoospores, androspores and the male gametes.

In asexual reproduction large multiflagellate zoospores are formed singly in each cell. Sexual reproduction is always oogamous with some species producing male gametes in special dwarf filament, the **nannandrium** which develops from special spores, the **androspore** produced inside **androsporangium**. Germination of zygote takes place after meiosis and four haploid zoospores are formed which develop into filament.

Family - Oedogoniaceae

Genus - *Oedogonium*

It is a common freshwater alga and is attached to various substrates including stone, wood, and on the stems and leaves of aquatic macrophytes. In young stage, the filaments are invariably attached to the substratum by a basal holdfast cell which lacks chlorophyll (Fig. 4.23 A).

Cells of the filament are short or elongate, cylidrical, may be nodulose, contain a parietal reticulate chloroplast with many pyrenoids, a large central vacuole or several small vacuoles and a prominent parietal nucleus (Fig. 4.23 B).

Cell division may take place from any intercalary cell of the filament except the basal cell. The process of cell division resulting in the formation of cap is unique (Fig.4.23 C-F). At the onset of cell division, the nucleus migrates from about the middle of the cell towards the distal end of the cell and divides mitotically. In early stages of mitotic division, the inner layer of cell wall near the distal end of the cell thickens so that a complete ring of thickened wall material is developed which later becomes grooved. At this stage ring appears U-shaped with thickened inner wall covered by outer wall (Fig.4.23 D). Finally outer layer ruptures

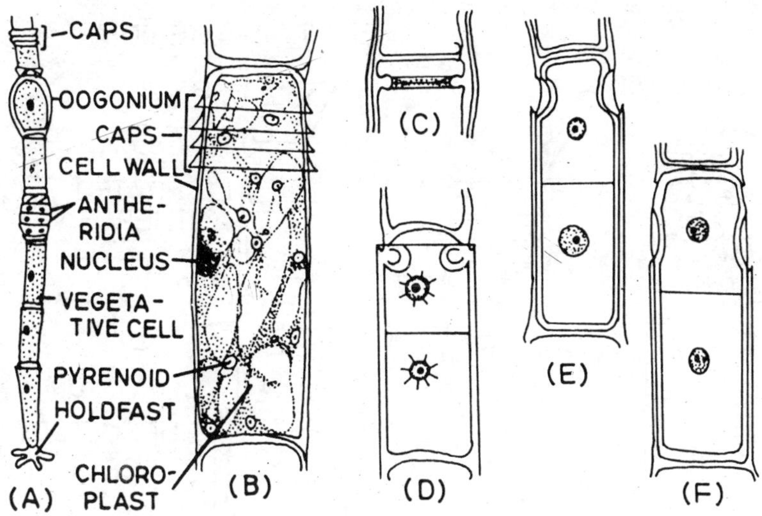

Fig.4.23. *Oedogonium* sp. (A) A portion of filament showing vegetative cells, caps, antheridia, oogonium and holdfast. (B) A single cell with structural details. (C -F) Stages in cell division.

so that the thickened portion becomes pulled out and cell elongates to about double its normal length (Fig.4.23 E) with the distal half composed almost entirely of new wall material. Nuclear division is followed by cytoplasmic cleavage and originally the septum occurs about mid - way along the original cell (Fig. 4.23 D) but later coincides with the point where the new material begins (Fig. 4.23 F). The remains of the original wall material near the distal end of the cell appears as a ring of thickened wall and successive divisions produce a series of rings.

Besides fragmentation of filament, reproduction is exhibited asexually by the formation of zoospores. Large multiflagellate zoospores are produced singly in vegetative cells which are then transformed into zoosporangia. In this process, the whole content of the cell gradually contracts to form a pear-shaped body with colourless spot at one side which develops into characteristic ring of flagella (Fig.4.24 A). Zoospores are liberated by rupture of the cell wall near the caps. After swimming for a while they settle on a suitable substratum and germinate.

Sexual reproduction in *Oedogonium* represents an advanced type of oogamy. Some species are monoecious or homothallic while others are dioecious or heterothallic. In dioecious species two different categories are found: (i) **macrandrous type** with antheridia produced on normal filament and (ii) **nannandrous type** with antheridia produced on dwarf male filament i.e., **nannandrium.**

In monoecious species, antheridia are formed in a series by successive transverse division of the vegetative cells. Contents of each antheridium divide into two which develop into multiflagellate antherozoids (Fig. 4.24 B-C). The oogonia are highly differentiated structures which arise by division of oogonial initial cell into a **supporting cell (suffultory cell)** and oogonium. The oogonium swells into round or oval outline becoming approximately double the size of the vegetative cell. A small colourless area appears at one point on the side of the oogonium through which swarming antherozoid reaches the immobile egg (Fig. 4.24 D-E). **Zygote (oospore)** is formed by the fusion of antherozoid and egg. The oospore develops a heavy wall and undergoes a period of rest. By meiotic division oospore divides into 4 parts which ultimately develop into 4 zoospores. In dioecious macrandrous species, out of these four, two zoospores form male filaments and the other two form the female filaments.

In nannandrous type of sexual reproduction, dwarf male plants are produced from a special type of

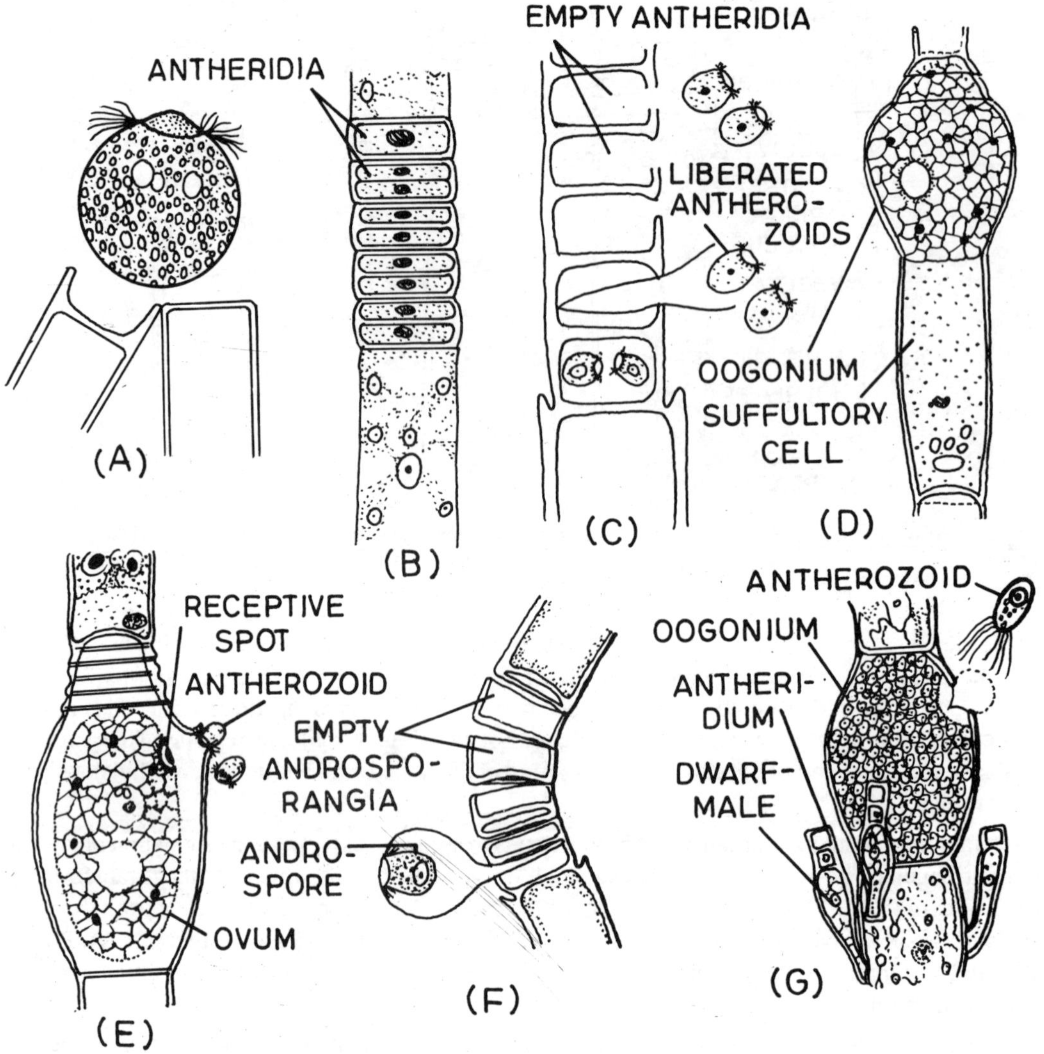

Fig.4.24. *Oedogonium* sp. (A) Liberation of zoospore. (B - C) Formation and liberation of antherozoids. (D) A part of filament showing oogonium and suffultory cell. (E) A stage in fertilization. (F - G) *Oedogonium* nannandrous type. (F) Liberation of androspore from an androsporangium. (G) Portion of a female filament bearing dwarf male filaments on suffultory cell.

zoospores known as **androspores** inside androsporangia. Each androsporangia is transformed into a single androspore having a size approximately intermediate between a normal zoospore and an antherozoid (Fig.4.24 F). Mostly the androsporangia are produced from the same filament which bear oogonia (**gynandrosporous type**). Sometimes they occur on distinct filaments (**idioandrosporous type**). After liberation, androspores germinate and get attached either to the oogonial wall or to the wall of the suffultory cell. Distal end of 2 or 3 celled dwarf male filament (Fig. 4.24 G) functions as the antheridium. Each antheridium produces two antherozoids bearing the characteristic crown of apical flagella. The antherozoids escape by a lid-like opening of the antheridium and enter through oogonial aperture (Fig. 4.25). Further process of germination of new haploid filaments is similar to the macrandrous species. Parthenogenesis is frequent in many species of *Oedogonium.*

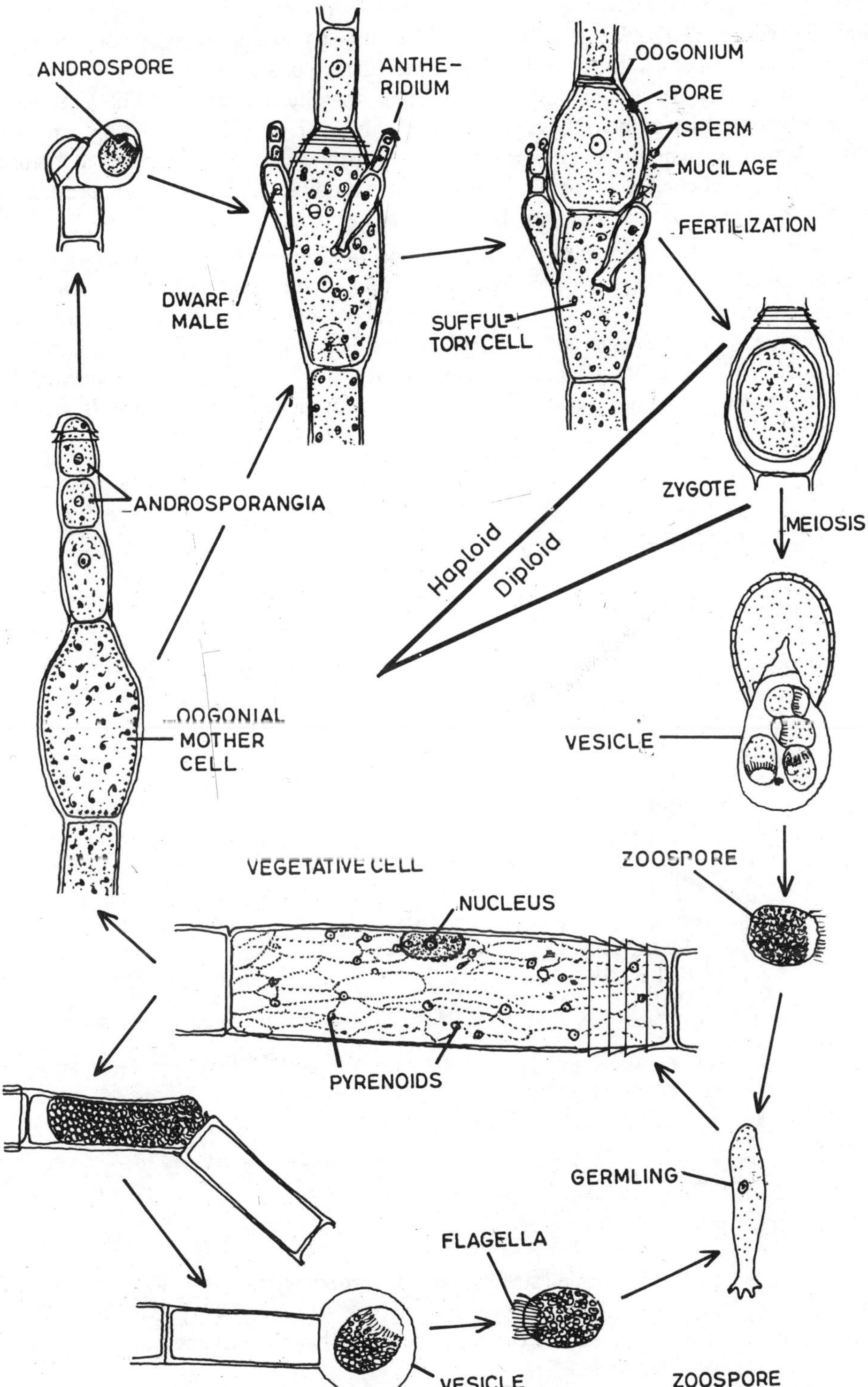

Fig.4.25. Life cycle of a nannandrous species of *Oedogonium*.

Genus - *Bulbochaete*

The branching filaments of *Bulbochaete* grow attached to stones and submerged wood and are mainly epiphytic on aquatic angiosperms and sometimes on other filamentons algae. Filaments are unilaterally branched and with the majority of cells bearing a long seta that is swollen at the base. Cells are broader at the upper end and tranverse striation of the wall is usually found only on the terminal cell of a branch (Fig. 4.26). Basal cell of the filament has a holdfast at its lower end. The bristles lose their protoplasts at maturity. The chloroplast is parietal and reticulate and contains a number of pyrenoids. Cell division is not restricted to the basal holdfast cell as earlier reported but intercalary divisions also occur.

Occurrence of vegetative multiplication from accidental breakage of the filament is rare as compared to *Oedogonium* and thus free floating mats of *Bulbochaete* are not found usually. Like *Oedogonium*, asexual reproduction is accomplished by zoöspores. Sexual reproduction is oogamous. Majority of the species are nannandrous with gynandrous androsporangia. A few species are macrandrous. The oogonia are always solitary and subtended by two suffultory cells. Pickett-Heaps (1974) has contributed significantly to the understanding of the organization, reproduction, cell division and bristle formation in *Bulbochaete* using electron microscopy.

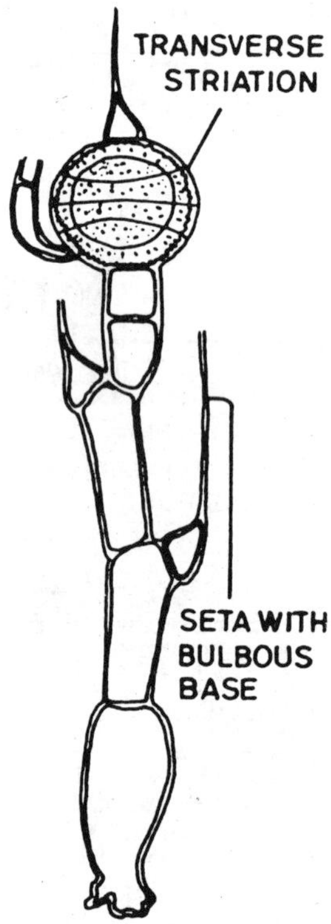

Fig.4.26. *Bulbochaete* sp. showing bulbous seta and basal hold fast.

ORDER 7 - CONJUGALES

The characters of this order are so specific and striking that they are sometimes assigned to a special class of algae, the Conjugatophyceae (Fott, 1971) or Zygnemaphyceae (Round, 1971). Members are mainly freshwater having wide distribution. This order contains both unicellular and filamentous forms and can be differentiated by three important features.

1. They show marked elaboration of chloroplast.

2. Flagellate stages are completely absent.
3. Algae reproduce sexually by fusion of amoeboid gametes during a specialised conjugation process.

The cells are uninucleate and characterised by prominent type of chloroplast (spiral, stellate or axile) with pyrenoids. Many of the members are capable of cell movement by excretion of mucilage. Cell wall is composed of three layers (Mix, 1975). The outer layer is mucous and amorphous while the two inner layers are fibrillar. Pores are present in the walls of the Desmidiaceae but not in other two families. Filamentous forms are included in the family Zygnemataceae and the unicellular species are divided between two families, Mesotaeniaceae and Desmidiaceae.

No asexual reproductive bodies are found in this order. During sexual union amoeboid gametes either fuse with the help of **conjugation papillae** and **conjugation tube** or the gametes can move from their parent cells into the medium and fuse.

Family - Zygnemataceae

The cells of this family are permanently united into unbranched filaments and the cell wall lacks pore. The cylindrical cells show great variety of chloroplast and the character is the basis of classification of the family.

1. **Single axile plate- like chloroplasts** occur towards the centre of the cells of *Mougeotia, Mougeotiopsis* etc. (Fig. 4.27 A, B). In those cells with flat axial chloroplast there is a marked chloroplast orientation in response to light intensity. In *Mougeotia* chloroplast presents a surface view under low light intensities whereas an edge view is presented under high light intensities (Haupt and Schonbohm, 1970). Phytochrome system is partly responsible for the chloroplast orientation.

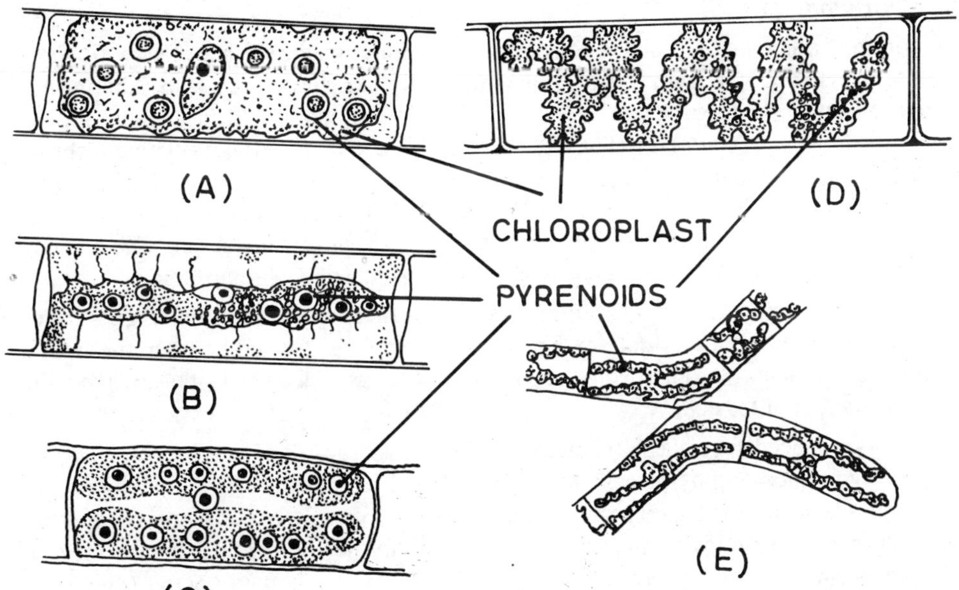

Fig.4.27. Chloroplast types is Zygnemataceae. (A - B) *Mougeotia* sp. showing single plate-like chloroplast. (A) Face - view. (B) Side - view. (C) Side - view of *Sirocladium* sp. showing two parietal plate - like chloroplast. (D) *Spirogyra* sp. showing spiral chloroplast. (E) *Sirocladium* sp. showing longitudinal bands of chloroplast.

2. **Two parietal plate-like chloroplasts** are found in *Sirocladium* (Fig. 4.27 C) which can be seen in side view.

3. **Stellate axile chloroplasts** are found in *Zygnema* (2 in each cell). In *Zygogonium* the chloroplasts tend to be more rounded and are not regularly stellate.
4. **Helical** or **spirally** twisted chloroplast are found in *Sirogonium* and *Spirogyra* (Fig 4.27 D). In *Spirogyra* the chloroplast is either narrow with an entire margin or is broad with a serrate margin which runs in an anticlockwise direction around the cell. In *Sirogonium* (Fig. 4.27 E) the bands of chloroplast run in a longitudinal direction. During septum formation a septum of pectic material is formed first and a layer of cellulose is later formed each side of it. Thus a middle lamella is formed which is superficially like the higher plants. The nucleus is usually in the centre of the cell in a band of cytoplasm which traverses the central vacuole. Nucleoli are large and complex and nucleolar organising chromosomes are involved in the organisation of nucleolus.

Fragmentation is the common mode of vegetative propagation. Akinetes and aplanospores are also reported. There are two types of conjugation, **lateral** and **scalariform**. In lateral conjugation, gametes of the same filament fuse while cells of different filaments form gametes in scalariform conjugation. When the gametes fuse in the conjugation tube, it is called as isogamous *(Mougeotia and Debarya)*. When only one gamete moves to fuse, it is anisogamous (*Zygnema* and *Spirogyra*). In the anisogamous processes, the female gametangia and the zygospores occur in the same filament. During conjugation, the filaments become attached to each other along their lengths with the help of mucilage. Papillae is formed first from one filament and later from the other. The wall between the papillae breaks down and a conjugation tube is formed. Nuclear fusion between the gametes occur immediately after cytoplasmic fusion. Zygospores germinate by undergoing meiosis and three of the four daughter nuclei abort. New plant is regenerated after the rupture and germination of thick walled zygospore.

Genus - *Zygnema*

This common freshwater alga is found in considerable quantities on the surface of slow flowing water. The genus is closely related to *Spirogyra*, differs in having two stellate chloroplasts with radiating protuberances and each with a central pyrenoid. Cells of *Zygnema* are short, cylindrical and the nucleus is located between the chloroplasts in a mass of cytoplasm (Fig. 4.28). The cells are surrounded with gelatinous sheath of varying thickness, depending on the species. Generally the filaments of *Zygnema* are few centimeters long in contrast to the much longer filaments of *Spirogyra*. The cell wall is composed of two cellulose layers. The characteristic pectose layer and the replication of cross-wall are practically absent. The nucleus appears to have a chromatin reticulum, a nucleolus and one nucleolar organizing chromosome.

Vegetative cells other than basal cell are capable of division. Nuclear division is followed by the division of chloroplasts and the cytoplasm. Transverse septum is formed between the two nuclei. Each daughter cell receives one chloroplast with the pyrenoid. Nucleus which lies in the lateral position migrates to the median position of the cell.

Fragments produced by accidental or mechanical injury are capable of forming new filament. A number of species survive desiccation by means of akinetes. Sexual reproduction by conjugation occurs in most species and both scalariform and lateral conjugation have been observed. In scalariform conjugation, both the amoeboid isogametes move towards each other and unite in the conjugation tube to form a zygospore (**physiological isogamy**). In some species, one of the two fusing gametes may be active (male) and may move upto female gamete bearing cell where it fuses with passive female gamete. This process is **physiological anisogamy**. In some species, **sexual dimorphism** has been found in which the cells producing passive gametes are longer and broader with larger chloroplasts and pyrenoids than the cells producing the active gamete. The wall of dormant zygotes (zygospores) are blue or brown, depending on the species. After germination, the zygospore gives rise to a single vegetative filament. A zygospore may be developed where conjugation does not occur.

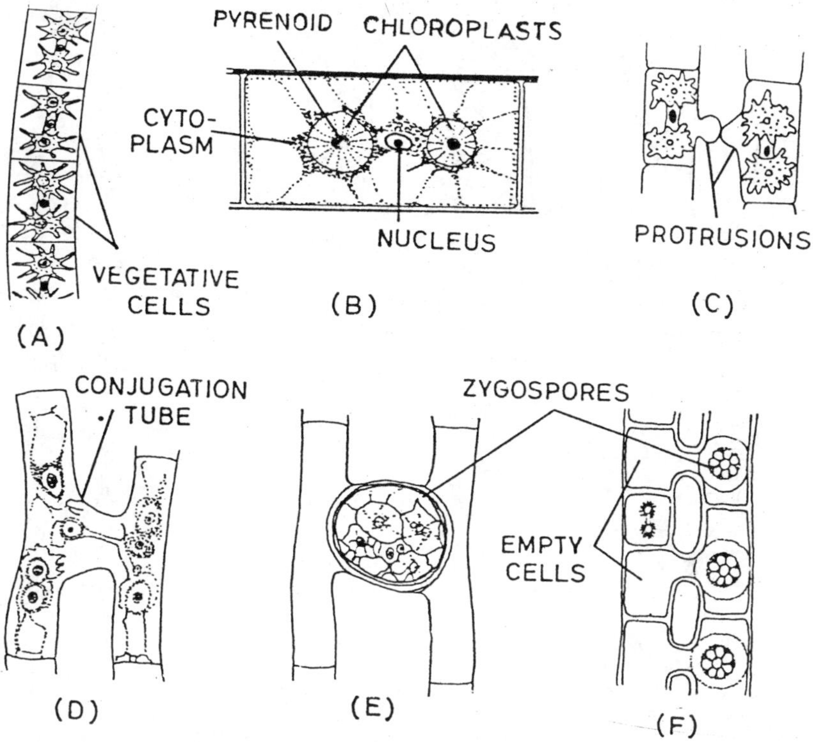

Fig.4.28. *Zygnema* sp. (A) A portion of a vegetative filament. (B) A cell with structural details. (C - F) stages of conjugation.

Family - Mesotaeniaceae (Saccoderm desmids)

Members of this family are basically unicellular, even though in some instances they are joined together to form filaments. The walls do not have pores. These are called as **saccoderm desmids** and unlike the **placoderm desmids** they do not have a new semicell formed after cell division. Chloroplast structures are similar to those of Zygnemataceae and the nucleus is centrally located in the cell. Flat axile plate like chloroplast with many pyrenoids is found in *Mesotaenium* (Fig. 4.29 A,B), parietal spiral band-shaped like *Spirogyra* in *Spirotaenia* (Fig. 4.29 C) and a pair of stellate chloroplast like *Zygnema* in *Cylindrocystis* (Fig. 4.29 D). Asexual reproduction is uncommon and sexual reproduction is like other two families of the order. These organisms are indicator of unpolluted water as they prefer to grow in low magnesium and calcium contents.

Family - Desmidiaceae (Placoderm desmids)

These **placoderm desmids** have two distinct halves or **semicells** by a median constriction, the **sinus** and joined by a connecting zone, the **isthmus.** Their cell walls contain pores. The cells may be solitary, joined end to end in filamentous colonies, or united in amorphous colonies. Desmids of this group are complex and very attractive (Fig. 4.29 E-J). Due to **polymorphism** (different shapes with the same organism), the taxonomy of many of these desmids become complicated.

Though members of this family show a high degree of specialisation of chloroplast but unlike different types found in other families only an axile chloroplast in each semicell is most common. Localised exudation of mucilage through the pores of cell wall might be related to gliding of desmids. The number of chloroplasts varies and each is associated with pyrenoid. The conspicuous nucleus lies at the equator of the cell in the region of isthmus. These desmids show different patterns of symmetry when they are viewed in polar aspect.

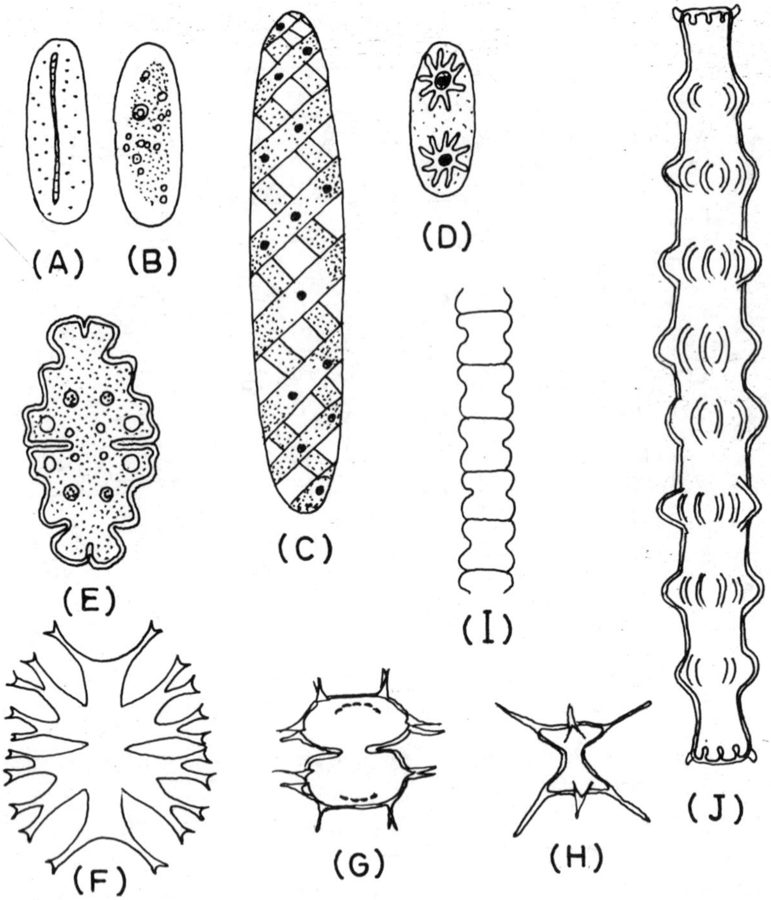

Fig.4.29. Members of Mesotaeniaceae (saccoderm desmid) and Desmidiaceae (placoderm desmid). (A - B) Axile plate -like chloroplast of *Mesotaenium* sp. showing high and low intensity orientation. (C) *Spirotaenia* sp. showing spiral chloroplast like *Spirogyra.* (D) *Cylindrocystis* sp. showing a pair of stellate chloroplast like *Zygnema.* (E) *Euastrum* sp. (F) *Micrasterias* sp. (G) *Xanthidium* sp. (H) *Staurastrum* sp. (I) *Spondylosium* sp. (J) *Pleurotaenium* sp.

Some have forms of beautiful complexity (e.g., *Micrasterias,* Fig. 4.29F).

The mode of cell division in unconstricted species resembles that of Mesotaeniaceae but the process is complex in constricted species having two semicells. Commonly isthmus elongates and 2 semicells are forced apart. A membrane and gradually a wall is developed across the isthmus. Thus one semicell consists of old cell wall material and the other of new.

Asexual reproduction is rare and sexual reproduction takes place between conjugating cells embedded in mucilage. Conjugation canal may or may not develop. In the filamentous members conjugation canal are characteristically present. In most of the species the amoeboid gametes unite in the conjugation canals and develop dark zygotes with ornamented or smooth walls. Nuclear fusion does not occur until immediately before germination. Germination begins with reduction division of the nucleus. Two of the 4 nuclei degenerate and the remaining develop into new individuals.

The desmids are good indicators of relatively unpolluted water, low in calcium and magnesium with a slightly acidic pH.

Genus - *Closterium*

This is a planktonic common desmid found in freshwater aquatic habitats. Cells are solitary, elongate with no median constriction, narrowed towards both poles and distinctly lunate or arcuate. The semicells contain single axile chloroplasts that may be radially ridged (Fig. 4.30A). A prominent nucleus lies between the two plastids at the equator of the cell. Near each pole of a cell, there is a vacuole containing one or more granules of barium sulphate (Brook, 1980). Mucilaginous material is secreted through the pores of the cell walls. According to Mix (1969), cell wall of *Closterium* is three- layered, an amorphous outer layer and a fibrillar primary and secondary wall. The pyrenoids are usually few in number and arranged in an axial series but sometimes they are numerous and irregularly distributed in chloroplast. Nucleus has a conspicuous nucleolus and a well-defined chromatin network.

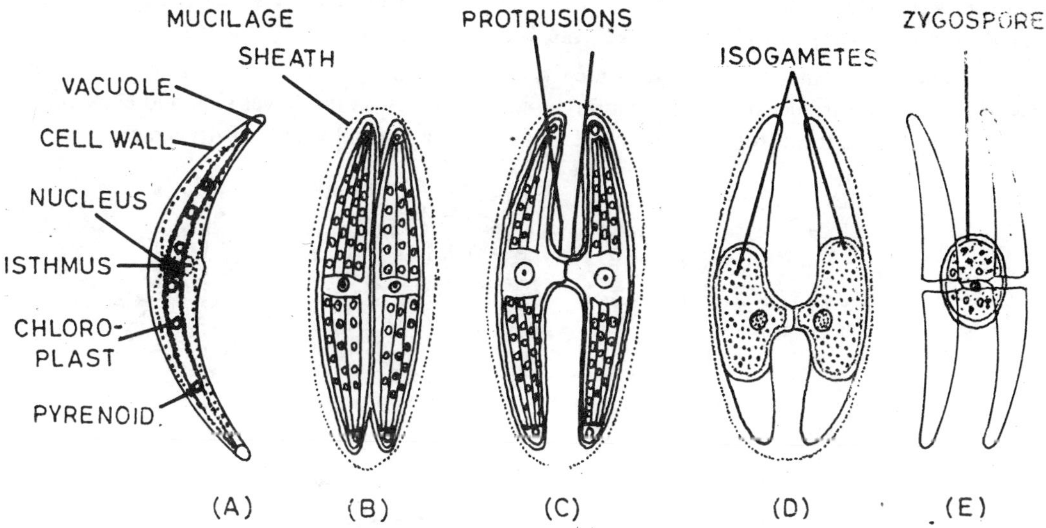

Fig.4.30. *Closterium* sp. (A) Individual cell with structural details. (B - E) Stages in sexual reproduction.

During cell division, the nucleus undergoes mitosis in the centre of the cell and this is followed by cytokinesis. The daughter nuclei migrate to the equator of new semicells in which each plastid divides into two and the nucleus assuming a position between the two halves. Meanwhile, each gradually develops a new semicell and these secrete new cell walls.

During conjugation, two compatible individuals become associated in pairs (Fig. 4.30 B-E) and are enveloped by mucilaginous sheath. They arrange themselves parallel to each other and develop conjugation papillae which merge and enlarge to form a broad canal. Two gametes fuse together forming zygospore which undergoes a period of rest by developing a thick wall. Zygospore germinates by meiotic division and form 4 haploid nuclei. Out of these, two degenerate and the remaining form two new cells.

ORDER 8 - SIPHONALES

Members of this order comprise coenocytic plant body with simple or branched filaments or aggregated to form some what elaborate structures. In the simpler forms, plant body is differentiated into rhizoidal and assimilatory portions. In some species, ring like thickenings at certain points divide the cavity into separate chambers. Small discoid chromatophores and diploid nuclei are many with or without pyrenoids. Many genera have two characteristic xanthophylls, **siphonein** and **siphonoxanthin** which are not found in other green algae. Cytoplasm forms a thick lining layer and surrounds a continuous large central vacuole filled with a cell sap. Many members have abundant deposition of calcium carbonate and are responsible for calcareous deposits in oceans and for many fossil remains. Though members are extremely varied with regard to size and external form, they are all fundamentally similar in that the entire plant body is single multinucleate cell with numerous discoid chloroplasts.

Reproduction is both asexual and sexual. Asexual reproduction is achieved by zoospores or aplanospores and sexual reproduction varies from isogamy to oogamy. Gametes are formed in undifferentiated or well developed gametangia.

Members are mainly found in marine habitat of tropics but some are freshwater also. Different phycologists have classified this order in different ways. According to Fritsch (1935), however, there are 9 families in Siphonales. These are : Protosiphonaceae, Caulerpaceae, Derbesiaceae, Dasycladaceae, Codiaceae, Valoniaceae, Chaetosiphonaceae, Phyllosiphonaceae and Vaucheriaceae. The genus *Phyllosiphon* of Phyllosiphonaceae is an endophytic alga which occurs in leaves and petioles of Araceae.

Most common members of this order are *Caulerpa, Bryopsis, Codium* etc. *Caulerpa* has a creeping rhizome like portion from which arise small rhizoidal branches on the lower side and erect branches of complex and variable form from the upper side (Fig. 4.31 A). Central vacuole is traversed by a number of

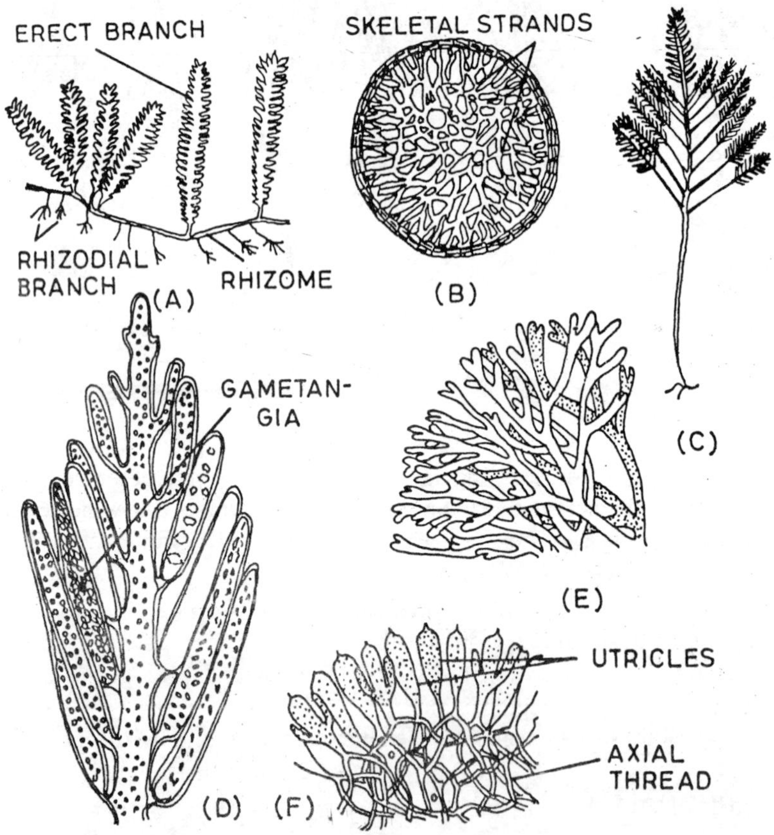

Fig.4.31. Members of Siphonales. (A - B) *Caulerpa* sp. (A) Showing creeping rhizome, rhizoidal and erect branches. (B) Section of rhizome showing skeletal strands. (C - D) *Bryopsis* sp. (C) Showing bipinnate branches. (D) A portion of apex showing gametangia. (E - F) *Codium* sp. (E) Showing pseudoparenchymatous habit. (F) Differentiation of axial threads and utricles.

projections from the cell wall. These longitudinal and transverse skeletal strands are responsible for increasing the rigidity of the plant body (Fig. 4.31B). *Bryopsis* has a rhizoidal portion with erect system showing extremely regular branching in a precise pinnate or bipinnate manner (Fig. 4.31 C,D). The lower region of the axis is naked and the upper apart bear pinnae of variable length. Each pinna is constricted at the point of origin. In *Codium,* large elaborate pseudoparenchymatons thallus is strengthened by interwining of the filaments and tight packing of utricles (Fig. 4.31 E,F). Central medulla consisting of axial threads frequently having no chloroplasts and cortex are usually differentiated. In the cortical region the axial threads turn at right angles and the tips swell to form the utricles.

Family - Dasycladaceae

Genus - *Acetabularia*

The genus with the distinct appearance (*L.acetabulum,* vinegar cup) presents an example of highly specialized thallus organisation of the family. It is commonly known as "mermaid's wine glass". It is very well studied genus because of its great role in the studies of morphogenesis.

The upper umbrella-like structure is composed of many gametangia attached together (Fig. 4.32 A). These gametangia are connected to the primary branch by a constriction at the base which form **corona.**Some vertical sterile hairs are present in the middle of the umbrella (Fig. 4.32 B). Structures placed between the

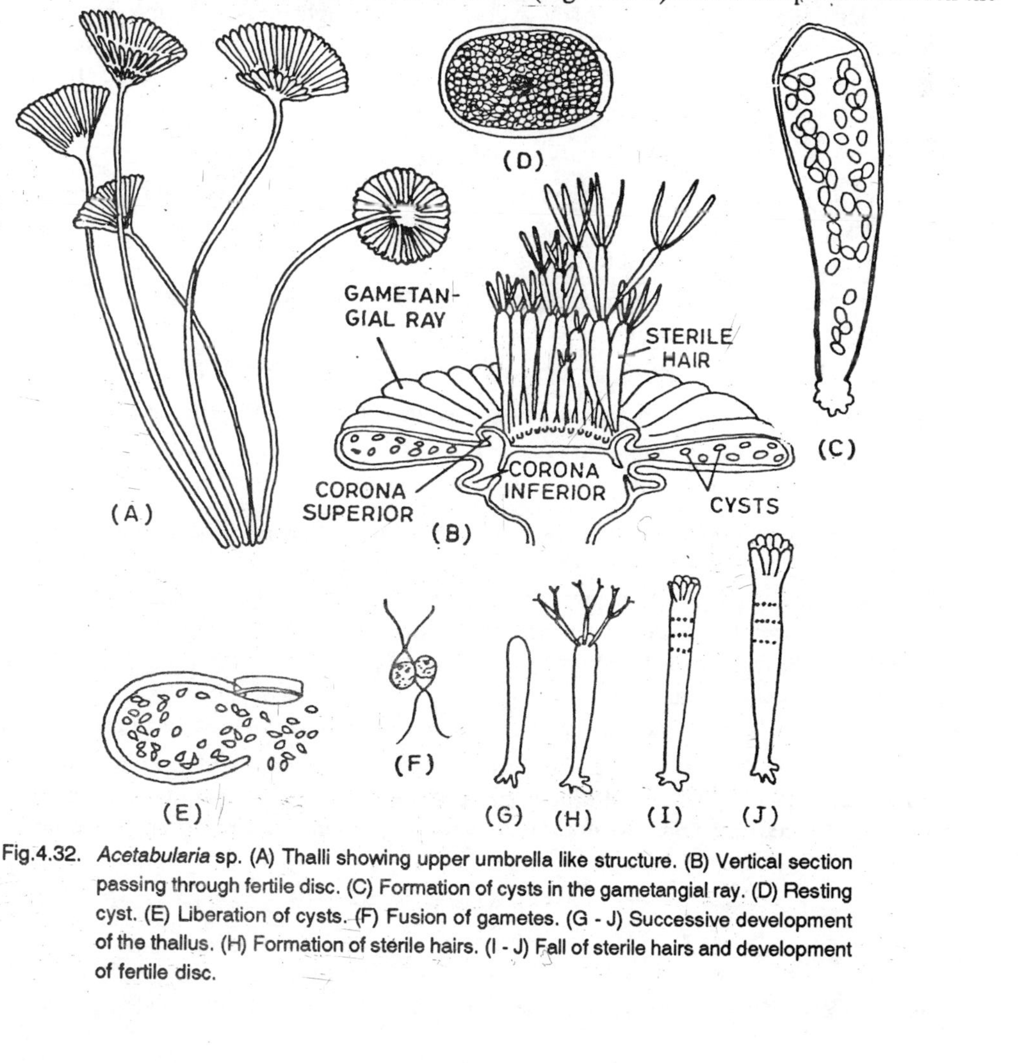

Fig.4.32. *Acetabularia* sp. (A) Thalli showing upper umbrella like structure. (B) Vertical section passing through fertile disc. (C) Formation of cysts in the gametangial ray. (D) Resting cyst. (E) Liberation of cysts. (F) Fusion of gametes. (G - J) Successive development of the thallus. (H) Formation of sterile hairs. (I - J) Fall of sterile hairs and development of fertile disc.

gametangia are termed as **'corona superior'** and **'corona inferior'**. All these structures are in continuation with main axis without having any septa. The whole thallus has a single diploid nucleus located in one of the lobes of the rhizoid and during cyst formation it may divide repeatedly and reach the upper gametangia by the protoplasmic currents. Chloroplasts have no starch but many carotene granules are found.

During cyst formation basally located diploid nucleus divides mitotically and migrate into upper gametangia where the protoplast gets fragmented into many uninucleate structures surrounded by a thick wall. The so formed cysts (Fig. 4.32 C,D,E) have the upper lid. The nuclei undergo meiosis with or without having a rest. The resulting nuclei produce biflagellate isogametes (Fig. 4.32 F). The resulting zygote germinates to form a rhizoidal system (Fig. 4.32 G-J).

The two species that have been used most extensively for graft experiments are *A. crenulata* and *A. mediterranea.* If a nucleate portion of one species is grafted to an enucleate portion of the other species, a cap is formed that shows a greater resemblance to the species of the portion containing the nucleus, although their is still some indication of the influence of the species of the enucleate portion. If that first cap is removed prior to maturation and a second cap is allowed to develop, this second cap will be a pure expression of the species of the portion containing the nucleus (Puiseux-Dao, 1970). These support the contention that morphogenetic materials of nuclear origin reside in the cytoplasm and influence the formation of a cap of a particular shape.

From the time of germination of the zygote to the time right before final maturation of the cap, a single nucleus is maintained in the rhizoidal portion and this so-called primary nucleus increases greatly in volume. Lampbrush-type chromosomes have been observed in it and there is some evidence that it is polyploid.

ORDER 9 - CHARALES

Because of their highly developed vegetative and reproductive features it is widely believed that members of this order should be elevated to the rank of either class (Charophyceae) or division (Charophyta). Members differ from the Chlorophyceae in following ways.

1. The plant body consists of an erect axis differentiated into nodes and internodes with laterals of limited growth at the nodes.
2. The reproductive organs are complex and are surrounded by a layer of sterile tissue.
3. The motile male gamete, antherozoid is an elongated biflagellate spiral structure unlike other Chlorophyceae.
4. The zygote germinates into a protonemal stage.

Thus the members commonly known as **stoneworts** or **brittleworts** are unique group of nonvascular hydrophytes occupying an isolated position between the Chlorophyceae and the Bryophyta (Grambast, 1974). The morphology of the members is similar to Chlorophyceae and particularly with *Draparnaldiopsis* (Chaetophorales) while the structure of sex organs, sperm and germination of zygote have resemblance with Bryophytes. The presence of chlorophyll **a** and **b** and storage starch in their plastids and their totally aquatic habitat have led some botanists to assign them as an order or class of Chlorophyta. Although there are eight genera in this order but *Chara, Nitella* and *Tolypella* (Fig. 4.33 A-C) are the most common ones.

Members are generally present in freshwaters but some are reported from saline waters. Many are heavily calcified with growths on the bottom of lakes leading to the formation of **marl** ($CaCO_3$ and $MgCO_3$ deposits) and hence the common name, the stoneworts. They are anchored to sandy or muddy substrates with the help of well developed rhizoids.

Family - Characeae

Genus - *Chara*

Chara is a submerged aquatic alga consisting of an erect branched axis attached to the substratum by an

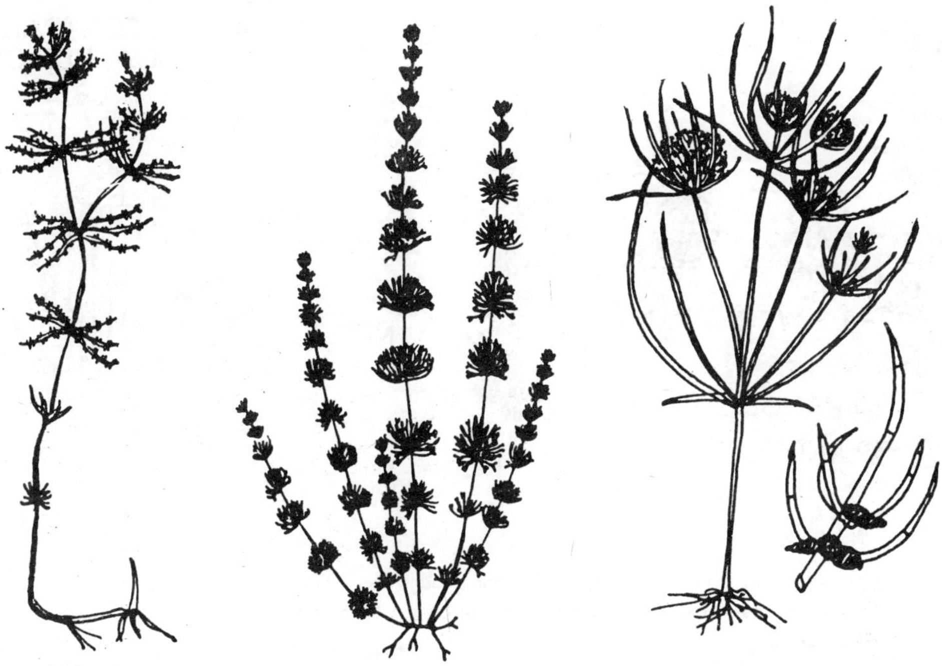

(A) (B) (C)

Fig.4.33. Common members of Charales showing their habits. (A) *Chara* sp. (B) *Nitella* sp. (C) *Tolypella* sp. with details of branchlet and position of sex organs.

elaborate system of multicellular rhizoids. Plant body is erect, slender and flexuous. It may attain a length of 30 cm or more and resembles in appearance to a miniature horse-tail. The erect system is differentiated into well-defined nodes and internodes. In both *Chara* and *Nitella* the nodes consist of several small isodiametric cells in a cluster of two central cells surrounded by 6-20 peripheral cells. In *Chara* the single central internodal cell is surrounded by a **cortex** of smaller elongated cells, half of which are derived from the node above and half from the node below while in *Nitella* the internode consists of a single elongated cell.

From each node, a whorl of short **laterals**, often spoken as "leaves" arise from one of the peripheral cells (Fig. 4.34 A). The number, degree of branching and length of these laterals are characteristic to particular genus and species. The laterals of *Nitella* are simple or branched uniseriate filaments whereas the struture of laterals resemble that of the main axis in *Chara*. Lateral branches having nodes and internodes just like main axis may arise axillary to the leaves. Internodal cell is several times longer than the node. The multicellular rhizoids with oblique septa developed from the lowest node are not differentiated into nodes and internodes. Their functions are in anchoragement and absorption.

Growth of the main axis and its branches is initiated by a single, large, dome-shaped apical cell which cuts off segments toward the posterior or bottom face of the cell (Fig. 4.34 B). Each segment divides transversely into a biconcave upper nodal initial cell which divides and redivides to give rise to a node and a biconvex lower internodal initial cell which remains undivided but elongates considerably to form the internode.

Cytoplasm in the nodal cell is dense and without conspicuous vacuole. Nucleus is single, located towards the centre of the cell and a number of small discoid chloroplasts without pyrenoids are distributed throughout the cytoplasm. In the internodal cell the central region is occupied mainly by a large central vacuole with the small chloroplasts arranged in well-defined longitudinal series in the peripheral cytoplasm. It divides amitotically to form a large number of nuclei in the lining layer. The lining layer of cytoplasm is

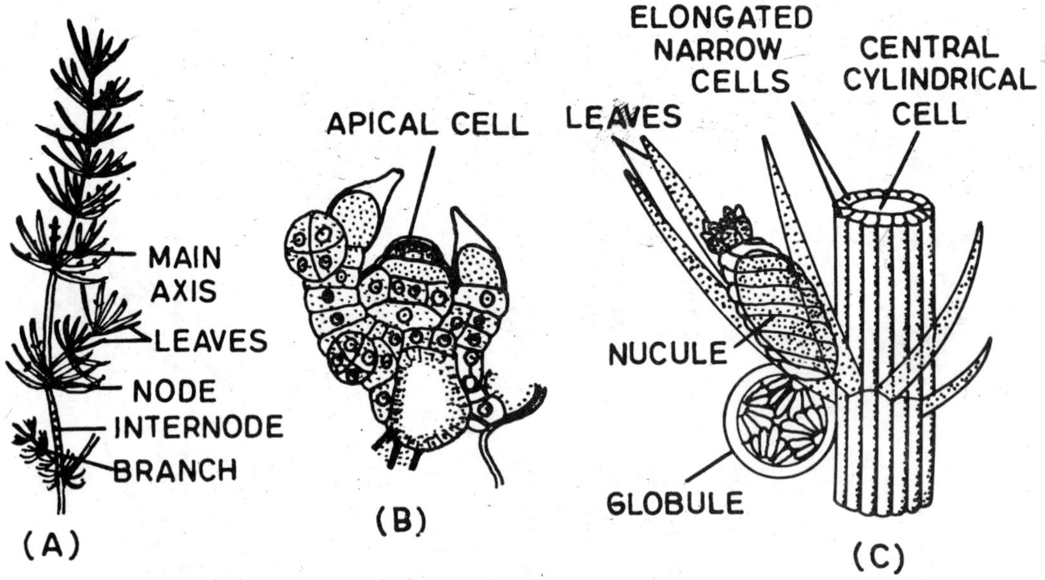

Fig.4.34. *Chara* sp. (A) Habit. (B) Vertical section through apex of main axis showing methods of segmentation. (C) Structure of nodal portion of main axis showing globule and nucule.

made up of two distinct areas, a stationary **exoplasm (ectoderm)** and an **endoplasm (endoderm)** which undergoes conspicuous streaming movement. Jarosch(1960) suggests that there are protein fibrils of the cytoplasm fixed to the cell wall and these fibrils undergo regular contraction and expansion which produce transverse wave and subsequently the streaming of cytoplasm. Only the endoderm is capable of streaming. According to Kersey and Wessels (1976) bundles of 50 to 100 microfilaments (not microtubules) are arranged on the inner face of the ectoplasm and propagation of oscillatory waves along the bundles of microfilaments causes the endoplasm to flow.

Different modes for vegetative propagation are (1) star-shaped aggregated cell filled with starch which are developed near lower nodes (**amylum stars**), (2) tuber like structures formed on the rhizoids (**bulbils**) and 3) protonema like outgrowths originating from the rhizoids. In addition, sexual reproduction is also known which is oogamous. Members do not form zoospores. The reproductive structures are borne at the nodes and their detailed arrangement is dependent on species. The mature male and female structures are called as **globule** and **nucule** respectively. Except for a few, most species of *Chara* are homothallic bearing globule and nucule at the nodes with a definite orientation, the nucule always lying above the globule (Fig.4.34 C). A mature globule is spherical, stalked large structure having bright yellow to red colour. The wall of globule is composed of 8 curved plates, the **shield cells** which are joined end to end giving pseudocellular appearance (Fig. 4.35 A). Each shield cell bears in its centre a rod-shaped cell - the **manubrium** from where **primary capitulum** are developed at the distal end in form of rounded cells. Each primary capitula produces 6 **secondary capitula**, each of which bears two long, whip like threads- the **antheridial filaments.** An antheridial filament consists of single row of cells ranging between 100 to 200 in number. One elongate biflagellate antherozoid is formed from each cell of this (Fig. 4.35 B-D).

The ontogeny of the male structure is more complex than that of female. Development takes place from the apical cell of the lateral branch which cuts off one or two discoid cells at its base and then becomes spherical. By longitudinal and transverse division, the spherical cell develops into **octants** which undergo further division. By enlargement the inner segments become separated from one another. 8 shield cells are formed from 8 peripheral cells and from the centre of each shield cell, manubrium, primary and secondary capitula and antheridial filaments develop gradually.

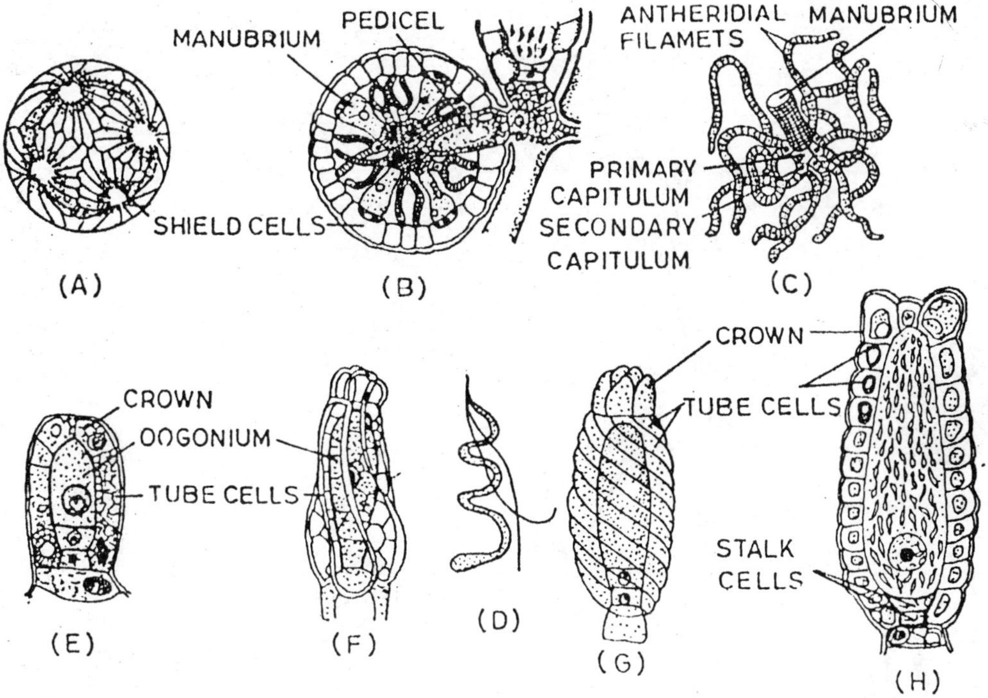

Fig.4.35. *Chara* sp. (A) A globule showing shield cell. (B) Internal structure of globule. (C) Manubrium with capitula and antheridial filaments. (D) A mature antherozoid. (E -G) Stages in devolopment of nucule. (G) Section of nucule showing structural details.

Nucule, the mature female structure consists of a central oogonium surrounded by spirally arranged cortical cells (Fig. 4.35 H). Envelope of 5 long spiral filaments with twisted appearance provide protection to enclosing oogonium. Each filament terminates in a small erect cell. Thus 5 erect cells of the 5 filaments form **corona** or **crown** capping a mature nucule. Single large uninucleate ovum filled with oil globules and starch grains is formed inside the oogonium. The adaxial cell of the basal node acts as nucule initial which divides twice giving rise to a row of three cells, the upper develops into oogonium proper, the lower forms the stalk and the middle cuts off 5 peripheral cells which elongate into filamentous structures forming **tube cells.** The tube cells envelop the oogonium in a spiral manner, each of which cuts off coronal cell at the apex (Fig. 4.35 E-H). When the egg cell is ready to be fertilised, the tube cells usually split a way from the corona and fertilization takes place through these slits. The details of corona structure are important taxonomic features used to separate *Chara* and *Nitella. Chara* has 5 - celled single-tiered corona whereas *Nitella* has 10-celled two tiered corona.

After attaining maturity, antherozoids enter in the oogonium, fuses with ovum and resulting zygotic nucleus migrates to the apical pole and there divides into 4 nuelei. It suggests that the nuclear division at this point is meiotic although there is no direct evidence. One of the 4 nuclei is cut off by a cell wall resulting into two unequal cells-upper uninucleate small **lenticular cell** and the lower **basal cell**,containing three nuclei and filled with food reserve. Basal cell soon disintegrates and small lenticular cell divides vertically to form a rhizoidal initial and a protonemal initial (Fig. 4.36). The former develops into a colourless rhizoid and the latter divides to form a small filamentous primary protonema. The protonema soon becomes green and subsequently develops into the aerial portion of a mature plant.

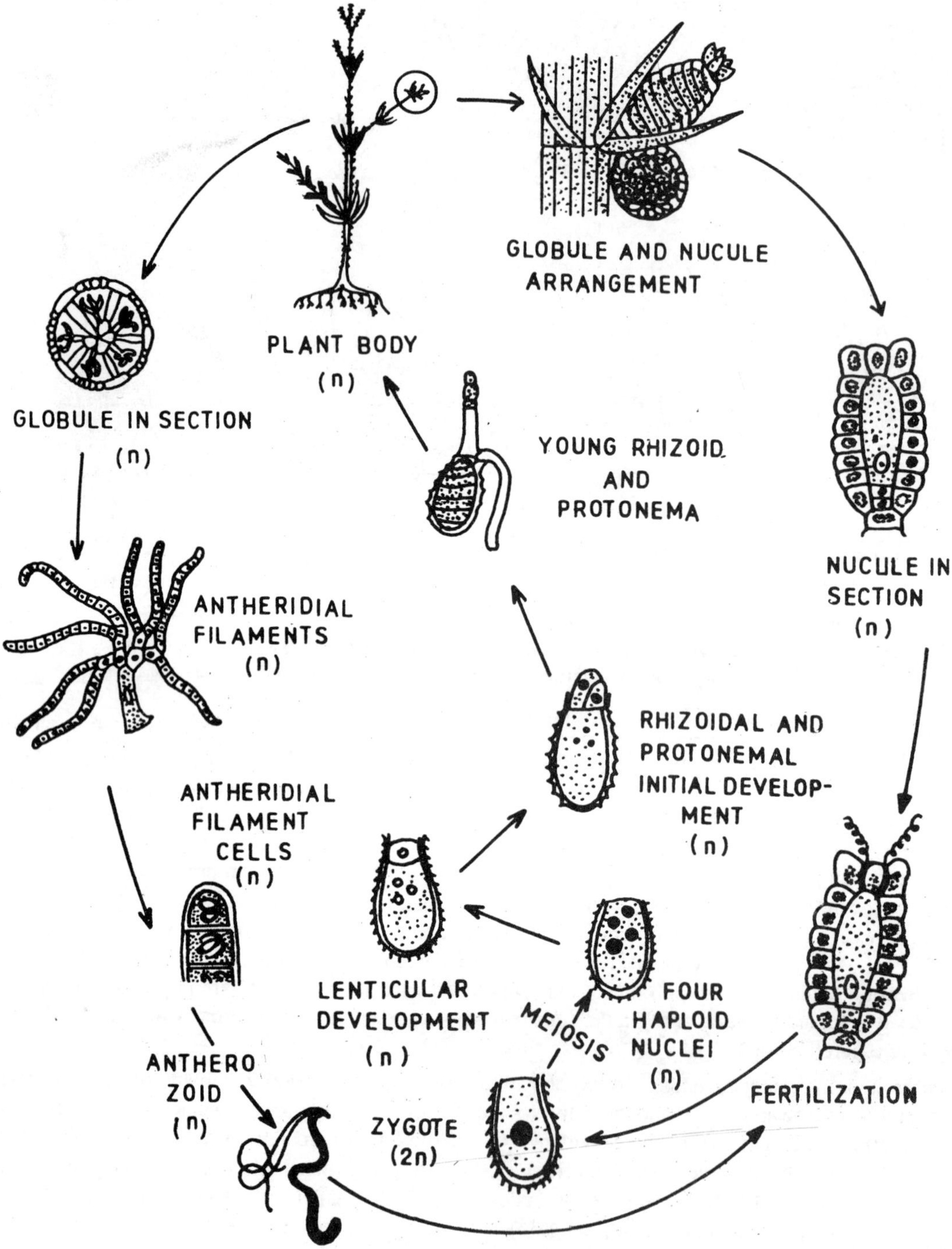

Fig.4.36. Sexual reproduction and life cycle of *Chara* sp.

Genus - *Nitella*

This genus resembles *Chara* in various respects. It also constitutes an extensive subaquatic vegetation in small water systems but unlike *Chara* it grows in much deeper water. Plant body in *Nitella* is comparatively more branched, mucilaginous and does not have calcareous deposition (Fig. 4.37). Vegetative body and sexual structures are smaller than *Chara.* Members are **ecorticated** (without cortical cells), without stipulodes and usually with furcated branchlet (**dactyles**).

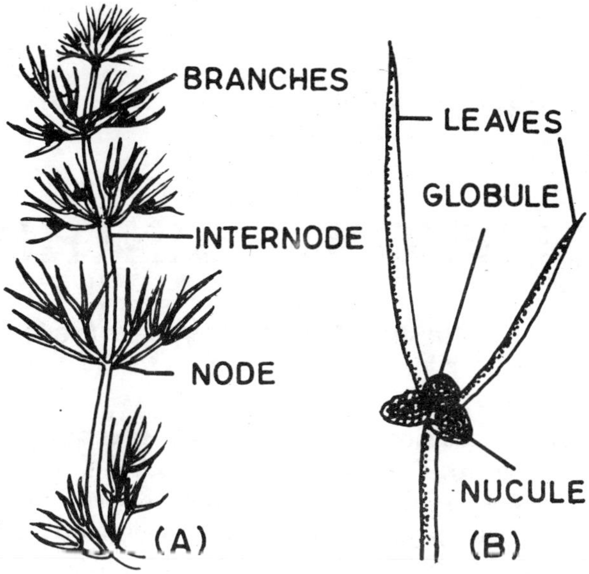

Fig.4.37. *Nitella* sp. (A) Habit showing nodes, internodes and pattern of branching. (B) Arrangement of nucule and globule.

The mode of sexual reproduction is similar to *Chara.* The globule and nucule are juxtaposed, the globule being directed upwards (terminal) and the nucules downwards (lateral). Both organs appear on lateral branches of limited growth. The nucule is relatively less elongated than *Chara* and has 10-celled corona arranged in two tiers which are again smaller than *Chara.* At an early stage of development, a single cell is cut off at the base of oogonium in *Chara* while in *Nitella,* three successive cells are cut off, the first two cells are longitudinal and third one is transverse.

As calcification occurs in most genera of Charales, the group is well represented as fossils. The female fructifications are calcified frequently and these are termed as **gyrogonites.** The earliest fossils of Charophyta occur in the uppermost Silurian.

5

DIVISION - XANTHOPHYTA

Class - Xanthophyceae (Yellow - green algae)

GENERAL FEATURES

The members of Xanthophyceae are commonly known as yellow-green algae. Because of unequal flagella they are classified in the sub-group **Heterokontae**. The species of this class are characterised by the following features. (1) There is an excess of carotenoids as compared to chlorophylls and they lack accessory pigment fucoxanthin. Chlorophyll **e** has been recorded only in two genera of Xanthophyceae i.e., *Tribonema* and zoospores of *Vaucheria.* (2) Out of the two unequal flagella, the longer is pantonematic (forwardly directed) while the shorter one is acronematic (posteriorly directed). (3) The food reserve products are normally oil and fat, starch has never been observed to get accumulated. (4) Cell wall may be frequently absent and when present it generally has a higher concentration of pectic materials. Silicification of the cell wall is common and it appears to be composed of two halves. (5) The eyespot in motile cells is always in the chloroplast and the chloroplasts are surrounded by two membranes of chloroplast E.R. The outer membrane of the chloropast E.R. is usually continuous with the outer membrane of the nucleus (Fig. 5.1).

Due to absence of chlorophyll **b** and accessory pigment fucoxanthin, members of Xanthophyceae appear greenish or yellowish-green and are often confused with chlorophycean algae. Uniformly simple shape of the chloroplast, usually discoid and parietal, differentiates it with green algae. The major xanthophyll found in the Xanthophyceae is **diadinoxanthin** (Whittle and Casselton, 1975). In some members statospores are formed which are unlike those found in Chrysophyceae as they consist of two parts of about equal size which overlap.

OCCURRENCE AND HABITAT

Members are primarily freshwater with only few marine representatives. As compared to Chlorophyceae this is a small group having mainly the planktonic forms. Some forms are epiphytic and are found on different plant parts. Some species grow in terrestrial habitats or in subaerial situations like trunk of trees, damp walls etc.

CELL STRUCTURE

Cell wall : Although cellwall is chiefly composed of pectic substances but in some members cellulose is the main constituent. In *Vaucheria,* cellulose comprises 90% of the wall with the remaining portion

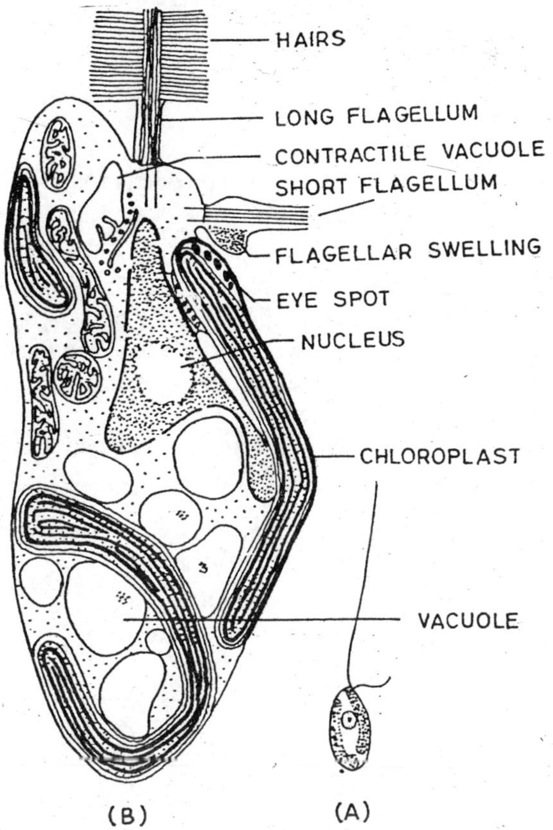

Fig. 5.1. (A) Light microscopic and (B) Electron microscopic diagram of a zoospore of a typical member of the Xanthophyceae (*Mischococcus sphaerocephalus*). (After Hibberd and Leedale, 1971).

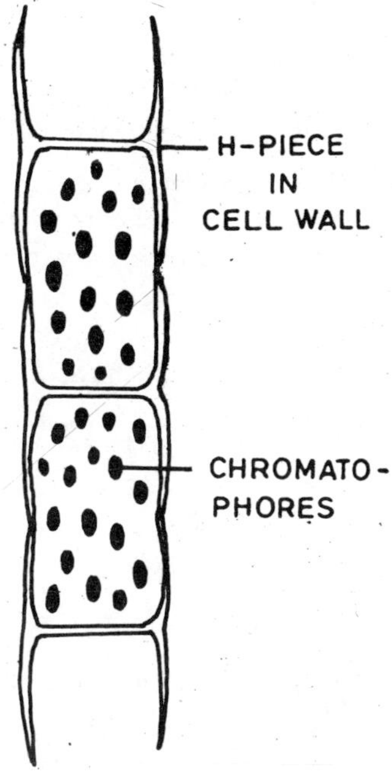

Fig. 5.2. A part of filament of *Tribonema* showing H-pieces in cell wall and discoid chromatophores.

comprising of amorphous polysaccharides composed primarily of glucose and uronic acids. The oomycetous fungi have probably evolved from *Vaucheria* like ancestor as is indicated by resemblances of their cell wall compositions.

In many members like *Ophiocytium majus* cell walls are composed of two overlapping halves. Some genera, mainly the filamentous forms (*Tribonema*) have walls that are made up of **H-shaped pieces.** These overlap with each other alternately so that each protoplast is enclosed by the halves of two successive H-pieces (Fig. 5.2).

Chloroplasts and food reserve : The thylakoids are grouped into bands of three and in many genera there is a pyrenoid in the chloroplast (Marchant, 1972). The eyespot consists of a layer of globules beneath the chloroplast envelope at the anterior end of the chloroplast (Fig.5.1). Flagellar sheath is dilated into flagellar swelling at the base of short flagellum. Chlorophyll **a** is present in the chloroplasts with be major carotenoids being diadinoxanthin, heteroxanthin and vaucherioxanthin ester. Chlorophyll **c** has only been reported from *Vaucheria geminata* (Caubergs and de Greef, 1971).

RANGE OF VEGETATIVE STRUCTURE

The morphological diversity of this class shows parallelism with Chlorophyceae. Primary division of the Xanthophyceae into orders has been based using the vegetative structure as the major criterion. Like green algae, the plant body of Xanthophyceae may be unicellular, palmellooid, coccoid, filamentous or siphonaceous. Yellow-green algae lack highly developed pseudo-parenchymatous or truly parenchymatous thalli as is observed in advanced members of green algae.

REPRODUCTION

Vegetative and asexual reproduction are of common occurrence while sexual reproduction is recorded only in few genera. The colonial and filamentous members may multiply by fragmentation into two or more parts which develop into new plants.

Asexual reproduction is achieved with the help of zoospores and aplanospores. Zoospores are formed in majority of genera which are naked and usually pear-shaped (pyriform). They possess one or more contractile vacuoles, a few chromatophores, rarely an eye-spot and two anteriorly attached flagella. The forward tinsel flagellum is usually four to six times longer than the shorter, posterior whiplash flagellum (Fig. 5.3). Normally one to 16 zoospores are produced inside each cell which are released by dissolution or rupture of the sporangial walls. Zoospores have a massive chloroplast, elongate nucleus and a pyrenoid.

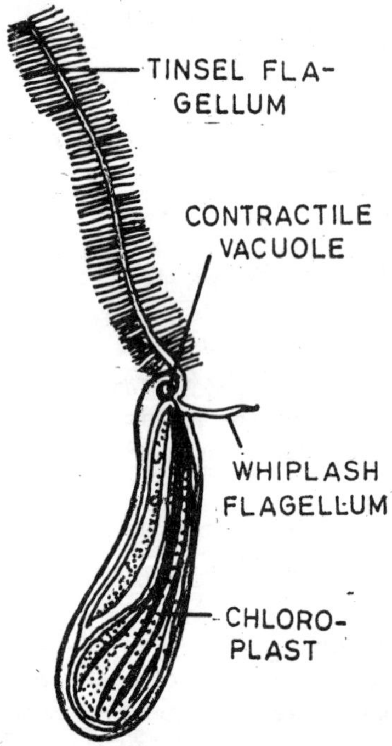

Fig. 5.3. A zoospore of Xanthophyceae showing forward tinsel flagellum 4-6 times longer than. posterior whiplasth flagellum.

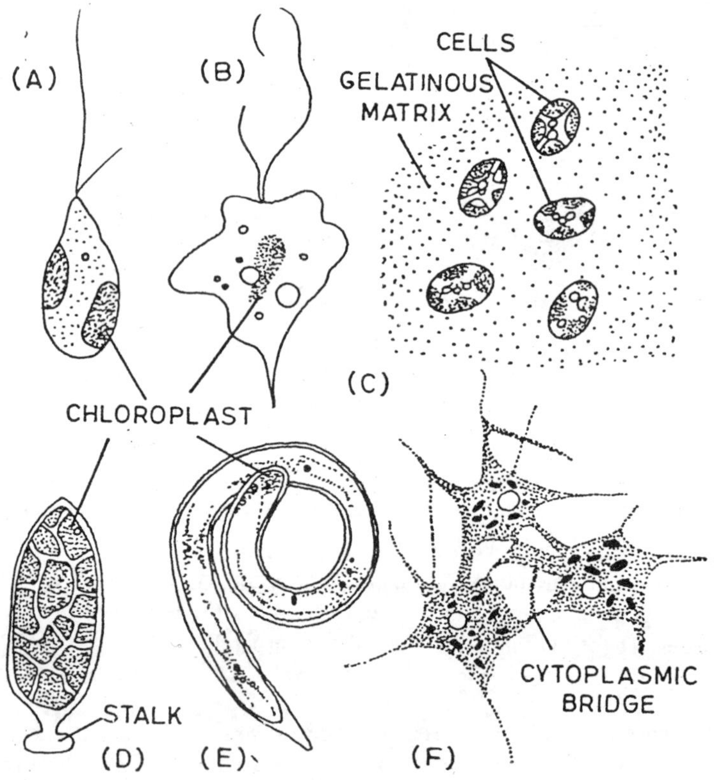

Fig. 5.4. Members of Xanthophyceae. (A) *Heterochloris* sp. showing two parietal chloroplast located at the side of cell. (B) *Chloromeson* sp. showing single central chloroplast. (C) A portion of thallus of *Gloeochloris* showing the irregular distribution of cells inside colourless gelatinous matrix. (D) *Characiopsis* sp. showing short stalk for attachment to the substratum. (E) *Ophiocytium* sp. showing elongated cells. (F) *Chlorarachnion* sp. showing cytoplasmic bridge.

Under unfavourable conditions **aplanospores** are produced in place of zoospores. Entire protoplast may produce a single aplanospore or divide into a number of parts, each of which becomes an aplanospore. An aplanospore liberated from a parent cell can grow directly into a new plant or it may give rise to zoospores which develop into new plants. Under certain conditions vegetative cells can also change directly into spore like resting stages with much thicker wall and abundant food reserves. Spore wall is not distinct from the parent wall in these. These spore like cells are usually found in filamentous genera and are called akinetes. Cysts and statospores are also produced by some members. Their formation starts with an internal delimitation of protoplast which is separated from the peripheral portion of the mother cell's protoplast by a membrane. The endogenously differentiated protoplast subsequently secretes a wall with two overlapping halves.

Sexual reproduction has been authentically reported in some genera. In *Botrydium,* the two fusing gametes are motile and may be isogamous or anisogamous. In *Tribonema* the male gamete is motile and female one is non-motile while in *Vaucheria* sexual reproduction is distinctly oogamous.

CLASSIFICATION

Although Fritsch (1935) classified this class into 4 orders, viz., Heterochloridales, Heterococcales, Heterotrichales and Heterosiphonales but most of the recent classifications distinguish the class Xanthophyceae into six orders. The vegetative tendencies of first five orders are parallel to those of the Chlorophyceae. These are Heterochloridales (motile), Heterogloeales (palmelloid), Mischococcales (coccoid), Tribonematales (filamentous) and the Vaucheriales (siphoneous). The counterparts of these are Volvocales, Tetrasporales, Chlorococcales, Ulotrichales and the Caulerpales respectively of the Chlorophyceae. The sixth order is the Rhizochloridales with amoeboid forms exhibitng a vegetative type not found in the Chlorophyceae.

The important features of these orders are:

ORDER 1 : HETEROCHLORIDALES

In this order organisms are flagellated, unicellular and motile but unlike the Volvocales of the Chlorophyceae the motile colony is unknown. Most of the species are naked and in many members cytoplasmic membrane is not rigid showing amoeboid tendency (*Heterochloris,* Fig. 5.4 A). In *Heterochloris,* the chloroplasts are parietal, two in number and are located at the sides of the cell. *Chloramoeba* represents the form of chloroplast most common in Xanthophyceae i.e., 2-6 discoid chloroplasts located towards the periphery of the cell. The cells of *Chloromeson* have a single central chloroplast (Fig. 5.4 B). Some species show a marked tendency to lose their photosynthetic pigments viz., cells of *Chloramoeba* become colourless when grown under heterotrophic conditions.

ORDER 2 : HETEROGLOEALES

This comprises palmelloid forms, solitary or colonial, permanently provided with or without gelatinous envelope. The immobile vegetative cells have the ability to return directly to the motile condition. The immobile vegetative cells of *Gloeochloris* are irregularly distributed throughout a colourless gelatinous matrix. The cells remain in colonies that are spherical and free-floating, or attached to submerged aquatics (Fig. 5.4 C).

ORDER 3 : MISCHOCOCCALES

Algae of this order are immobile, coccoid, solitary or colonial. Walls are distinct, smooth or sculptured in one or two pieces. Vegetative cells do not have the capacity to return directly to a motile condition.

This is the largest order in the class and the number of genera and species in this order exceeds the collective number in the other orders.

Members have two main tendencies, the trend towards the multinucleate condition and some species prefer to get attached to the substratum. The attached form *Characiopsis* consists of elongated cells attached to the substratum by a short stalk (Fig. 5.4 D). Most species have a number of discoid chloroplasts and some are multinucleate, mainly before the zoospore formation. The genus *Ophiocytium* shows distinct multinucleate condition with elongated cells attached to the substratum (Fig. 5.4 E).

ORDER 4 : TRIBONEMATALES

This order includes mainly multicellular unbranched filamentous genera. Cylindrical cells are uniseriately arranged end to end to form branched or unbranched filament. *Tribonema*, the most common genus of this order has strong morphological resemblance to *Microspora*, a member of green algae. Barrel - shaped cells are usually 2-5 times longer than the width. The wall is composed of two H-shaped pieces overlapping in the middle of the cell. When the filaments become dissociated, the H-pieces are detected and the ends of filaments appear as the open ends of a cylinder. Aplanospores are produced more frequently than the zoospores and are released by the pulling apart of the two portions of the cell wall.

ORDER 5 : VAUCHERIALES

The members of this order are multinucleate, siphonaceous (coenocytic) and comprise two families. In Botrydiaceae, cells are differentiated into an aerial portion with chloroplast and a colourless subterranean portion. Sexual reproduction is iso - or anisogamous. Vaucheriaceae represents tubular thalli with oogamous sexual reproduction.

ORDER 6 : RHIZOCHLORIDALES

This order includes those species which have an amoeboid form. In order Heterochloridales, some of the species become amoeboid temporarily but in the present order the species are permanently amoeboid. Many of the forms have tendency to form a large mass of cells through the connection of cytoplasmic strands (*Chlorarachnion*, Fig. 5.4 F). Like other amoeboid cells of Chrysophyta, the members of this order are holozoic. This sole method of reproduction is through zoospores apart from vegetative division of the cells.

PHYLOGENY AND INTERRELATIONSHIPS

Three major evolutionary lines may be traced among the Xanthophyta like Chlorophyta. They are :

1. Development of nonmotile solitary or colonial forms from motile unicellular ancestors,
2. Development of tubular or siphonaceous forms and
3. Development of multicellular filamentous types.

Due to strong morphological similarities with Chlorophyceae, many phycologists advocate on the theory of **parallel evolution** and similarity in the origin of the group from the flagellates. Since the members in Xanthophyceae lack elaborate thalli as found in Chlorophyceae, this sequence of parallel evolution did not go too far. Fritsch (1935) believed that the group is still in process of evolution while some consider it a reduced stock.

Due to similarities in food reserve, composition of cell wall and in other physiological and biochemical features, members of Xanthophyceae resemble with those of Chrysophyceae. Presence of chlorophyll **e** and other features have led modern phycologists to propose an independent status for Xanthophyceae as a class or even as a division Xanthophyta.

LIFE HISTORIES OF SELECTED MEMBERS

Order - Heterosiphonales

Family - Botrydiaceae

Genus - *Botrydium*

It is of widespread occurrence on damp soil near banks of streams and pools. The vesicular thallus is unicellular, multinucleate consisting of usually globose aerial portion with chloroplasts and a colourless rhizoidal portion which penetrates the soil (Fig. 5.5). It bears a strong resemblance to *Protosiphon*. The vesicular part has a large central vacuole and a peripheral layer of cytoplasm containing numerous nuclei and discoid chloroplasts. The shape of the aerial part is influenced by environmental conditions. It is usually elongate when growing in shaded habitats and spherical when growing in bright light. The rhizoidal portion may be simple or branched and has dense or vacuolated cytoplasm with many nuclei without plastids.

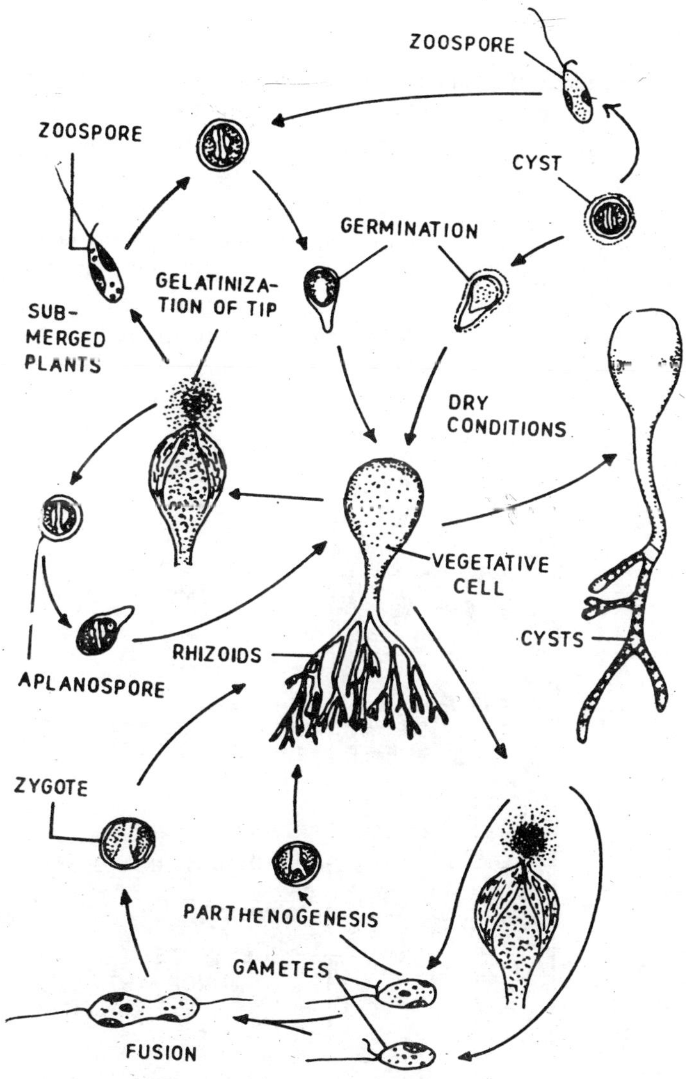

Fig. 5.5. Life cycle of *Botrydium* sp.

Reproduction is carried out by zoospores or aplanospores. The formation of aplanospores or zoospores is initiated at night. Zoospore production normally takes place if the thallus is flooded with water. The sac is transformed into large sporangium and after cleavage biflagellate pear-shaped uninucleate zoospores with two chloroplasts each are released. When exposed to dry conditions the protoplasmic contents of the vesicle migrate into the rhizoids and divide to form thick-walled globose to ellipsoid coenocyts, known as **cysts**. These cysts either germinate directly or produce zoospores.

Sexual repooduction has also been reported occasionally which are performed by motile iso- or anisogametes (depending on the species) with the cells being either homothallic or heterothallic. The gametes become apposed with their anterior ends when uniting in pairs to form the zygote. Parthenogenetic germination into thalli has been seen if the gametes do not fuse. A germinating zygote develops directly into vegetative thallus after undergoing reduction division.

Family - Vaucheriaceae

Genus - *Vaucheria*

The genus *Ectocarpus* shows many of the simplest features in the Phaeophyceae and is one of the best-known brown algae containing many species. It is a marine alga of wide occurrence in the temperate and tropical seas. This alga usually grows as an epiphytic or epilithic tufts in the littoral, sublittoral and supralittoral zones.

This branched filamentous alga has tubular coenocytic axis with septa being laid down only when gametangia or sporangia are formed. It has relatively thin cell wall within which the cytoplasm is restricted to the periphery of the coenocyte with a large central vacuole (Fig. 5.6). Numerous ellipitical chloroplasts with pyrenoids are present outside and nuclei are present in the center of the cytoplasm. In the apex large number of vesicles, mitochondria and dictyosomes are present where the growth of the

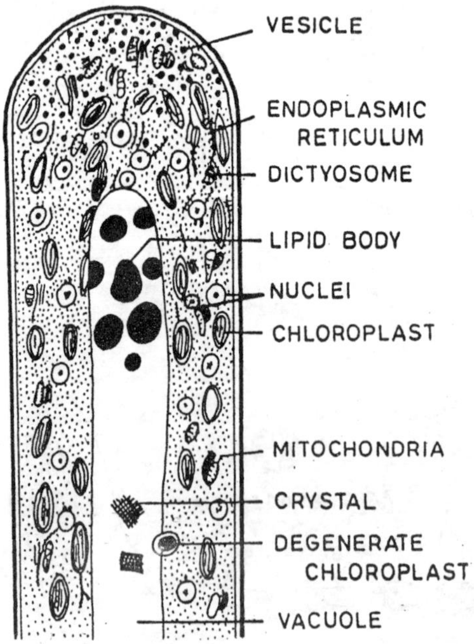

Fig. 5.6. Diagrammatic representation of the tip of a vegetative filament of *Vaucheria dillwynii*. (After Ott and Brown, 1974 a).

filament is restricted. Nuclei and chloroplasts are lacking in this apical zone. Large central vacuole possess lipids, degenerated chloroplasts and crystals. In the region of central vacuole cytoplasmic streaming takes place which involves nuclei, mitochondria and dictyosomes. The chloroplasts do not have definite streaming but have a random movement. The nuclear envelope remains intact during mitosis surrounding the spindle apparatus. Such "closed" type of nuclear division is known in some fungi and Protozoa which is regarded as a primitive condition.

Both asexual and sexual reproduction are found in *Vaucheria* besides accidental breaks of the filament which later regenerate. Zoosporangia are produced at the tip of the filament during asexual reproduction. The protoplasm is then metamorphosed into a single large zoospore (Fig. 5.7) with numerous pairs of slightly unequal flagella covering the surface. This multinucleate zoospore is regarded

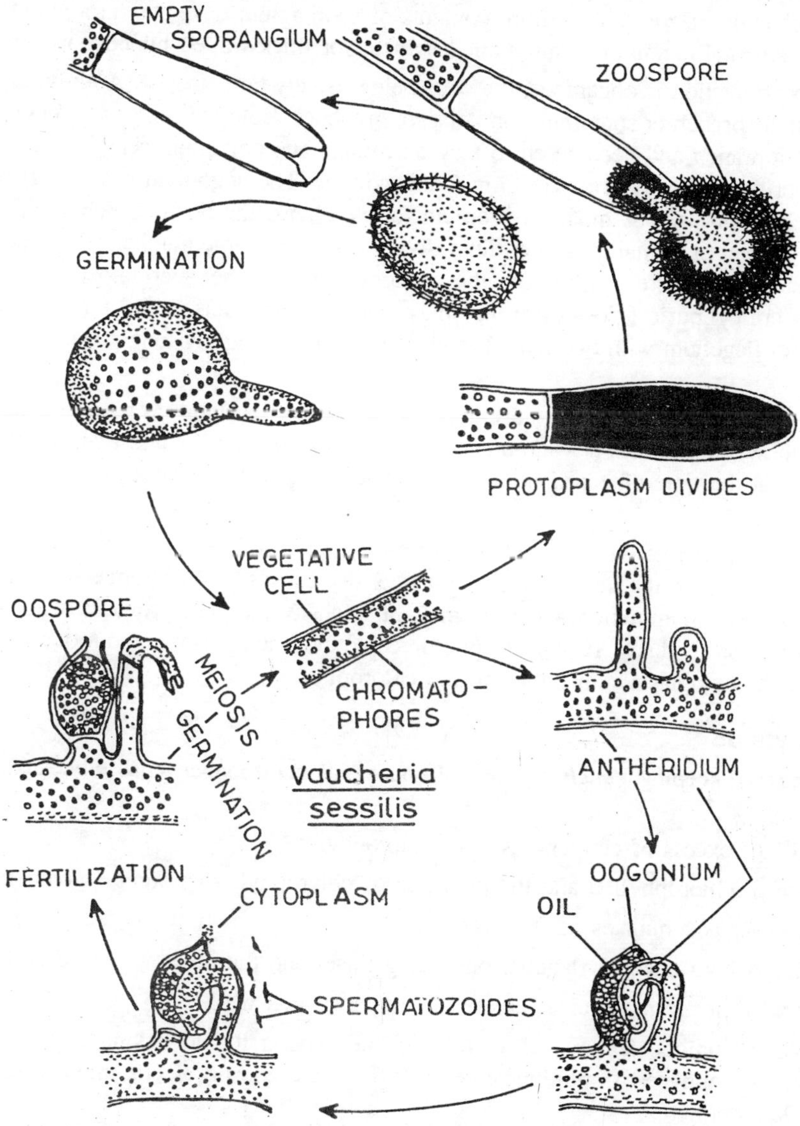

Fig. 5.7. Life cycle of *Vaucheria* sp.

as **compound zoospore or coenozoospore** as it is a composite structure of many nuclei with flagellar pairs. In the production of zoospores, large number of chloroplasts and nuclei stream into the tip of the filament, the central vacuole decreases in size and the tip appears dark green. The process of zoosporogenesis is controlled by an endogenous circadian rhythm. The pair of centrioles associated with each nucleus initiates two flagella which are located in an internal vesicle. The flagella of several nuclei become associated within common vesicles and are called "flagellar pools". The vesicles fuse with plasma membrane and numerous pairs of flagella are thus located on the surface of the cytoplasm within the mature zoosporangium. By the gelatinization of the apical tip of the zoosporangium, zoospores are released. There is no eyespot or pyrenoid in the zoospore. After swimming for about 15 minutes, these multiflagellate zoospores settle down, flagella get withdrawn and a wall is secreted around. Germination follows almost immediatedly by the development of tubular outgrowths. In some terrestrial species, instead of production of zoospores, entire contents of sporangium develops into an aplanospore. Under specific environmental conditions, thin-walled akinetes or thick-walled hypnospores may be produced.

Sexual reproduction is oogamous, the cells being usually homothallic. Antheridia and oogonia are borne on the same branch or sometimes on adjacent branches. Both the sex organs develop as a swelling on the side of a branch and become cut off by a septum. The oogonium remains as an ovoid swelling while the antheridium usually grows to form a hooked tube. The oogonium has a single egg and is filled with oil and chloroplasts. The mature oogonium forms a beak, the tip of which gelatinizes forming an aperture. The protoplast of mature antheridium divides to form a large number of uninucleate biflagellate spermatozoids. An aperture appears in the antheridium and the spermatozoids are released. The spermatozoids are cylindrical posteriorly having a flattened proboscis in the anterior region. A forward projecting tinsel flagellum with two lateral rows of hairs and a slightly longer trailing smooth flagellum is present.

For fertilization one spermatozoid enters the oogonium through aperture and fuses with the egg protoplast. The zygote secretes a wall and oil droplets fuse to form a small number of central droplets. Meiosis occurs at the time of germination of the zygote and new siphonaceous filament emerges from the zygote wall.

Vaucheria is one of the algae which exhibit chloroplast orientation movements in the light and distribution is random in the dark. Feeble light intensity causes the chloroplasts to move into position to receive maximum illumination and in high light intensity they are provided with device to receive minimal illumination (Nultsch, 1974). The photoreceptors of the light-oriented movement of chloroplasts are present in the cytoplasm rather than in the chloroplast.

Taxonomic status

The earlier view of keeping *Vaucheria* in Chloropnyceae is no longer tenable because of the following factors:

1. Presence of excess of carotenoids over chlorophylls,
2. Absence of chlorophyll **b** and the presence of chlorophyll **e** in some members,
3. Starch never accumulates as food reserve,
4. Presence of two unequal flagella, one being tinsel and the other whiplash and
5. Presence of pectic material as chief component of cell wall.

Some of the unique features of *Vaucheria* i.e., spermatozoid having a proboscis as in *Fucus* (a member of Phaeophyceae), multinucleate multiflagellate zoospores etc. make it distinct from other members of Xanthophyta.

6

DIVISION - BACILLARIOPHYTA

CLASS - BACILLARIOPHYCEAE (DIATOMS)

GENERAL FEATURES

The members of Bacillariophyceae are generally referred to as diatoms. These are mainly unicellular but occasionally may be colonial also. The cells are surrounded by a rigid two-part box like cell wall called the **frustule**. The diatoms are characterized by following features:

1. The cell wall is silicified and shows characteristic secondary structures.
2. The photosynthetic pigments include chlorophyll **a**, $\mathbf{c_1}$ and $\mathbf{c_2}$ with the major carotenoid being the golden-brown fucoxanthin giving the cells their characteristic colour.
3. Food storage products include fats and chrysolaminarin.
4. The motile stages possess a single pantonematic flagellum. An unusual feature of this single flagellum is the presence of nine peripheral doublet tubules but no central pair.

OCCURRENCE AND HABITAT

Diatoms are ubiquitous in their distribution and are found practically in every aquatic habitat (freshwater, marine and brackish waters) as free-living photosynthetic autotrophs, colourless heterotrophs or photosynthetic symbiotes. They may occur as plankton, epiphyte or periphyton on substrates such as rocks or aquatic plants. Members are also found in terrestrial or subaerial habitats receiving atleast periodic moisture. Epizoic and endozoic existence have also been recorded for the pennate diatoms. Some diatoms secrete a common tubular envelope and the cells glide within it. A large number of fossil diatoms is also recorded, the earlier types comprising marine species dating form the Cretaceous era (Patrick and Reimer, 1966). Diatoms play a very important role in the aquatic habitats as they provide basic food for aquatic fauna. The diatoms are increasingly being used as indicators of water quality.

CELL STRUCTURE

The cell of the diatom (frustule) consists of two parts which surround the protoplasm with a cetral nucleus suspended in it.

Cell Wall: Frustule is composed of two overlapping almost equal halves, the smaller fitting into the larger one like a pair of petri-dish (Figs. 6.1 B, 6.5). The outer and inner halves (the lid) are known as **epitheca** and **hypotheca** respectively. Each half consists of the main surface, the **valve (epivalve** and **hypovalve)** and the overlapping **connecting bands (epicingulum and hypocingulum)**. The epivalve and epicingulum comprise epitheca and hypovalve and hypocingulum comprise hypotheca. The two connecting bands, one attached to each valve are called as **girdle.** Sometimes the connecting bands themselves are called as girdle bands. Sometimes there are one or more additional bands between the valve and the girdle which are known as **intercalary bands**. The diatom cell, thus can be viewed from two directions, the **girdle view** and the **valve view** (Fig. 6.1 A, B). The axis between the middle of the two valves is **pervalvar axis** and the axis at right angles to this is **valvar plane** (Fig. 6.1 C). The girdle bands are often furnished with minute teeth which hold the valves together by their edges and there is the deposition of pectinaceous film. If this film is destroyed, the valve and the girdle bands get separated.

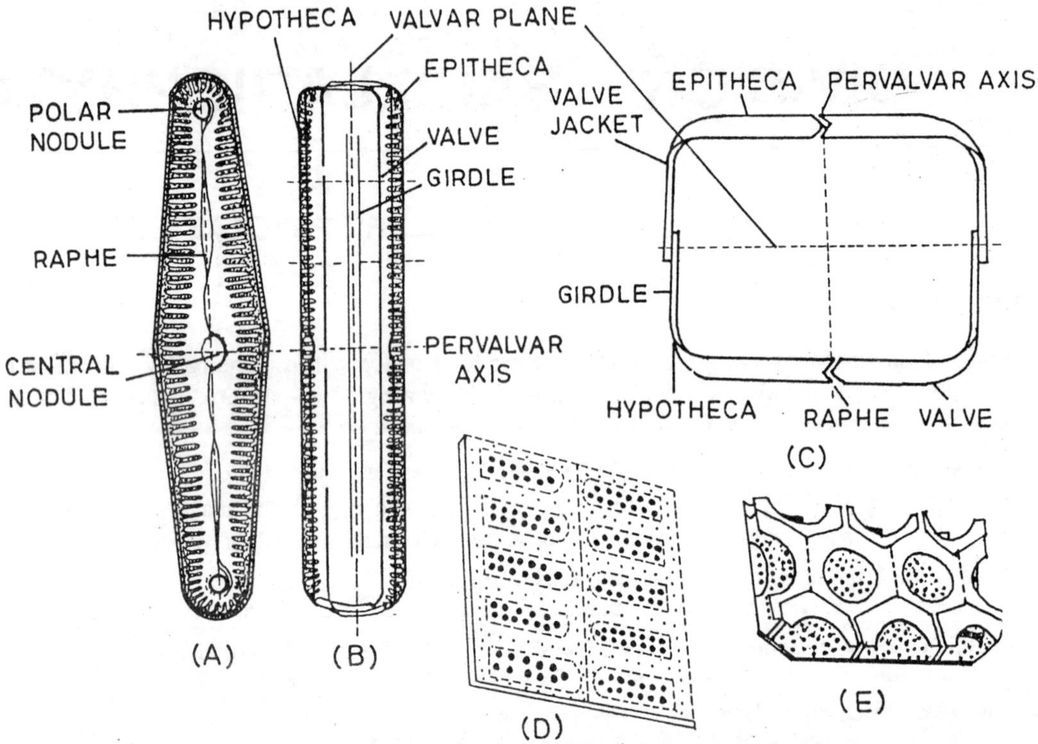

Fig. 6.1. Bacillariophyceae. (A) Frustule of *Pinnularia* in valve - view. (B) Girdle - view. (C) Transverse section of frustule of *Pinnularia*. (D) Laminar wall showing single silicified layer. (E) Locular wall showing two parallel layers with a complex reticulum of cross walls between them.

The diatom wall is ornamented due to regular deposition of siliceous material in defferent patterns. Ornamentation has been divided into four types by Hendey (1964). These are: (1) **centric** and radial where it is arranged with reference to a central point (*Coscinodiscus*, Fig. 6.2 A), (2) **trellisoid** where it is arranged uniformly over the surface without reference to a point or a line (*Eunotia*, Fig. 6.2 B), (3) **gonoid**, in which the structure is dominated by angles (*Triceratum*, Fig. 6.2 C) and (4) **pennate** where the structure is balaterally symmetrical (*Navicula*, Fig.6.2D, 6.8). In some pennate diatoms **raphe system** is present. Raphe i.e., longitudinal slit in the **theca** (valve) is divided into two parts by the **central nodule** (6.1 A). Each half of the raphe terminates in a swelling of the wall called the **polar nodule**. In some

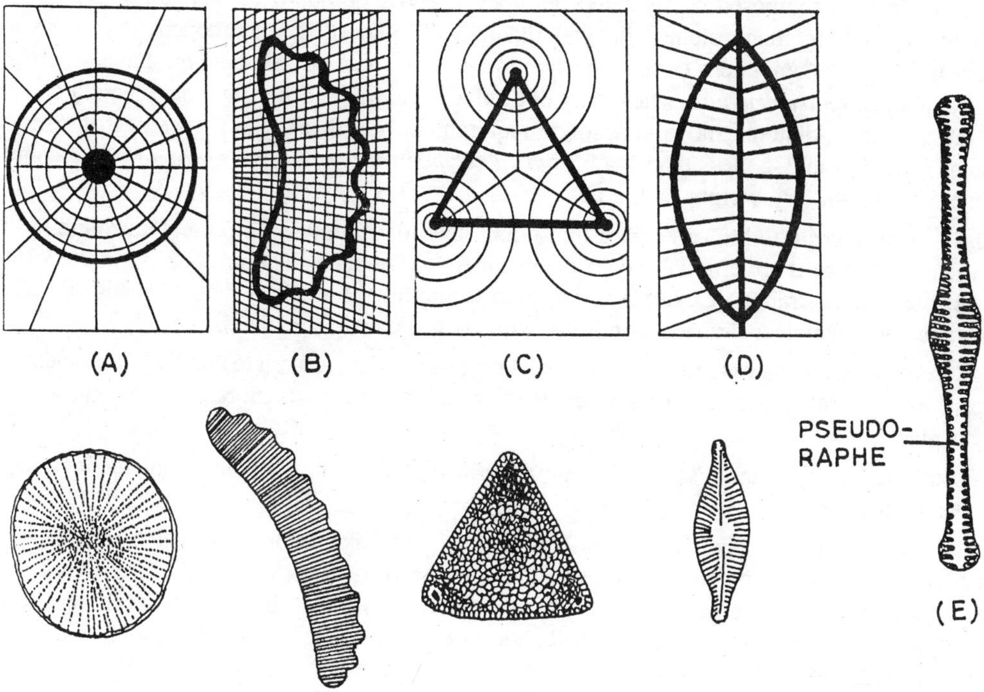

Fig. 6.2. (A -D) The basic patterns of ornamentation in the Bacillariophyceae. (A) Centric and radial showing a central point. (B) Trellisoid showing no reference to points. (C) Gonoid showing angles. (D) Pennate showing bilateral symmetry. (E) A cell of *Tabellaria* sp. showing pseudoraphe.

pennate forms instead of raphe system, an unornamented area running down the center of the valve (**pseudoraphe**, Fig. 6.2 E) is present.

Over the valve surface of the diatoms, many secondary structures are found. The four basic types of such structures are common.

1. **Punctae** or **striae** are small perforations arranged in regular lines or series.
2. **Aerolae** are relatively large, depressed box like structures with a sieve membrane.
3. **Canaliculi** are thin, narrow, tubular channels in the valve walls.
4. **Costae** or **ribs** are formed due to heavy deposition of silica.

The wall types may be either **laminar** which consists of single silicified layer with a great variety of patterns on it (Fig. 6.1 D) or **locular** where wall consists of two parallel layer with a complex reticulum of cross walls between them (Fig. 6.1 E). The girdle bands in some diatoms have the same sculpturing as the valve, but in many others sculpturing on the girdle and valve is different. In few members the girdle bands have no sculpturing. In many diatoms special pores (slime or mucilage pores) are found through which mucilage is secreted.

In order to assist colony formation the valve surface can have extensions called **processes** which help in maintaining contact between cells. Depending upon the shapes they may be **cornutate** processes (horn like), **spinulae** processes (very small), **awn** processes (hollow and elongated) etc. If the outgrowths of the valves project beyond the valve margin these are termed as **setae** which are prominent in *Chaetoceros* (Fig. 6.7).

The frustule is composed of hydrated amorphous silica or quartzite that may also have small amounts of aluminium, magnesium, iron and titanium. The inorganic component of the frustule is enveloped by organic component comprising amino acids and sugars. All diatoms have silicified wall except *Phaeodactylum tricornutum* where only one valve is silicified. Since no other element can replace silicon, the growth of diatoms show an absolute requirement of silicon. The percentage of silica varies in different genera. It may be less than 1% in *Phaeodactylum tricornutum* or much silicified comprising more than 50% of the dry weight of the cell.

Protoplast : Just internal to the cell wall a thin lining layer of cytoplasm is present surrounding a large central vacuole. The cells are uninucleate. In the members of Pennales the nucleus is normally suspended in the vacuole within strands of cytoplasm while in Centrales the nucleus mostly occurs in the lining layer of cytoplasm. Nuclear division is mitotic with two interesting features.

1. Chromosomes are arranged as an equatorial ring rather than as a plate and an intranuclear spindle is found at metaphase. The ring arrangement of chromosomes is more common in animals than in plants.
2. Like Chrysophyta and Pyrrophyta, the spindle is in the form of an elongated cylinder with flattened poles.

The cells of Pennales have normally a large parietal plate-like chloroplast which may be lobed. In Centrales cell usually contains numerous discoid chromatophores. Naked pyrenoids are present in some diatoms. The chloroplasts are surrounded by two membranes of the chloroplast envelope, outside of which are two membranes of chloroplast E.R., the outer membrane being continuous with the outer membrane of nuclear envelope. The bands of three thylakoids are present inside chloroplast matrix. The chloroplasts contain chlorophyll **a**, $\mathbf{c_1}$ and $\mathbf{c_2}$ along with principal carotenoid fucoxanthin giving the cells their characteristic golden-brown colour. Some colourless or **apochlorotic** diatoms are also known which survive on decaying marine vegtation and on mucilages of large seaweeds.

Chrysolaminarin, the storage product of diatoms is located in the vesicles. The chrysolaminarin from a mixture of freshwater diatoms was found to consist of ß-1,3 and ß-1,6 linked glucose units in a ratio of 11:1 and average chain length of 12 glucose units. The cytoplasm also possess other eucaryotic organelles such as mitochondria, dictyosomes, endoplasmic reticulum etc.

RANGE OF VEGETATIVE STRUCTURE

The morphological appearance and shape of the diatom cells are variable which help in their preliminary indentification. Six main types of morphological elaborations according to Fritsch (1935) are:

(1) the flat discoid type as found in many centric diatoms (e.g., *Cyclotella*, Fig. 6.9), (2) the needle shaped as found in *Synedra* (Fig. 6.3 A), (3) the long and at times coiled filament like *Melosira* (Fig. 6.3 B), (4) the valves having elongated bristles (*Stephanodiscus*, Fig. 6.3 C) or horns (*Chaetoceros*, Fig. 6.7), (5) the star- shaped colonies like *Asterionella* (Fig. 6.3 D), and (6) colony with extensive production of mucilage (*Cyclotella planctonica*). Although unicellular tendency is most frequent in diatoms but in some members colonies may be formed by any of the following three methods:

1. By localised production of mucilage, cells may be joined as in *Asterionella*.
2. Many cells may get embedded in a common mucilaginous envelope. This envelope may be tubular in some genera.
3. Colony may be formed by the joining of cells through special outgrowth like spines as seen in *Chaetoceros*.

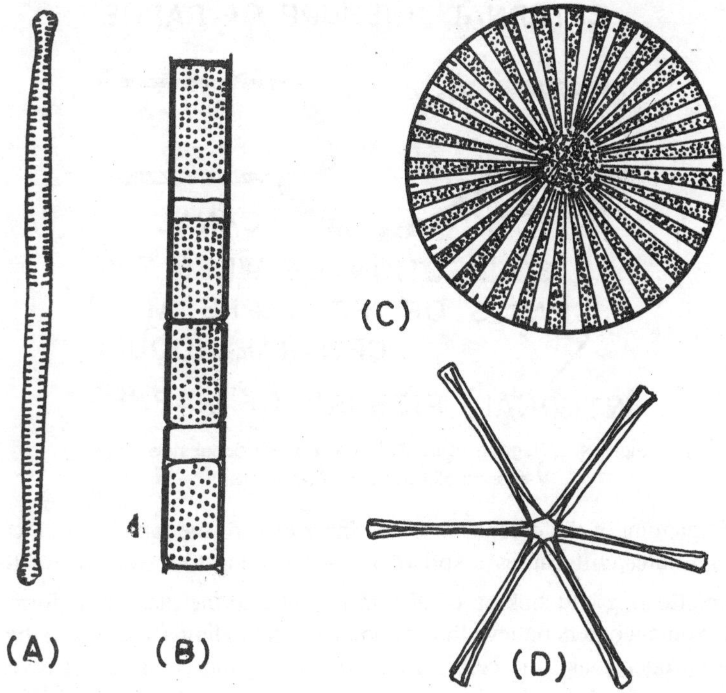

Fig. 6.3. Range of vegetative structure in diatoms. (A) Needle-shaped *Synedra*. (B) Filamentous *Melosira*. (C) *Stephanodiscus* sp. (D) Star-shaped *Asterionella*.

MOTILITY

In some of the pennate diatoms with one or two raphe systems, motility has been noticed. Many theories and interpretations have been put forward for this. One theory explains that the forward movement of diatoms is accomplished due to the secretion of mucopolysaccharide strand through the raphe which sticks to the substrate and by continuous production of the strands (Harper and Harper, 1967). Inhibition of motility can be caused by the addition of mucous dispersing agents.

Gordon and Drum (1970) proposed capillary mechanism but such particle movement along the raphe was demonstrated to be more jerky than the actual movement of diatoms (Harper, 1977). Nultsch (1956) distinguished three types of movement - (1) the *Navicula* type with a straight movement, (2) the *Amphora* type with a curved movement and (3) the *Nitzschia* type which always exhibits curved pathways with two different radii. Positive or negative phototaxis is demonstrated in these gliding diatoms, a positive reaction being expressed in white light by most diatoms. The observed rates of gliding in diatoms vary from 2 to 14 μm sec^{-1} at room temperature.

The rhythmic phenomenon noticed in *Hantzschia amphioxys* prevents the cells from being washed away by the tide. During low tide the cells are positively phototactic, so that they come to the surface of the mud and at high tide they are negatively phototactic, so that they descend in the mud. At high tide they also exude a mucilaginous material which causes them to stick to each other and to sand grains.

The detailed structure and connection between raphe with central and polar nodules explain **cytoplasmic cyclosis theory**. In transverse section, the raphe appears as "V" shaped slit (Fig. 6.4) which opens to the external environment by one limb and to the internal cytoplasm by the other. At the region of central nodule the external and internal clefts are joined by a loop canal which is in direct contact

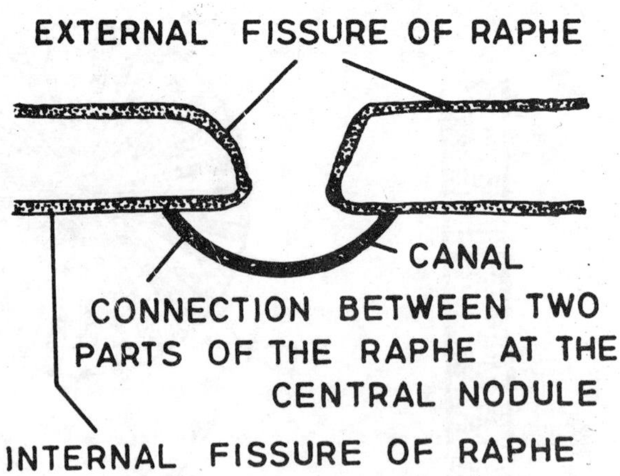

Fig. 6.4. Diagram of central nodule in sectional view showing V-shaped slit and structural details.

with protoplasm. Streaming of the cytoplasm in the inner cleft of the raphe and its contact with external environment through outer clift causes a sort of friction leading to movement of the cell.

The process of floating and sinking of diatom frustules in the planktonic forms has also not been understood clearly. Some workers believe that the increase in the lipids and oils in the cell contents along with deficiency in nitrogen make the cells lighter to float. If the condition is reversed, the cells sink down. Some workers have related this process with the release of metallic ions from the cell sap.

SILICIFICATION

Silicon is an absolute requirement for mitosis and frustule formation in diatoms. Silica deposition vesicles, either derived form Golgi apparatus or endoplasmic reticulum, are associated with silicification. The cytoplasmic vesicles fuse laterally and their common membrane is termed as **silicalemma** or **silica deposition vesicles**. The deposition of silica may be uniform over the valve or is initiated in the centre and proceeds centrifugally in some diatoms or initiates in the marginal region that proceeds centripetally in others. Volcani (1981) has reported all three patterns in the species of *Melosira.* Silicic acid is the form of silicon taken up by diatoms from their environment. The living diatoms do not leach out silica into the medium but that of dead cells does. Diatom frustules are quite resistant to natural degradation and their sedimentation over geological periods at the bottom of aquatic bodies is the significant feature. One of such most known deposits is in Lompoc, California which is known as **diatomaceous earth** or **diatomite**. This silicious deposition is mined and used for a variety of commercial purposes. The succession and periodicity of planktonic diatoms have been found to be correlated with the concentration of silica.

CELL DIVISION AND VEGETATIVE REPRODUCTION

Most common method of reproduction in diatoms is vegetative division of cell. Diatoms usually divide at night and the plane of division is parallel to the valve surfaces. Before cell division, the parent cell gets enlarged and two valves separate slightly. The nucleus divides mitotically into two daughter nuclei which is followed by cytokinesis. After the protoplasm has divided into two by the invagination of plasmalemma, two silica deposition vesicles are formed per cell to form two new hypovalves. The valves of the parent cell separate and the new silica valve becomes the hypotheca of each daughter cell. The daughter cell having the original hypotheca of the parent (now the epitheca of daughter cell) is smaller than the parent cell. Thus, there is progressive decrease in average cell size in a population of diatoms (Fig. 6.5). Though this reduction in size is found in most of the species, some species manage to maintain

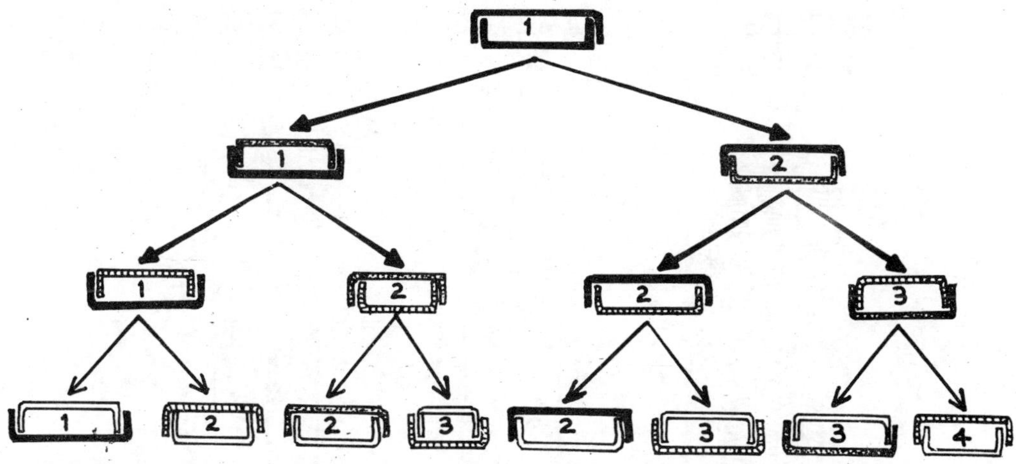

Fig. 6.5. Successive cell division in diatoms showing progressive decrease in cell size.

a constant size which may be due to plasticity of their cell wall. In some forms constant size is maintained during favourable growth conditions while there is diminution of cell size in unfavourable growth conditions. The prevalent method of attaining maximum size is by sexual process i.e., by the formation of auxospores.

AUXOSPORE FORMATION AND SEXUAL REPRODUCTION

Auxospores are large cells which devélop from zygotes, the product of gametic union. Thus, the process of sexual reproduction is intimately linked to reduction in cell size which varies from species to species. There seems to be a distinct size range (usually 30-40 percent of maximum) within which the cells start reproducing sexually. Sexual reproduction is oogamous in centric diatoms and isogamous in pennate diatoms. The product of sexual fusion i.e., **auxospores** (zygote) is not a resting stage but characteristically increases in volume immediately after gametic fusion. The vegetative cells of both Pennales and Centrales are diploid and gametes are formed after meiotic division. Gametes are amoeboid in Pennales while spermatozoids are flagellated in Centrales.

In pennate diatoms, cells aggregate in pairs i.e., **gamontogamy** (Fig. 6.6) and 1 or 2 nuclei develop into gametes out of 4 produced after meiotic division. The parent cells fuse **(cystogamy)** and the gametes generally derived from different cells fuse in a copulatory jelly. Majority of members are isogamous and the gametes move towards each other. In some species the gametes are morphologically similar but they show physiological anisogamy.

In the members of Centrales, oogonia are found as slightly extended cells with an elongated nucleus. One egg cell is enclosed inside an oogonium until fertilization. The male sex organ spermatogonium produces four uniflagellate spermatozoids. In *Melosira varians,* the vegetative cell divides to form 4-8 spermatogonia while in *Cyclotella tenuistriata,* the vegetative cell acts directly as a spermatogonium.

Immediately after gametic fusion, zygote begins to grow into an auxospore. After two mitotic division, auxospore ultimately gives rise to the vegetative cell which has the largest dimensions to the particular species. There are some exception to the general mode of sexual reproduction. In some Centrales like *Cyclotella meneghiniana* gametes have not been observed. In some species auxospore is surrounded by a fertilization membrane known as **perizonium.** In some members like *Eunotia,* copulation tube is formed during sexual reproduction and the contact is achieved by chemotropic growth. Sometimes two haploid nuclei from a single cell fuse to form the auxospore. Such variation is known

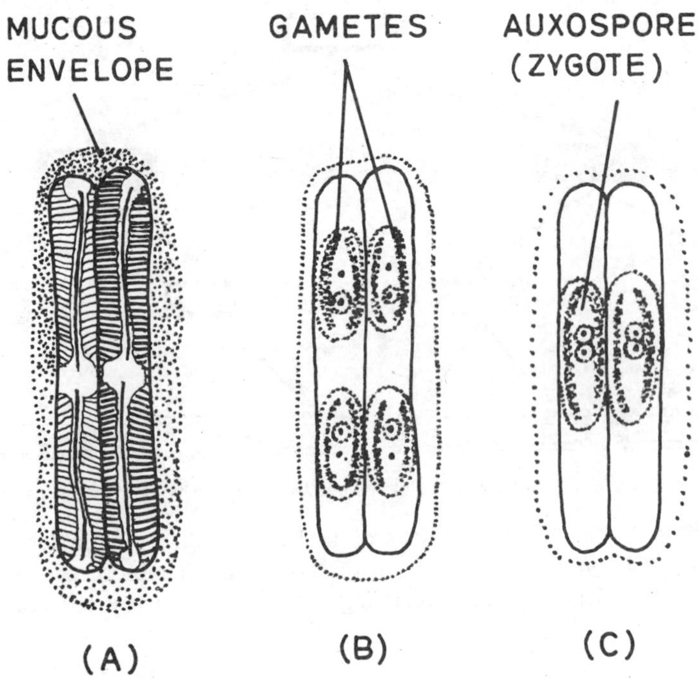

Fig. 6.6. Auxospore formation in *Pinnularia* sp. (A) Two cells lie next to each other. (B) Formation of two gametes in each mother cell (with one functional nucleus) after meiosis. (C) Formation of one auxospore in each mother cell.

as **autogamous** reproduction. By the process of apogamy, auxospores may also arise parthenogenetically, in which the nucleus of the parental cell undergoes division without any reduction in chromosome number.

RESTING SPORES AND RESTING CELLS

At different times in their life cycles, some diatom cells form thick, ornamented walls and become **resting spores.** Resting spores may be formed singly, in pairs, or in series of four. The protoplasm of the parental frustule darkens and is contracted which develops a very heavy siliceous wall of its own. If such cells are planktonic, they sink to the bottom and tide over unfavourable conditions. In some members like *Chaetoceros,* the valves of the resting spore are dissimilar to each other, one being much more convex in girdle view than the other (Fig. 6.7). During germination, resting spores elongate and divide forming a valve appearing at one end within 24 hours and a second valve, appearing at the other end after 48 hours. The new cells are much wider than the parent cells.

Resting cells differ from the resting spores in having same morphology as vegetative cells and do not form a protective layer. During formation of the resting cells large vacuoles decrease in size and many small ones develop, large lipid bodies are produced and the mitochondria become fewer. Such cells remain viable for atleast 2 months into deep water below the euphotic zone. Within 2 days the cells become active and begin to reproduce.

CLASSIFICATION

On the basis of symmetry and orientation of the secondary structures on the valve surface, diatoms have been divided into two orders, the Centrales and the Pennales. The members of Centrales exhibit radial symmetry with many discoid chromatophores while Pennales represent bilateral symmetry with relatively

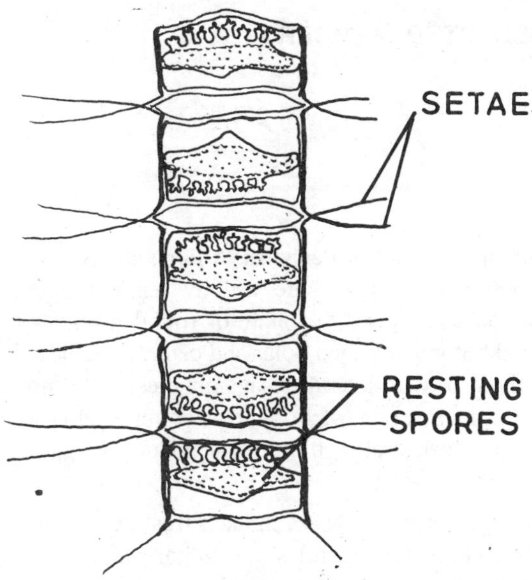

Fig. 6.7. *Chaetoceros* sp. showing a chain of cells containing resting spores and setae.

large chloroplasts in the form of plate or lobes. Other differences between the two orders are: 1. Centrales are more commonly planktonic and marine while Pennales are usually freshwater and frequently occur as epiphytes and as benthic forms 2. Centric diatoms are non-motile whereas many pennate species exhibit a gliding movement. 3. Sexual reproduction in Centrales is oogamous while in pennate forms it is isogamous.

Members of the Centrales are considered to be ancestral to the Pennales, the latter group having developed in late Cretaceous or early Tertiary period (Simonsen, 1972). On cell shape and on the presence or absence of particular processes, Centrales are divided into 3 groups - 1. disc - shaped with no particular processes (*Cyclotella*), 2. valve surface with various horns (*Chaetoceros*) and 3. elongated cells with no processes (*Rhizosolenia*). On presence, absence and development of raphe, Pennales is also divided into 1. Araphideae - with no raphe *(Fragilaria),* 2. Monoraphideae - with one raphe (*Cocconeis*) and 3. Biraphideae - with two fully developed raphes (*Navicula*). *Tabellaria* and *Asterionella* are the examples which only possess a pseudoraphe.

PHYLOGENY AND INTERRELATIONSHIPS

Phylogenetic evolution of diatoms has been traced as the record of large number of fossil diatoms is available due to their siliceous skeleton. The members of Centrales are considered to be most primitive, appearing first in the Jurassic. The pennate forms evolved only from Tertiary period. From the available evidences it appears that pennate forms were derived from the centric forms. The representatives of Pennales with no raphe appeared first in course of evolution. This view finds support that the Centrales are predominantly marine while Pennales inhabit freshwater habitats mainly.

Composition of pigment i.e., presence of chlorophyll **a** and **c** and absence of chlorophyll **b**, dominance of carotenoid, accumulation of oil and leucosin as reserve food, bivalved cell wall, deposition of silica on the wall and occurrence of cysts in many members suggest the similarity and close relationship of diatoms with Xanthophyta and Chrysophyta. Because of these common features many earlier phycologists placed all the three in single division Chrysophyta. Apart from similarities many differences are very sharp which led to possibility of their parallel evolution rather than from divergent lines. Pigment composition and presence of fucoxanthin suggest some resemblances to Phaeophyta.

LIFE HISTORIES OF SELECTED MEMBERS

Order - Pennales

S.order - Biraphideae

Family - Naviculaceae

Genus - *Navicula*

Navicula is one of the common examples of Pennales. A large number of species of this genus are found in freshwater. Frustules are symmetrical and are usually without intercalary bands. The valves are attenuated towards the pole and usually with capitate or rounded apices (Fig. 6.8). The raphe is axial, distinct and straight with small but well defined polar and central nodules. The axial area where the raphe lies is narrow and either without or with lateral expansions near polar nodule or central nodule. Lateral to the axial field, parallel rows of striae or rows of punctae are present. These are either strictly transverse or some what radiate in the region lateral to the central nodule. In girdle view, frustules appear to be small and rectangular.

Usually there are two large chromatophores in each frustule but in few forms there is only one or four or eight chromatophores. Generally frustules are solitary and free -. floating but in some marine species they lie parallely along their long axis within repeatedly branched tubes.

Sexual reproduction is initiated after cell divisions have reduced the cell to a certain critical size.

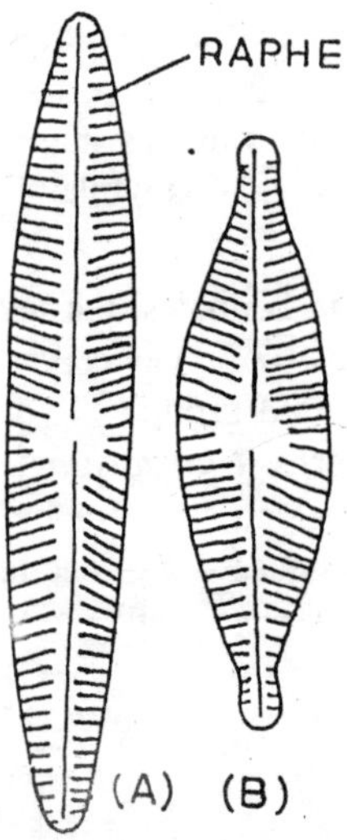

Fig 6.8. (A-B) *Navicula* spp. showing symmetrical frustules and rounded or capitate apices.

Two cells then come together and invest themselves in a mucous envolope. The nucleus in each cell then undergoes two meiotic divisions followed by cytokinesis into two cells. Two gametes are formed in each mother cell which fuse in pairs to form two auxospores, one within each mother cell (Fig. 6.6). These auxospores become greatly enlarged and ultimately new siliceous frustules of maximum size are produced within each auxospore.

Order - Centrales

S.order - Discoideae

Family - Coscinodiscaceae

Genus - *Cyclotella*

The genus *Cyclotella* is primarily marine, though some species are also found in fresh, brackish and salt water. The discoid cells of *Cyclotella* are generally circular in valve view. A few species have elliptical valves. Majority of the species are free-floating and solitary.Bymeans of common gelatinous envelope, cells are united in straight or spirally twisted filaments in some species. The ornamentation of the valve shows two concentric regions - an inner smooth region and an outer peripheral zone with radial straie or punctae (Fig. 6.9). Girdles of the frustules are smooth and intercalary bands are not present between them.

Numerous discoid chromatophores are present in the protoplast. Auxospores are produced singly in each cell and the mode of formation is autogamous.

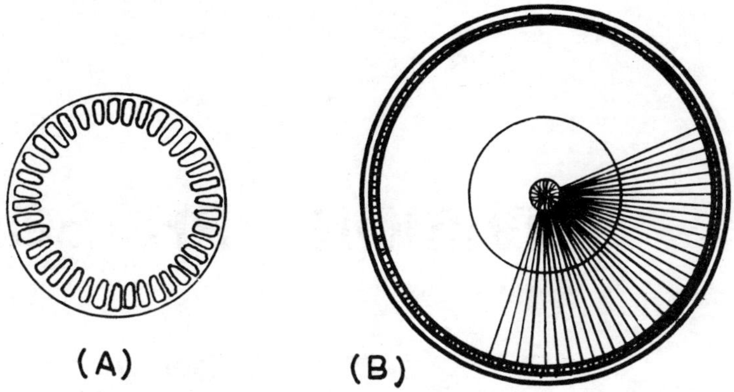

Fig 6.9. (A-B) *Cyclotella* spp. showing circular valve.

7

DIVISION - PHAEOPHYTA

Class - phaeophyceae (Brown algae)

GENERAL FEATURES

The members of the division Phaeophyta constitute a distinct taxonomic group due to their morphological, anatomical and physiological complexities. Phaeophyta comprise about 265 genera and over 1500 species. They are characterised by following features : (1) The photosynthetic pigments include chlorophyall **a** and **c**, **ß-** carotene, violaxanthin and fucoxanthin with occasional presence of diatoxanthin and diadinoxanthin. (2) Food reserve products are in form of soluble carbohydrate called laminarin and an alcohol called mannitol. (3) Characteristic constituents of the cell wall are alginic acid and fucinic acid. (4) The motile stages are pear-shaped with 2 flagella (1 acronematic and 1 pantonematic) arising from the side of the cell. (5) There are two membranes of chloroplast E.R which are usually continuous with the outer membrane of the nuclear envelope.

The methods of growth in different members may be **apical, diffuse** or **trichothallic**. There are no unicellular or colonial members in this class and the members are basically filamentous, pseudoparenchymatous or parenchymatous. The most elaborate type is the simplest type in the Phaeophyceae. The methods of reproduction and life-histories show high degree of specialisation.

OCCURRENCE AND HABITAT

Members are almost exclusively marine except 6 genera which have been reported from freshwater habitat. These are *Heribaudiella, Sphacelaria, Pseudobodanella, Lithoderma, Pleurocladia* and *Porterinema*. A number of marine forms penetrate into brackish water where they often form an important part of the salt marsh flora. These brackish water plants have totally lost the ability to reproduce sexually and they propagate by vegetative means only. Most of the phaeophycean members grow in the intertidal region and the upper littoral region. They are dominant in colder waters particularly in the northern hemisphere. Some species are reported from 100-150 meters deep water (sub-littoral zone), lower littoral zone and the supra-littoral zone. The littoral zone of cold temperate seas is popularly known as sea forests comprising a large number of brown algae i.e., **sea-weeds**. In the tropics the only region where large number of Phaeophyceae are encountered, is the Sargasso sea of the Atlantic. Large sized brown algae such as *Macrocystis* and *Nereocystis* occur in much deeper waters of Pacific Ocean. These algae have long stipes and a number of air bladders occur at the base of their blades. Both the plants are perennial and can reach length upto 60 meters (*Macrocystis pyrifera*) and upto 100 meters in case of *Nereocystis* species. Although some members are epiphytes or endophytes but most of the other genera are lithophytes being attached to rock or stone with the help of well developed holdfasts.

CELL STRUCTURE

In many respects the cell structure of Phaeophyceae is similar to Chrysophyceae, Bacillariophyceae and Xanthophyceae.

Cell walls: The cell walls have an inner cellulose layer and an outer gelatinous layer of pectic material containing chracteristic mucilages such as alginic acid and fucinic acid (fucoidin). **Alginic acid** is made up of ß-1,4 linked mannuronic acid units that have a variable amount of guluronic acid units attached through C-1 and C-4 linkages. **Fucoidin** is primarily composed of α-1,2 linked sulfated fucose units with a lesser amount of α-1,4 linked sulfated fucose units. The relative amount of alginic acid and fucoidin vary between different species, different parts of the same plant as well as in different environments. The alginates (salts of alginic acid) have both structural and ion-exchange roles. They are also helpful in reducing dessication. Alginic acid may constitute upto 50% of dry weight of the alga. Due to their emulsifying or stabilizing properties, the alginates are used in a variety of commercial purposes.

In some species of *Padina*, calcification of the wall occurs where calcium carbonate is deposited as meedle-shaped crystals of aragonite in concentric bands on the surface of the fan like thallus (Fig. 7.9A). In parenchymatous members, plasmodesmata or pores are present between the cells and are bounded by the plasmalemma.

Flagella and Eye spot : Motile cells of the Phaeophyceae have a long anterior pantonematic flagellum with tripartite hairs and a shorter posteriorly directed whiplash (acronematic) flagellum. An exception is found in the Fucales where posterior flagellum of the spermatozoid remains longer than the anterior flagellum. The eyespot consists of 40 to 80 lipid globules arranged in a single layer between the outermost band of the thylakoids and the chloroplast envelope.

Chloroplast : Chloroplasts may be present singly or many in each cell and their number is a dependable taxonomic criterion. Apart from distinct stellate type *(Bachelotia)* they may be discoid, platelike or ramified. The greatest variety of chloroplast form in vegetative cells occur in the Ectocarpales. The chloroplasts have three thylakoids per pand and are surrounded by the chloroplast envelope and two membranes of the chloroplast E.R. (Fig. 7.1). The outer membrane of the chloroplast E.R. is generally continuous with the outer membrane of nuclear envelope in the Ectocarpales but discontinous in the members of Dictyotales, Laminariales and Fucales. Girdle lamellae are present lying just within the chloroplast envelope. Just beneath the girdle lamella in *Sphacelaria*, microfibrils of DNA or **genophore** are present. The DNA microfibrils are both linear and circular and are attached to the thylakoid membranes. The plastids contain chlorophyll **a**, $\mathbf{c_1}$, and $\mathbf{c_2}$ with the major carotenoid being fucoxanthin.

Pyrenoids are single stalked type and they project outward from the chloroplast but are continuous with the chloroplast matrix. They contain a granular substance and are not traversed by thylakoids. Pyrenoids are surrounded by polysaccharide sheath which presumably contains the reserve product.

Other inclusions: In many brown algae, cells contain various small vesicles and small colourless highly refractive vesicles (**fucosan vesicles**). Their abundance has been correlated with the sites of high metabolic activity like dividing cells. These vesicles contain a colourless, highly refractive acidic fluid (**phaeophycean tannins**) that stains red with vanillin and hydrochloric acid. They are readily oxidized in the air leading to formation of brown or black pigment, **phycophaein**. Fucosan vesicles originate in

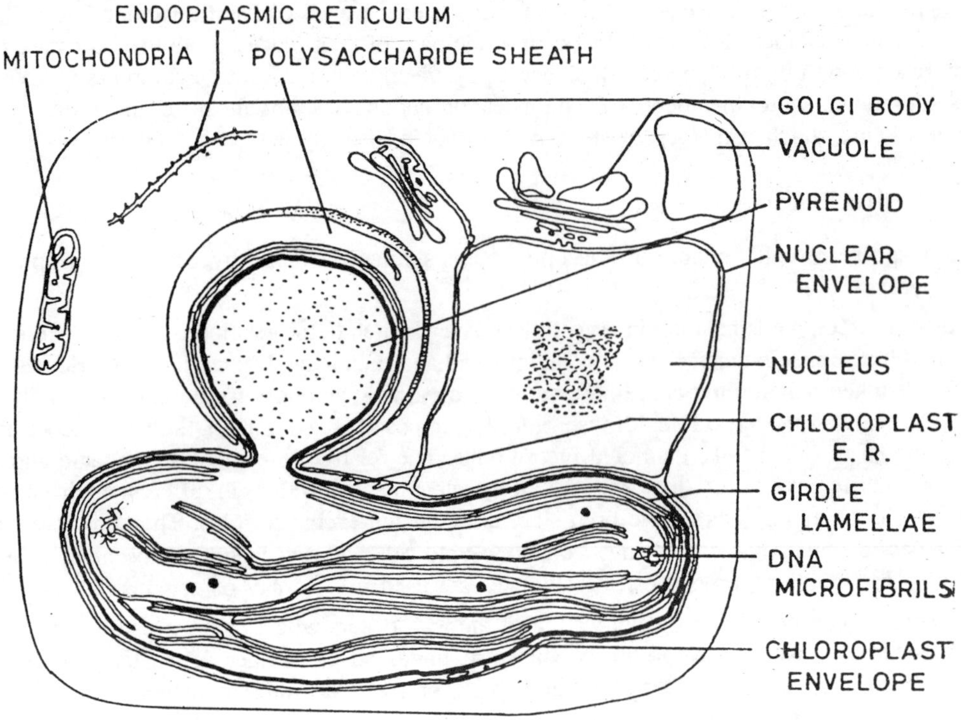

Fig. 7.1. Diagram of a brown algal cell to show the relationships of the organelles and various membranes.

the plastids where the tannins are produced and are then extruded from the plastid area into the cytoplasm in a vesicle.

Nucleus : A prominent membrane-bound nucleus is present inside the cells which contains one or few nucleoli. The resting nuclei do not normally appear to contain chromatin. Several algae possess many feulgen-positive bodies of about 0.6 μ diameter. These are the **chromocentres** and can be seen arranged along the chromosomes. The number of chromocentres is variable and has no relationship with the cytological state of the cell. They are not found in any other algal group and resemble those of some Angiosperms.

GROWTH

In Phaeophyceae, growth may occur by various methods. In most Ectocarpales and many Chordariales, growth is **diffuse**. In **trichothallic** type of growth, cell divisions are localized at the base of one or several filaments. This is found in the members of Desmarestiales, Cutleriales and some Chordariales. **Apical** growth is exhibited by the members of Sphacelariales, Dictyotales and Fucales. A single apical cell, groups of apical cells or marginal apical cells cut off segments proximally. An intercalary meristem is present in the Laminariales. A **meristoderm** which consists of a layer of superficial meristematic cells is also found in Laminariales and Fucales.

RANGE OF VEGETATIVE STRUCTURE

Members of Phaeophyceae do not represent unicellular, colonial, coccoid, dendroid, palmelloid and simple filamentous forms. The simplest form is heterotrichous filament. The thalli in many of the members are composed of pseudoparenchymatous or true parenchymatous cells. In the majority of forms the thallus is differentiated into a basal holdfast and an erect unbranched or branched, solid or hollow, tubular or compressed part. In many Laminariales, the holdfast resembles like root and the erect part bears leaf-like blades and stem like stipes. The plant body presents a close morphological similarity to the vascular plants. *Macrocystis pyrifera* (Fig 7.2 A) attains a maximum length of 60 metres. Plants of *Lessonia flavicans* (Fig. 7.2 B) look like a miniature tree. *Postelsia palmaeformis* resembles a palm tree in appearance, hence it is commonly known as palm tree (Fig. 7.2 C). The examples of different types of thallus construction are illustrated in the following self-explanatory diagrams.

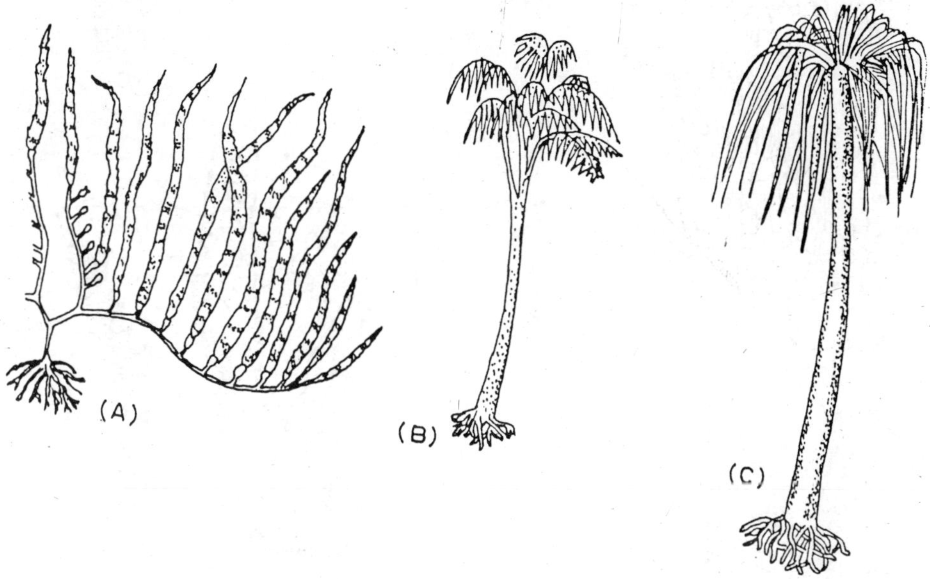

Fig. 7.2. Members of Phaeophyceae. (A) *Macrocystis pyrifera.* (B) *Lessonia flavicans.* (C) *Postelsia palmaeformis.*

1.	Heterotrichous type	-	*Ectocarpus* (Fig. 7.6 A)
2.	Discoid prostrate type	-	*Ascocyclus* (Fig. 7.3 A)
3.	Crusts or Cushions	-	*Ralfsia* (Fig. 7.3 B)
4.	Elaborated erect system	-	*Desmarestia* (Fig. 7.3 C)
5.	Compact pseudoparenchymatous type (uniaxial)	-	*Spermatochnus* (Fig. 7.3 D)
6.	Compact pseudoparenchymatous type (multiaxial)	-	*Castagnea* (Fig. 7.3 E)
7.	Foliose parenchymatous type	-	*Punctaria* (Fig. 7.3 F)
8.	Tubular parenchymatous type	-	*Asperococcus* (Fig. 7.3 G)

Besides these types, endophytic type of association is also found in *Streblonema* (Fig. 7.3 H).

ANATOMY

Anatomical elaborations in the members of Phaeophyceae are also distinct. A transverse section of *Dictyota* shows three layers (upper, central and lower). Lower and upper layers are densely occupied

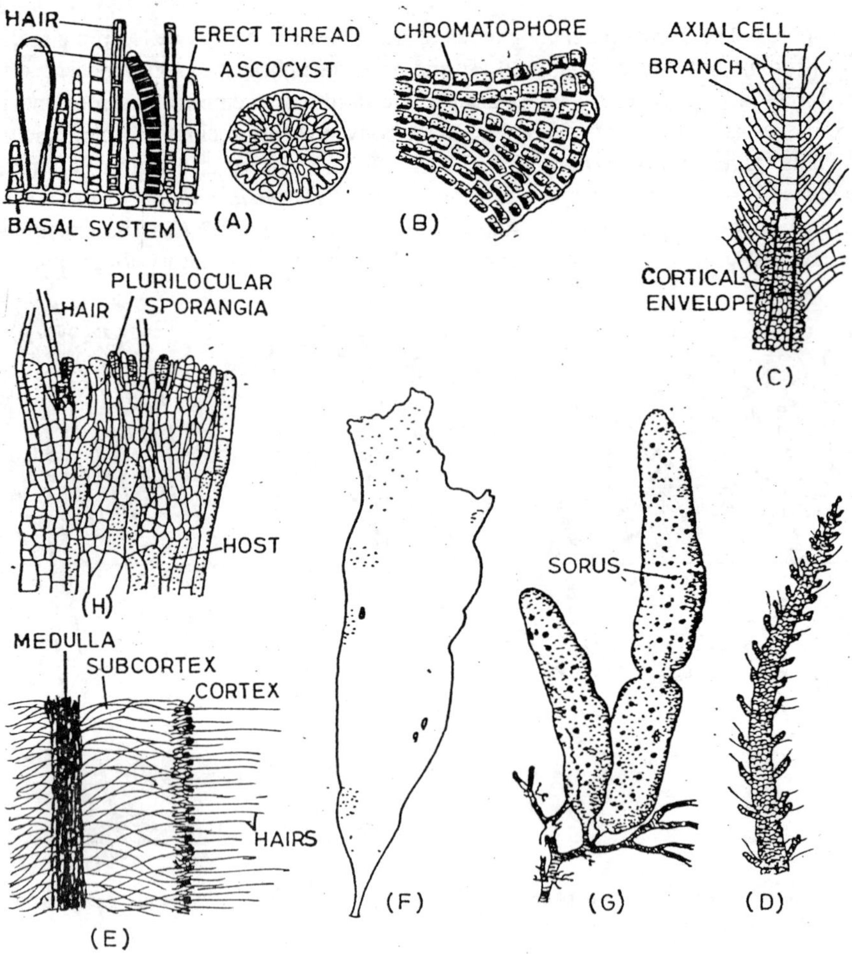

Fig. 7.3. Members of Phaeophyceae showing range of vegetative structure. (A) Section of mature disc of *Ascocyclus orbicularis* and a developed disc. (B) A part of basal stratum of *Ralfsia clavata*. (C) Apex of growing thallus of *Desmarestia aculeata*. (D) External view of *Spermatochnus paradoxus*. (E) Longitudinal section of part of thallus of *Castagnea crassa*. (F) Habit of *Punctaria plantaginea*. (G) Habit of *Asperococcus bullosus*. (H) *Streblonema*, endophytic in *Fucus*.

with chromatophores while central layer is relatively less dense and is filled with fucosan vesicles and food reserve (Fig. 7.4 A). Greater internal differentiation is exhibited by the members of Laminariales and Fucales. In *Laminaria,* the anatomy of stipe and lamina show one or more layers of small cells (**meristoderm**) covered externally by mucilage, the **outer cortex, inner cortex** and **medulla**. Cells of the outer cortex is much wider and elongated with pointed ends while the cells of the inner cortex have square ends. There is the mass of tangled thread in medulla (Fig. 7.4 B) which is characterised by the presence of cells modified into elongate hypha-like structures (**trumpet hyphae**). The ends of transverse cell wall swell out to form bulbs (Fig. 7.4 C). The transverse wall is perforated resembling and behaving like sieve- plate. Some evidences suggest that these hyphae behave as the conducting elements for transportation of water and nutrients. There is a system of anastomosing mucilage ducts which are particularly confined to the stipe. The trumpet hyphae of *Macrocystis* may be compared with the sieve-tubes and sieve-plates of higher plants. However, the former does not have companion cells.

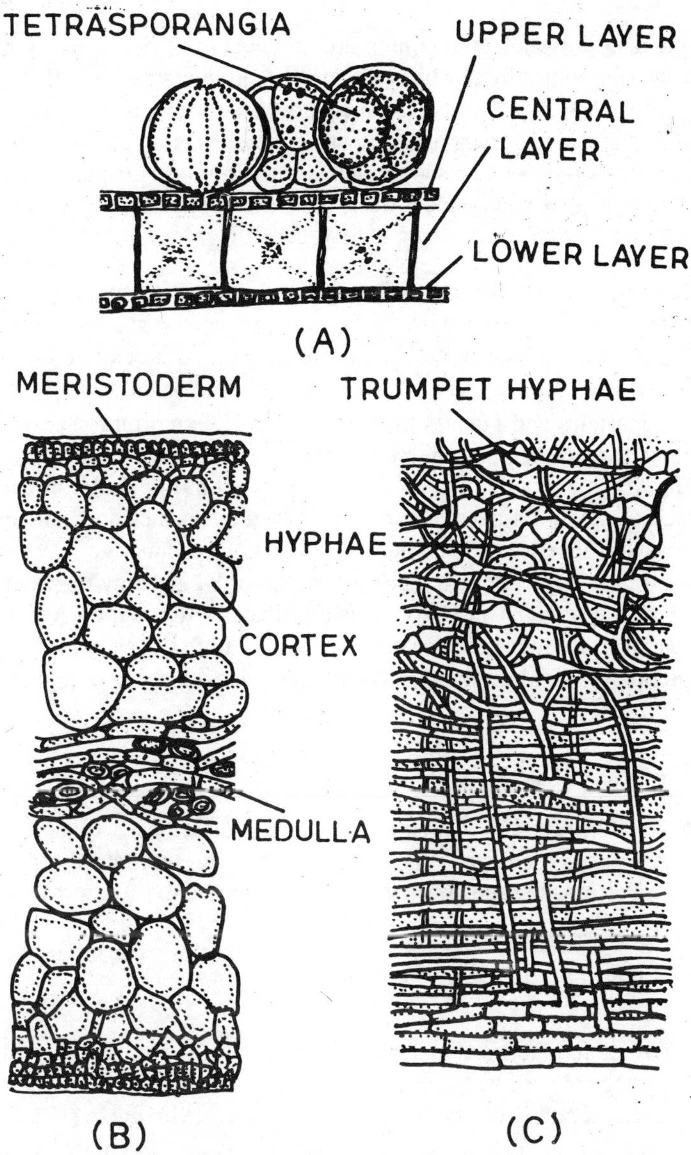

Fig. 7.4. (A) Anatomy of *Dictyota* sp. showing three layers and tetrasporangia. (B) Section of *Laminaria* lamina showing meristoderm, cortex and medulla. (C) Section of central portion of stipe of *Laminaria* showing trumpet hyphae.

REPRODUCTION

Vegetative : Fragmentation of thallus and regeneration of fragmented part is a common feature in many brown algae. In *Sphacelaria* the thalli propagate by the formation of **propagules** or **gemmae.** These are borne on lateral branches and when get detached they float on the surface of water which subsequently germinate into a new plant.

Asexual reproduction : This is performed by typical zoospores which are biflagellate having two laterally placed unequal flagella. Anterior flagellum is longer and pantonematic with a double row of mastigonemes while the shorter acronematic flagellum is directed posteriorly. Two types of sporangia

(multilocular or **plurilocular** and **unilocular)** are involved in the production of zoospores. The protoplast of unilocular sporangia divides by meiotic division and subsequently by mitosis forming 64 or 128 uninucleate fragments. Each fragment metamorphoses into a zoospore which form haploid gametophytic plant. Plurilocular sporangia germinate either on gametophytic plant or on sporophytic plant. On sporophytic plant, plurilocular sporangium forms diploid zoospores or neutral spores which germinate to form the diploid sporophytic plant while plurilocular sporangium present on the gametophytic plant behaves either as male gametangium or female gametangium.

Sexual reproduction: Gametes are produced from plurilocular gametangium which are borne on the gametophytic plants. The opposite sex gametes from different gametangia fuse in pairs. In most members of the orders Ectocarpales and Sphacelariales, the gametes are isogamous while anisogametes are found in Cutleriales. Sporophytic plant is developed after the germination of zygote.

Members of Laminariales and Fucales exhibit oogamous sexual reproduction. Female gametes are large, nonflagellate and nonmotile while the male ones are small and flagellated. The male gametes are produced in antheridia which are unicellular in Laminariales but multicellular in Dictyotales (in sori) and Fucales (in conceptacles). About 64 antherozoids are formed in Fucales and each of which bears a short anterior and long posteior flagellum. Oogonium, the female reproductive organ is always one celled which forms usually a single large nonflagellate egg but 2,4 or 8 eggs may be produced in Fucales. Eggs are discharged out from the oogonia in Dictyotales and Fucales while they remain attached at the tips of oogonia in Laminariales. Parthenogenetic development of unfused gametes may also occur in some genera. In the Tilopteridales, a single, large protoplast is released per sporangium and it is referred to as **monospore**.

ALTERNATION OF GENERATION

The alternation of generation refers to a situation in which a plant has two somatic phases- the haploid or gametophytic and the diploid or sporophytic which regularly alternate during the life cycle. Three different types of life cycles are found in Phaeophyceae which exhibit alternation of generation:

1. Isomorphic,
2. Heteromorphic and
3. Cytological.

In isomorphic alternation of generation, the alternating diploid sporophytic and haploid gametophytic generations are morphologically alike. The members of Ectocarpales, Culteriales (except *Cutleria),* Tilopteridales, Sphacelariales and Dictyotales exhibit such type of life cycle (Fig. 7.5). In heteromorphic type, the two alternating individuals in the life cycle differ morphologically. The thallus is either pseudoparenchymatous **(haplostichous)** formed by aggregation of filaments or parenchymatous **(polystichous)** formed by early longitudinal division of the cells. This type of life cycle is exhibited by genus *Cutleria* of Culteriales and the members belonging to Chordariales, Sporochnales, Desmarestiales, Dictyosiphonales and Laminariales. The tendency for the sporophyte to become increasingly elaborate whereas the gametophyte becoming more reduced culminates in the Fucales in which there is no gametophytic stage. Haploid phase is represented only by gametes and in this type of life cycle only cytological alternation of chromosome number takes place.

CLASSIFICATION

The basic principle in the classification of Phaeophyceae is the pattern of life history, although manner of growth, thallus construction and presence or absence of pyrenoids with the chloroplasts have also been taken into consideration. Kylin (1933) classified brown algae into three series and 11 orders. Plants with

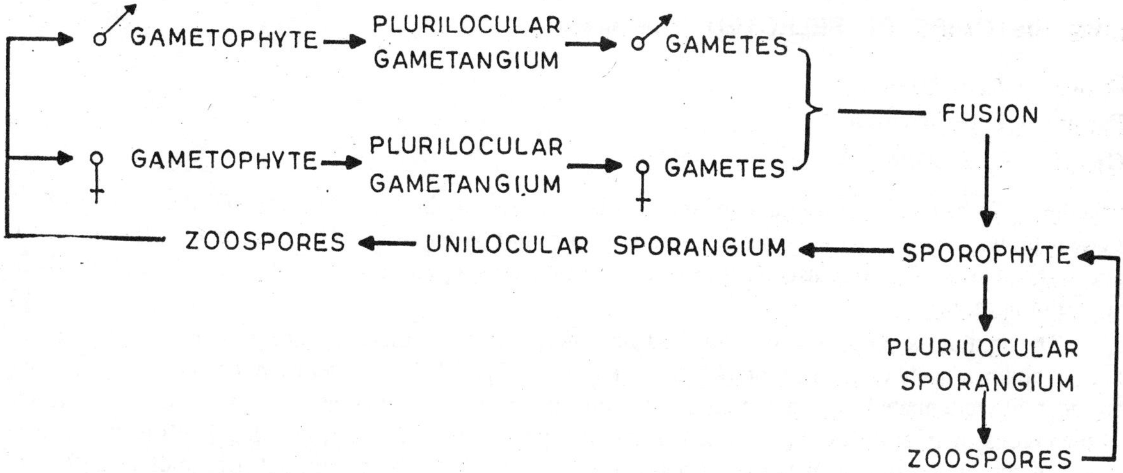

Fig. 7.5. Life cycle pattern in *Ectocarpus* sp.

an isomorphic alternation of generation were placed in **Isogeneratae**. Members having heteromorphic alternation were assigned to **Heterogeneratae** which was subdivided into **Haplostichineae** (with pseudoparenchymatons thallus) and **Polystichineae** (with true parenchymatous thallus). Plants without any morphological alternation of generation were placed in **Cyclosporae** which comprised the order Fucales only.

PHYLOGENY AND INTERRELATIONSHIPS

Due to their well developed thallus construction, Phaeophyceae occupy a distinct status. They are sometimes considered parallel to Chlorophyceae in the nature of plant body and method of reproduction but the presence of multicellular reproductive organs and develoved anatomical features do not justify such conclusion. The nature of swarmer suggests that the Phaeophyceae possibly evolved from some unicellular flagellate organisms. Since the brown algae lack jacket layer around reproductive organs, xylem for conduction, a cuticle for protection against evaporation and mode of reproduction is simple, the Phaeophyceae did not receive any higher position.

Phaeophyceae resemble Cryptophyceae, Dinophyceae and Bacillariophyceae due to presence of chlorophyll **c**. As compared to Chlorophyceae, all these classes have excess of carotenoids over chlorophylls.Fucoxanthin, the main xanthophyll of Phaeophyceae also occurs in Chrysophyceae. Occurrence of oil and saturated fat is a common character for Phaeophyceae, Xanthophyceae, Chrysophyceae and Bacillariophyceae. The sum of the evidence indicates that the brown algae evolved from a prechrysophycean stock. However, differentiation of epidermis, cortex and medulla in the thallus anatomy, development of sieve tube cells and in some cases development of cambium like regions make them distinct from other classes.

Due to motionless tetraspores of Rhodophyceae and supposed immobility of the spermatozoids in Dictyotales, some relationship with red algae has also been suggested.

Fossil records resembling *Fucus* and other members have been recorded from early Paleozoic rocks. Among the few fossil types described, the most striking is *Prototaxites* from the Silurian and Devonian but there is no satisfactory evidence about its systematic position. As the fossil records are not sufficient and the members are not well preserved, these are of not much help in determining phylogenetic relationships.

LIFE HISTORIES OF SELECTED MEMBERS

Order - Ectocarpales

Family - Ectocarpaceae

Genus - *Ectocarpus*

The genus *Ectocarpus* shows many of the simplest features in the Phaeophyceae and is one of the best-known brown algae containing many species. It is a marine alga of wide occurrence in the temperate and tropical seas. This alga usually grows as an epiphytic or epilithic tufts in the littoral, sublittoral and supralittoral zones.

The thallus is composed of branched uniseriate filaments differentiated into irregular prostrate system and profusely or sparsely branched erect system (Fig. 7.6 A). The terminal ends of the branchlets are generally attenuated to an acute point. In some species prostrate system is not well defined. Growth of the erect system is diffuse but that of the prostrate portion is apical. The individual cell is uninucleate and contains ribbon like or plate like or band-shaped chloroplast with or without pyrenoids (Fig. 7.6 B). The end branches bear long, tapering, hyaline and vacuolated cells without chromatophores. In the prostrate portion branched rhizoids are found which help to fix the thallus with the substratum.

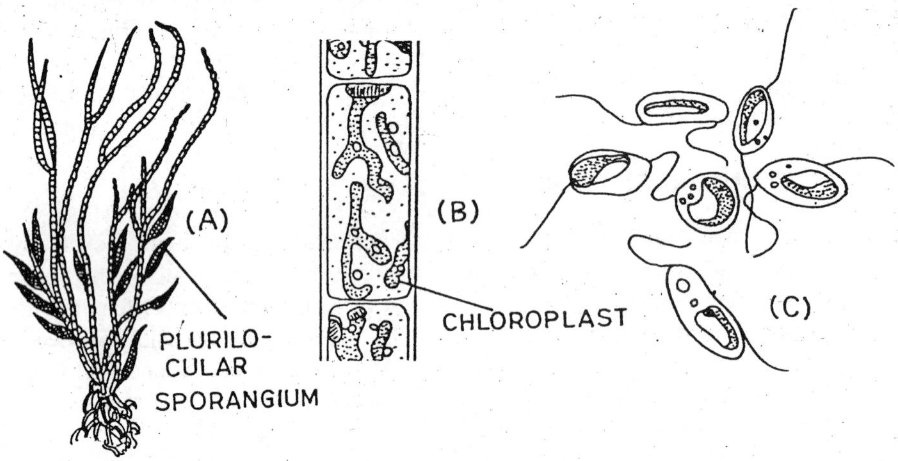

Fig. 7.6. *Ectocarpus* sp. (A) Filaments bearing plurilocular sporangia. (B) Vegetative cells containing ribbon-shaped chloroplasts. (C) Cluster of male gametes around a settled female gamete.

Gametophytic plant of *Ectocarpus* bears **plurilocular gametangia** which are elongated structures and develop from terminal cells of lateral branchlets (Fig. 7.7). The protoplast of each cell is transformed into a single pear - shaped swarmer with two laterally inserted unequal flagella. Swarmers are released through a pore on the cell wall. These swarmers are of same size and are morphologically alike behaving as gametes. Some gametes are relatively less active and behave as female gametes. The more active ones cluster around the relatively passive female gamete and fix themselves to it by their forwardly directed flagella (Fig. 7.6 C). The actual fusion takes place rapidly between the passive gamete and one of the more active male gametes resulting into zygote. This method of fusion and clustering of the active gametes around the passive one is known as **clump formation**.

Sporophytic or diploid plant is produced after the germination of zygote which resembles morphologically with the haploid gametophytic plant. Sporophytic plant bears two types of sporangia - **plurilocular** and **unilocular** (Fig. 7.7). Plurilocular sporangia are elongated, multilocular and are sometimes known as **neutral sporangia** while unilocular sporangia are oval and unilocular. Both develop

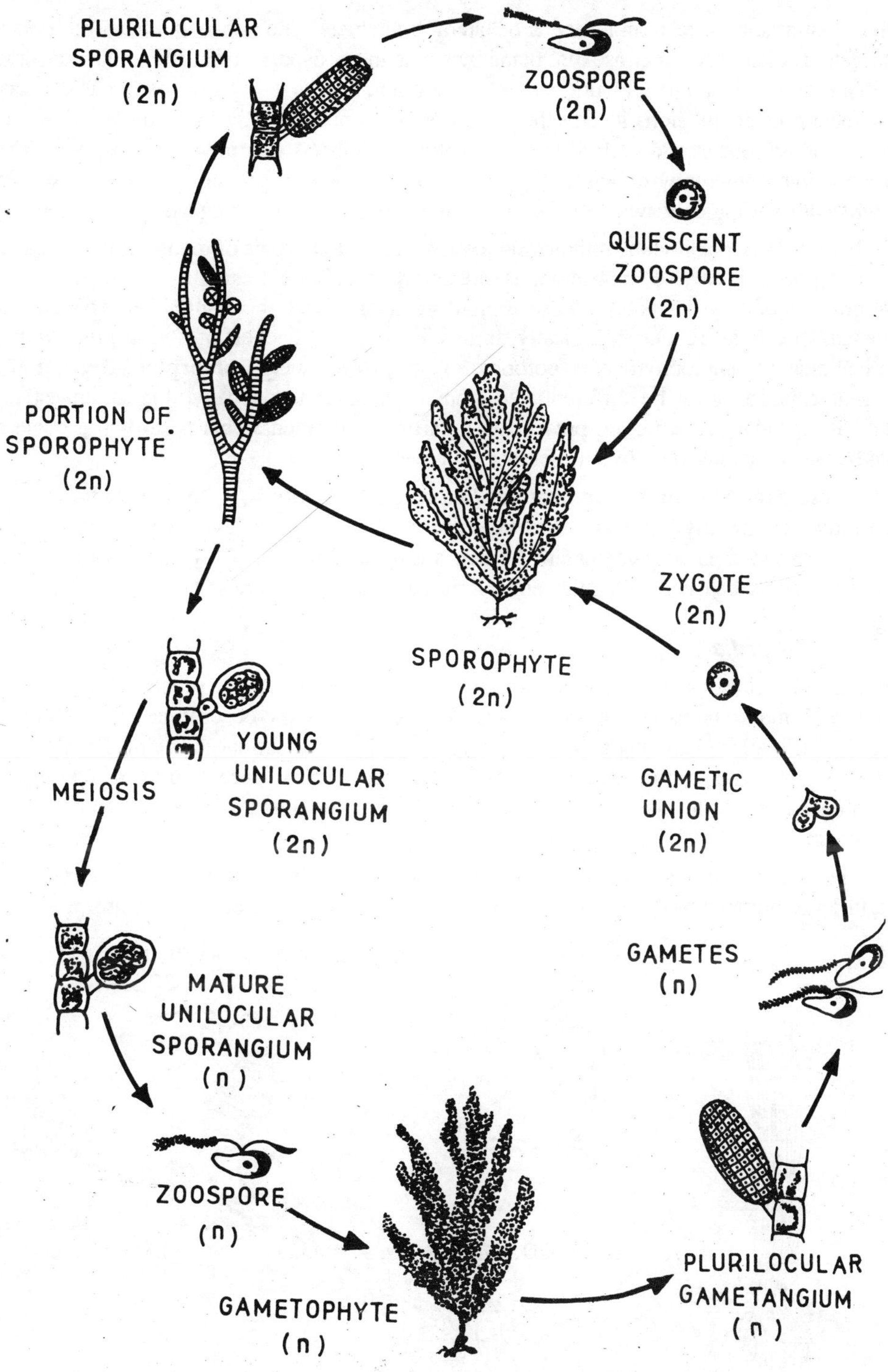

Fig. 7.7. Life cycle of *Ectocarpus* sp.

from the enlargement of terminal cells of lateral branchlets. The zoospores from the plurilocular organs germinate directly into other asexual plants whereas the zoospores from the unilocular organs germinate into haploid sexual plants. In unilocular sporangium, the first division of nucleus is reductional and subsequent mitotic divisions lead to the formation of 64 nuclei. After the completion of nuclear division, cleavage of protoplast takes place which eventually form 64 swarmers or zoospores. These zoospores germinate into gametophytic plants (Fig. 7.7). The plurilocular sporangia borne on the sporophytic plant resemble morphologically with the plurilocular gametangium produced on the gametophytic plant.

Each mature plurilocular sporangium consists of vertical rows of many small cubical compartments. The protoplast of each compartment is metamorphosed into a single biflagellate zoospore without undergoing reduction division. These diploid zoospores and the haploid zoospores produced in the unilocular sporangia are morphologically similar. Thus in the life cycle of *Ectocarpus* two morphologically identical plants (gametophytic and sporophytic) alternate showing **isomorphic alternation** of generation. The filaments can only be differentiated from each other through cytological investigations and the nature of reproductive structure present on a particular filament. This type of life cycle shows a great resemblance with that of *Ulva,* a member of green algae.

Investigations have shown that in *Ectocarpus siliculosus,* environmental factors probably temperature commonly inhibits the development of one of the two generations. Haploid filaments producing gametangia or gametes are formed during warmer part of the year while during colder part diploid filaments bearing unilocular and plurilocular sporangia are formed.

Genus - *Giffordia*

Although most species of Ectocarpales exhibit an isogamous sexual process but in *Giffordia secunda* there is well marked anisogamy. In genus *Giffordia,* there is the occurrence of small-loculed and large-loculed plurilocular gametangia, smaller being antheridia and the larger being female gametangia (Fig. 7.8 C). A third type of plurilocular organs of intermediate size have also been reported. Male gametes are produced from the small-loculed gametangia and the female ones are formed in large-loculed gametangia.

This alga is composed of small to large tufts of much branched uniseriate filaments in which cells contain large number of discoid chloroplast (Fig. 7.8 A,B). Mass of rhizoids are present at the base of

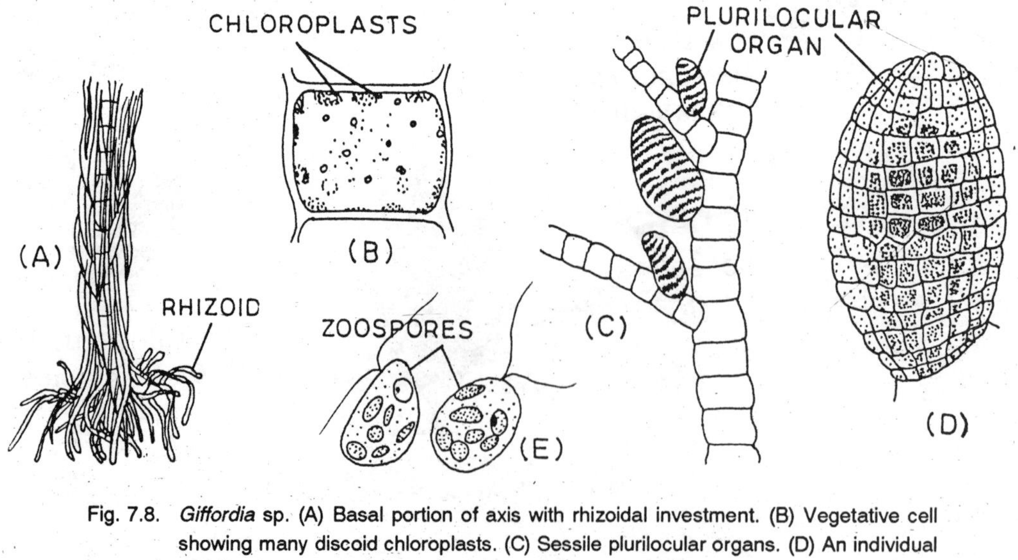

Fig. 7.8. *Giffordia* sp. (A) Basal portion of axis with rhizoidal investment. (B) Vegetative cell showing many discoid chloroplasts. (C) Sessile plurilocular organs. (D) An individual plurilocular organ. (E) Zoospores showing several chloroplasts.

erect threads for attachment with the substratum. There may be heterotrichous organisation with the development of prostrate filament. Plurilocular organs are usually sessile and depending on species may be asymmetric or cylindrical in shape (7.8 D). **Plurispores** released from plurilocular organs have an eyespot and several chloroplasts (7.8 E). In some species the plurilocular organs and unilocular organs remain in close, unilateral series. There is no evidence of sexuality in most of the species (Clayton, 1974). Sometimes the contents of female gametangia are capable of parthenogenetic development.

Order - Dictyotales

Genus - *Padina*

Padina is the only calcified genus of the Phaeophyceae. It is a common and easily recognisable alga found in tropical and subtropical seas throughout the world.

Plants are typically fan-shaped with lacerated proliferations (Fig. 7.9 A) and occur in clusters. The growing edge is circinately rolled protecting the margin with its apical initials (Fig. 7.9 B). Growth of the mature thallus is achieved by a marginal meristem. The stalk of each frond is the upward continuation of a branch of the prostrate perennial rhizome which is richly branched and attached to the substratum by tufts of rhizoids. A distinctive feature of this genus is the deposition of limestone over the surface, the amount of which varies from species to species. In *P. pavonica* calcification is relatively heavy on the upper blade surface but lighter on the under surface.

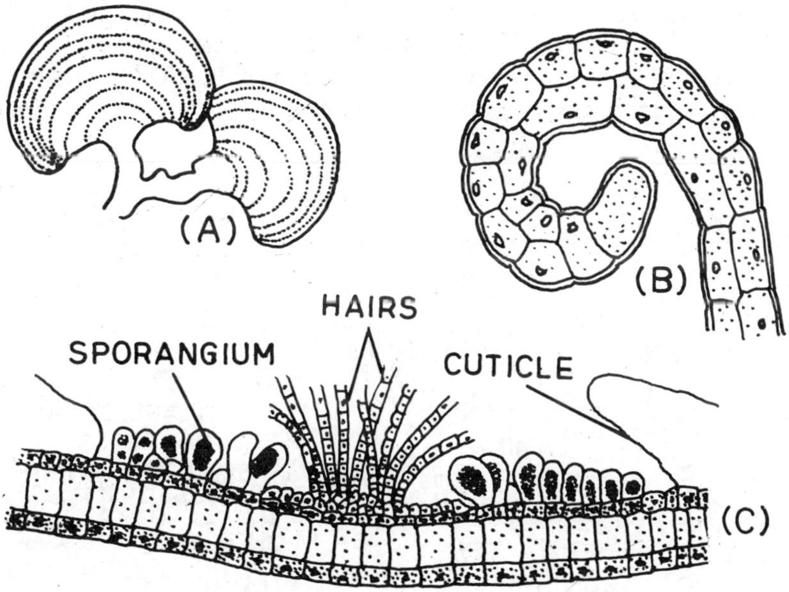

Fig. 7.9. *Padina* sp. (A) Typical fan-shaped plant. (B) Vertical section of margin showing a single initial from the marginal row of apical cells. (C) Section through sporangial sorus

Presence of 4-8 concentric rows of hairs form a conspicuous zonation on the fronds. The zones of hairs occur alternately on the two surfaces but strongly developed on the upper surface. These concentric hairs are prominent in plants exposed to strong light. The reproductive organs (**tetrasporangia, antheridia and oogonia**) are formed on either side of these zones (Fig. 7.9 C) although during their development the hairs are shed. Most of the species are dioecious. Fagerberg and Dawes (1973) have made ultrastructural investigation of gametophytic and sporophytic phases and have found no substantial differences between the two isomorphic phases except minor differences in growth activity. In the cultural studies of *P. pavonica,* a succession of diploid plants was observed, showing **apomeiosis** in the sporangia.

Nonmotile spores produced in pear-shaped tetrasporangia on the diploid plant are distinct. They are relatively larger and sometimes are termed as **tetraspores.** Meiotic division takes place before spore formation. Out of 4, two tetraspores produce male gametophyte and 2 female. Thus, there is genotypic determination of sex at division. Antheridia and oogonia are developed from the surface cells in zonate bands. There is some evidence of a correlation between their development and spring tidal cycles.

Order - Laminariales

Family - Laminariaceae

Genus - *Laminaria*

Laminaria forms an important component of marine flora and is widely distributed in colder as well as temperate and tropical waters. Most species prefer quieter water.

Members are commonly known as **kelps** and are a good source of food, fertilizers and iodine. The sporophytic thallus of most species is differentiated into basal discoid or branched holdfast, a stipe and a blade (lamina) lacking a midrib. The blades of some species are always undivided (Fig. 7.10 A) whereas those of others are divided (Fig. 7.10 B). Growth occurs at intercalary meristem usually placed between lamina and the stipe (Fig. 7.10 C). Although some annual species are known but most of the species are perennial. The blade erodes away above the transition zone and only the holdfast and stipe persist during the nongrowing season. With the onset of new growing season, a new blade is developed from the tip of the stipe. The sporophytes are very durable and may attain even six feet or more in length. It can withstand great deal of tidal stress without getting fragmented. In a few species branched rhizome is established which gives rise to new crops of stipes and blades.

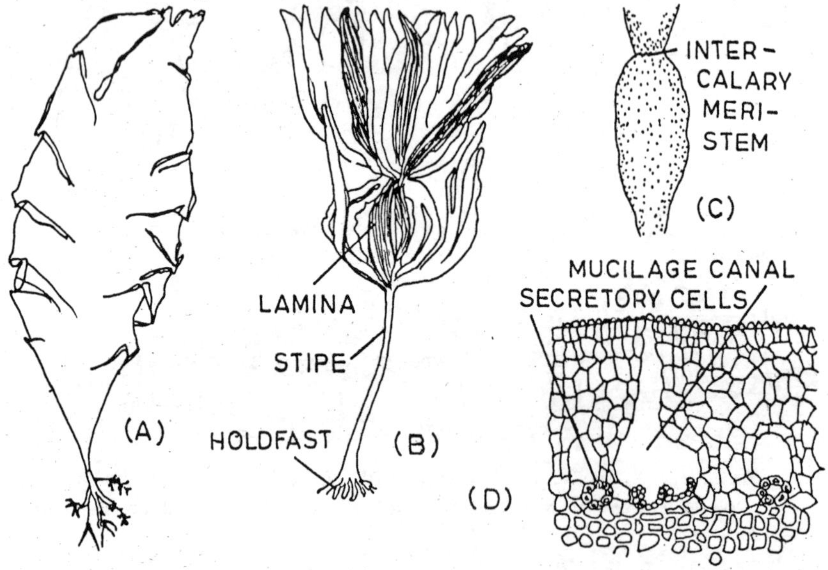

Fig.7.10. *Laminaria* sp. (A) Undivided blade of *L. saccharina.* (B) Thallus showing holdfast, stipe and divided lamina. (C) A portion of the thallus showing intercalary meristem. (D) T.s. of stipe showing mucilage canal and secretary cells.

There are three different tissues in the sporophyte - the central medulla, the cortex and the epidermis. The hold fast (**haptera**) lacks a medulla but all three tissues are present in the stipe and blade. The blade and stipe have the same anatomy, the only difference being the cylindrical to elliptical shape of the stipe and the flattened shape of the blade. At the thallus surface photosynthetic meristematic cells (**meristoderm**) are present which add to the girth of the organ. The meristoderm is composed of small cells (Fig. 7.4 B,C) that cut off daughter cells to the inside which in turn form the cells of the outer cortex. The meristoderm is usually covered with a layer of mucilage. Larger cells of **outer cortex, inner cortex** and **medulla** are differentiated inside the meristoderm. Another type of cells are the hyphae which originate as outgrowths of the cells of the cortex and develop into slender, often branched cells of considerable length that grow into the mucilage of the medulla (Fig. 7.10 D). The medullary cells are characterized by the presence of **trumpet hyphae.** These are also termed as **sieve cells** because there are sieve plates with pores separating the cells.

The sporophyte bears only unilocular sporangia which are produced in **sori** along with sterile **paraphyses.** 32 biflagellate zoospores are formed by repeated division of diploid sporangial nucleus after undergoing meiosis and cleavage of the protoplast. Zoospores are released through the thickened apex of the sporangium and upon germination form two types of mioroscopic gametophyte (Fig. 7.11). The

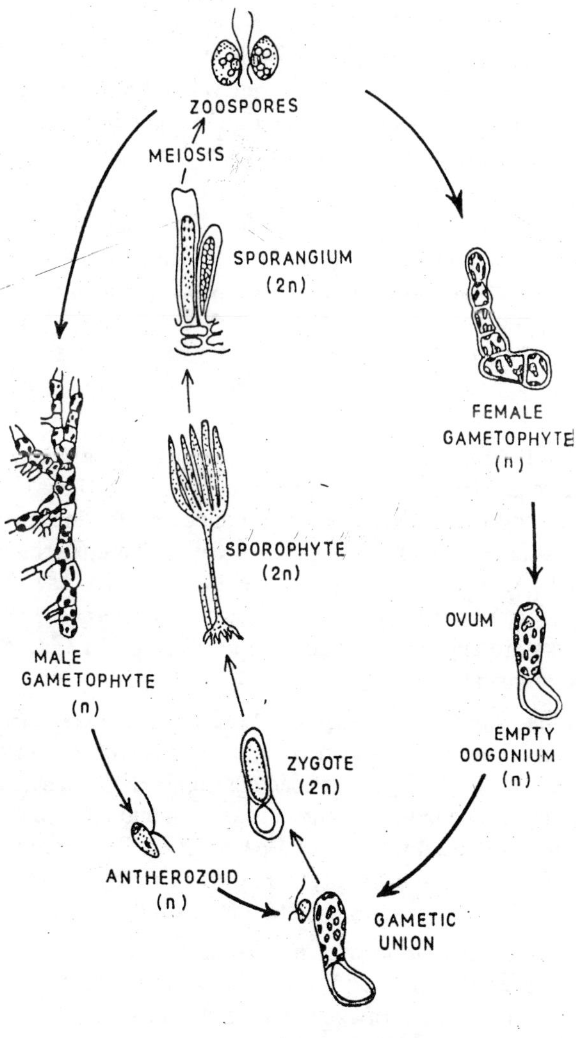

Fig.7.11. Life cycle of *Laminaria* sp.

male gametophyte has smaller cells and is more branched than the female. The antheridia are produced singly or in groups at the terminal cells of the projecting threads or as lateral outgrowth. Single pear-shaped biflagellate spermatozoid or antherozoid is liberated through a pore in the apex. After the release of antherozoid, the male gametophyte disintegrates. Elongated oogonium is produced from any cell of the female gametophyte. A single egg is released from the oogonium which remains attached to the oogonial wall and in this position fertilization takes place. Without undergoing any resting period the zygote germinates to form a flat proembryo and subsequently into the mature sporophyte.

The life cycle in *Laminaria* exhibits **heteromorphic alternation** of a large sporophyte with a microscopic gametophyte. Retention of egg and zygote on oogonial wall represents a transitional stage between primitiveness and advancement. Environmental factors particularly light and termperature control the life cycle of *Laminaria* to a great extent.

Order - Fucales

Family - Sargassaceae

Genus - *Sargassum*

Sargassum is most conspicuous brown alga in tropical and subtropical waters ranging from midlittoral to sublittoral zones. It is commonly known as **gulf weed**. Some species of *Sargassum* grow attached and others float in tremendous masses in the Atlantic Ocean, west of Africa (the Sargasso Sea). It is also very common along the east and west coasts of India growing attached to rocks below low-tide level.

Plant body of *Sargassum* resembles the seed plants as its bushy body is differentiated into irregular to solid holdfast, cylindrical stem like main axis, flattened sterile leaflike laterals. Solitary spherical **air bladders** are developed in the axils of laterals. The main axis is perennial and slow growing giving rise to annual crops of radially arranged primary laterals. These laterals are flattened, with or without a midrib and serrated margin (Fig. 7.12 A,B). The fertile branches are usually finger like and are termed as **receptacles**. It is more condensed and with narrower and shorter segments. Flask-shaped **conceptacles** are embedded in these branches.

Anatomically the thallus is differentiated into mucilaginous **cuticle**, epidermis and peripheral layer of compact, columnar **cortical** cells which have abundant plastids (Fig. 7.12 C). Inner to cortex is the zone of **medulla** with loosely arranged cells. Hyphae are produced by the inner cortical cells but there are no trumpet hyphae. There is an orientation of organelles within the epidermal cells. They have an outer layer of alginic acid vesicles with a basal nucleus and chloroplast. The cap of alginic acid vesicles may shield the chloroplasts and nucleus from intense illumination, especially at low tide when the plants are usually exposed. Air bladders are apparently filled with gases (Fig. 7.12 D). Growth of the thallus is accomplished by a three-sided apical cell.

In floting species, the thallus reproduces by fragmentation. The members of Fucales do not produce spores and the sexual reproduction is oogamous. Antheridia and oogonia are embedded inside conceptacles. The plants are either monoecious or dioecious. In monoecious forms the antheridia and oogonia can be in the same or different conceptacles. The male receptacle is smooth and the female one appears spinous in dioecious plants. The antheridia are usually formed on paraphyses. The first two division of primary nucleus are meiotic and then four mitoses lead to the formation of 64 nuclei inside a mature oval antheridia. Each of which is transformed into biflagellate antherozoids.

The oogonia are usually sessile and embedded in the wall of the conceptacle (Fig. 7.12 E). The oogonial cell undergoes three nuclear divisions and form 8 haploid nuclei. Only one nucleus survives and the others degenerate. The entire content of oogonium metamorphoses into an uninucleate ovum. After emergence through the ostiole, the mature oogonium remains attached to the conceptacle wall by means of long gelatinous strands (Fig. 7.12 F).

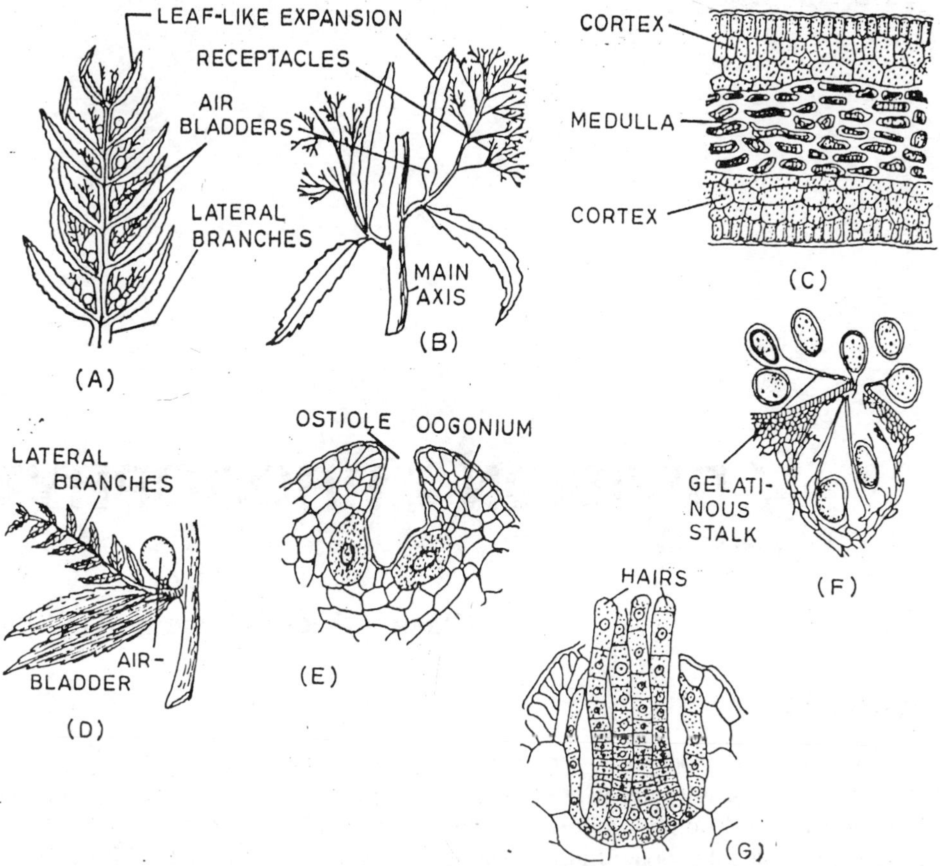

Fig.7.12. *Sargassum* sp. (A) A branch showing leaf-like expansions, air-bladders and receptacles. (B) Small portion of the main axis showing structural details. (C) Section of a portion of thallus. (D) Lower portion of the main axis showing air bladder and lateral branch. (E) Section of a young female conceptacle. (F) Female conceptacle showing liberated oogonia and one oogonium remains attached to the conceptacle wall by gelatinous stalk. (G) Section of a cryptoblast.

Antherozoids after liberation into sea water swim towards female conceptacle and many of them surround each oogonium. Fertilization between one of the attached antherozoid and ovum takes place within the oogonium which is still attached to the conceptacle wall. Without undergoing a period of rest zygote starts germination having no meiotic division. Ultimately by the gelatinization of the oogonial wall, the zygote is liberated out which rests on some suitable object. Through transverse division upper and lower cells are produced germinating into erect portion and rhizoids respectively.

Some species of *Sargassum* have sterile flask-shaped conceptacle like structures (**cryptoblasts**) which are present on primary branches and laterals (Fig. 7.12 G). These cryptoblasts do not bear sex organs but unbranched hairs arise from the floor of the cryptoblasts which secrete mucilage and project in tufts through the openings of cryptoblasts (**cryptoblamata**).

There is no morphological alternation of generation in *Sargassum.* Morphology and anatomy of sporophytic thallus is complex. At the time of gamete formation, meiosis takes place. Thus the gametes are the only representatives of haploidic phase. Only there is alternation between haploidic chromosomal complement with that of diploid sporophyte. The diploid plant instead of producing sporangia and spores forms gametangia and gametes. Thus there is only **chromosomal alternation** of generation.

8

DIVISION - RHODOPHYTA

Class - Rhodophyceae (Red algae)

GENERAL FEATURES

The members of the division Rhodophyta can be distinguished from other eucaryotic algae by the following characters. (1) There is complete absence of any flagellated stage. (2) Their photosynthetic pigments include the characteristic **phycobilins** (R - phycocyanin and R- phycoerythrin) together with chlorophyll **d** and the carotenoid **taraxanthin.** (3) The food storage products are **floridean starch** and the galactoside **floridoside**. (4) There is the presence of nonaggregated photosynthetic lamellae or thylakoid within the chloroplast and there is lack of chloroplast E.R. (Fig. 8.1). (5) Polysulphate esters are characteristic components of the cell wall. (6) They exhibit a highly specialised process of sexual reproduction involving characteristic female cells **(carpogonia)** and the male gametes **(spermatia)**. However, sexuality is apparently absent in some members.

The red algae are probably one of the oldest groups of eucaryotic algae as evidenced by their characters. One of the striking features of most of the red algae is the presence of a pore like opening **(pit-connections)** in the cell wall through which cytoplasmic connection is maintained from cell to cell. Among the carotenoids, α-and β- carotenes are found and besides taraxanthin other xanthophylls are lutein, zeaxanthin, violaxanthin, neoxanthin etc. Majority of the members exhibit triphasic life cycle.

OCCURRENCE AND HABITAT

Although marine red algae occur at all latitudes, there is a marked shift in their abundance from the equator to colder seas. A majority of seaweeds are from red algae having about 400 genera and 4,000 species. As compared to brown and green algae, they prefer deeper waters where they recieve only blue-green wavelengths of light. Few species are also present in polar and subpolar regions where brown and green algae predominate.

According to their geographical distribution, the average size of the plants differ. The larger species of fleshy red algae occur in cool-temperate areas while the members are mostly filamentous and small (except massive calcareous forms) in tropical seas. Red algae can live as deep as 200 metre. This ability

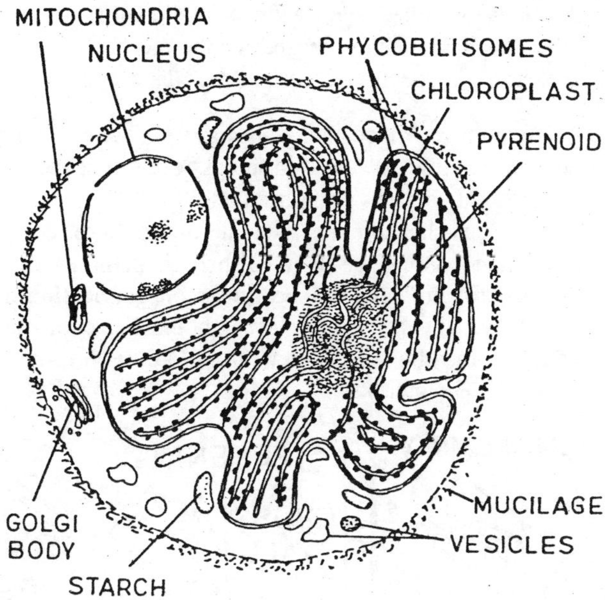

Fig. 8.1. Diagram of a cell of *Porphyridium cruentum* showing structural details. (After Gantt and Conti, 1965).

is due to the presence of accessory pigments. About 200 species are found in freshwater and preferably in moving water attached on rocks or other substrata. Some calcareous members are either partial parasites or parasites on larger algae. *Porphyridium* has been recorded from moist soil.

Brilliant red pigmentation is well expressed in subtidal members or those present in shaded locations. Due to photodestruction of the phycoerythrin, many red algae do not appear reddish instead they look violet, purple, brown, black, yellow or green. This property of the algae in altering their proportion of various pigments in response to differing qualities of incident light is referred to as **chromatic adaptation** (Crossett et al., 1965).

CELL STRUCTURE

Cell of a red alga shows chloroplast with one thylakoid per band and no chloroplast E.R., floridean starch grains in the cytoplasm, no flagella, pit connections between the cells in filamentous genera and eucaryotic type of nucleus (Bronchart and Demoulin, 1977).

Cell Wall : The cell wall in red algae is composed of an inner rigid microfibrillar component and an outer more amorphous component consisting of mucilage or slime. Cellulose forms the microfibrillar framework in most red algal cell walls while in the Bangiales a ß-1,3 linked xylan performs this function. An outermost cuticle has been detected in Bangiales which was found to be primary protein in *Porphyra.* The mucilaginous, amorphous matrix of the wall is usually a sulfated galactan polymer of which agar, porphyran, furcellaran and carrageenan are the examples. These mucilages may constitute upto 70 % of the dry weight of the cell wall. These sulfated galactans are capable of forming gels under the appropriate conditions like alginate and extracted products are of commercial value.

Calcification : In the members of order Corallinales, cell wall has capacity to deposit calcium carbonate. Alginate, a wall component previously considered to be restricted to brown algae, has been obtained in coralline red algae. It is believed to play a key role in the calcification process. The crystals of calcium carbonate - **calcite** (in Corallinales) and **aragonite** (in Nemalionales) differ in their shape and other physical properties. Many theories have been proposed for the mechanism of calcification but these do not explain as to why calcification is specific to certain red algae only.

Pit connection : In some filamentous red algae, cell walls have **pits** which allow protoplasmic continuity between adjacent cells. Chemical analyses of these pit-plugs reveal that the core is made up of protein and the outer cap layer, when present, is partly composed of polysaccharides. The pit connection consists of a circular septal aperture and a lens-shaped plug fitted into the aperture. On the basis of their ontogeny (Fig. 8.2), two types of pit connections can be distinguished. A **primary pit connection** results from equal division of a cell into two progeny cells. In **secondary pit connection** neighbouring filaments or cells establish contact secondarily. The occurrence of high incidence of parasitism, often involving closely related genera of red algae, may be correlated with this ability to form secondary pit connections in order to facilitate transolcation of materials.

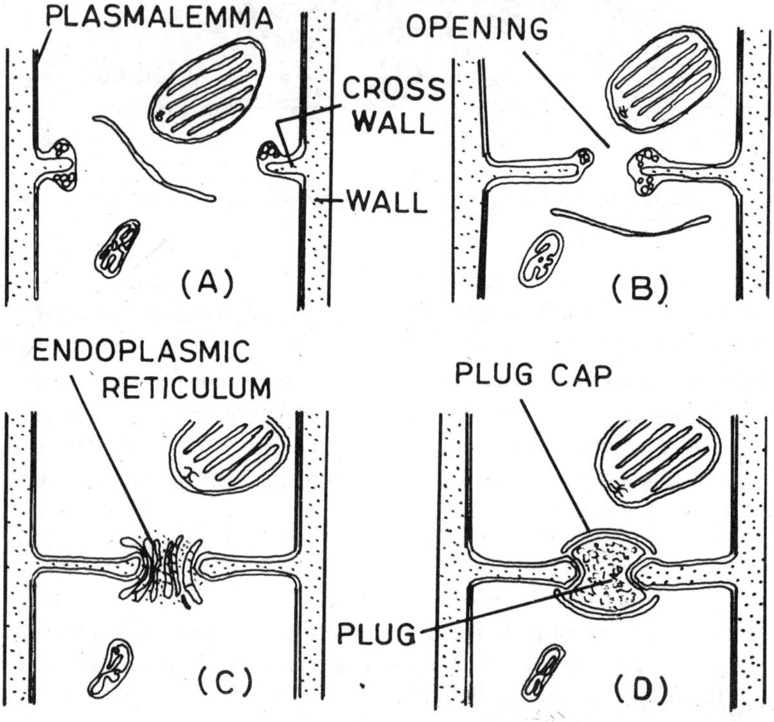

Fig. 8.2. Diagram to show the formation of pit connection in a red alga. (A) Formation of cross wall towards inner side. The wall precursors are found in vesicles derived from the cytoplasm. (B) Completion of cross wall septum leaving an opening in the center. (C) Accumulation of endoplasmic reticulum and condensation of electron-dense material across the opening. (D) Formation of pit connection comprising plug and plug cap with the plasmalemma continuous from cell to cell. (After Lee, 1971).

Chloroplast and storage product : In morphologically simple Rhodophyceae (Bangiophycideae) cells usually contain a single, more or less stellate axile chloroplast with a central pyreroid while in the Florideophycidae, chloroplasts are commonly discoid. Chloroplasts are surrounded by the two membranes of chloroplast envelope without chloroplast E.R. (Fig. 8.1). Inside the membrane bound chloraplast, homogenous matrix is presnt which is traversed by a number of bands and each band has single disc. This is quite in contrast to other algae and is considered to be of phylogenetic significance. The phycobilin pigments containing **phycobilisomes** (35 nm in diameter) are located on the surface of the thylakoids like Cyanophyceae.

Chylorophyll **d** is usually found in the chlorplast besides chlorophyll **a**. Out of the different carotenoids, **zeaxanthin** is found in largest quantity. The phycobiliproteins include **R-phycocyanin, allophycocyanin** and three forms of **phycoerythrin**. Phycoerythrin, the dominant accessory pigment, is responsible for imparting the red coloration often completely masking the presence of chlorophyll **a**. The three forms of phycoerythrin B-, R - and C- phycoerythrin refer to the taxonomic groupings: the order Bangiales, the division Rhodophyta (except Bangiales) and the division Cyanophyta respectively. In those species where phycocyanin predominates over phycoerythrin, the phycobilisomes appear discoid or cylindrical. If both phycoerythrin and phycocyanin are present, the phycobilisomes are spherical. The light absorbed by the phycobiliproteins is tranferred primarily to chlorophyll **a** involved in photosystem II. There are about 35 phycobiliprotein molecules in a phycobilisome (Gantt, 1969). With the change in light intensities, the relative amounts of phycobiliproteins and their ratio with chlorophyll vary.

The food reserve of the red algae is floridean starch, similar to amylopectin of higher plants. Other reserves recorded from red algae include sugars and glycosides such as trehalose, floridoside, isofloridoside, maltose and sucrose. The floridean starch granules are always found scattered in the cytoplasm and never within the chloroplast. In some red algae particularly in members of the Bangiophycidae and the order Nemalionales, pyrenoids are present in the chloroplasts.

Nucleus : The members of Bangiophycidae, Nemalionales, Cryptonemiales and Gigartinales of Florideophycidae have uninucleate condition whereas in other orders multinucleate condition is observed. The nucleus is bounded by well-defined membrane and usually with one or sometimes more than one nucleolus. Mitotic division has also been observed in the members.

Other inclusions : One or more vacuoles are found usually in the cells. In *Lomentaria* vacuoles are bound by a tonoplast membrane. Ultrastructure of *Lomentaria* and *Porphyridum* revealed the presence of dictyosomes and an endoplasmic reticulum. Mitochondria have also been seen in *Lomentaria*.

Growth : By means of an apical cell or a group of apical cells the growth takes place in most Rhodophyta. The apical cell appears as conspicuous dome-shaped structure cutting off segments proximally and this pattern is **uniaxial** mode of growth. There may be a number of filaments each terminated by an apical inital which cut off segments. This pattern is **multiaxial** system of growth.

Secretary cells : In some Rhodophyceae, secretary cells in form of **vesicles** or **glands** are found (Fig. 8.3). These cells are colourless at maturity and commonly have a large central vacuole. Their nature and composition are variable and they lack a nucleus and chloroplasts. The secretary cells in the family Bonnemaisoniaceae usually contain high concentrations of bromine or iodine. In Ceramiaceae and Rhodymeniales, secretary cells are not associated with the formation of halogens but are involved in the secretion of mucilage.

Other cells : There are variety of devices in the red algae for anchorage to the substrate. In many members single-celled rhizoids are formed with the terminal end swelling into a broad discoid attachment or branching into an elaborate unicellular hold fast (Fig. 8.3 C). Multicellular rhizoidal attachments are also produced in many members like *Porphyra, Nemalion* etc.

In many Florideophycidae, colourless hairs are commonly found emerging from the cortical cells. Their exact function is unknown. Some specialised hook like structures or **tendrils** are present in few members which help in attachment.

RANGE OF VEGETATIVE STRUCTURE

The thallus structure in Bangiophycidae is relatively simpler than that of Floridiophycidae. Species of

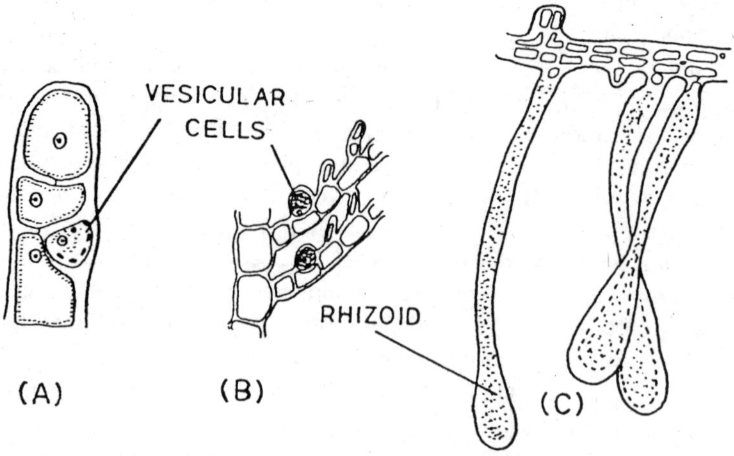

Fig. 8.3. Special cells of Rhodophyceae. (A) Vesicular cell of the *Trailiella* stage of *Bonnemaisonia.* (B) Vesicular cells of *Antithamnion* sp. (C) Unicellular rhizoids of *Polysiphonia rubrorhiza.*

Bangiophycidae are unicellular, palmelloid, simple or branched filaments and never show aggregation of filaments into a pseudoparenchymatous structure as is characteristic of the Florideophycidae. The members of Florideophycidae reveal either uniaxial or multiaxial construction. In uniaxial forms the main axis is composed of a row of usually large cells while in the multiaxial forms the main axis consists of more than one filament (axial filaments) arranged in a parallel manner. Some forms are held together merely by the production of copious mucilage. The largest size encountered among this class never approach those found among Phaeophyceae. The range of vegetative structure in red algae can be understood with the help of self-explanatory diagrams (Fig. 8.4).

1. Unicellular - *Porphyridium* (Fig. 8.1)
2. Palmelloid - *Asterocystis* (Fig. 8.4 A)
3. Filamentous - *Goniotrichum* (Fig. 8.4 B)
4. Heterotrichous type - *Acrochaetium* (Fig. 8.4 C)
5. Discoid prostrate type - *Erythrocladia* (Fig. 8.4 D)
6. Crusts/Cushions - *Hildenbrandia* (Fig. 8.4 E)
7. Elaborated erect system - *Batrachospermum* (Fig. 8.18)
8. Compact pseudoparenchymatous type (uniaxial) - *Lemanea* (Fig. 8.4 F)
9. Compact pseudoparenchymatous type (multiaxial) - *Nemalion* (Fig. 8.4 G)
10. Foliose parenchymatous type - *Porphyra* (Fig. 8.4 H)

Besides these types, parasitic adaptation can be seen in *Choreocolax* (Fig. 8.4 I). Red algae also include nonobligate epiphytes (*Acrochaetium-Rhodochorton* complex), obligate epiphytes (*Polysiphonia lanosa* on *Ascophyllum*) and semiparasites with photosynthetic pigments (*Gonimophyllum*).

ANATOMY

Though some cellular differentiation exists in red algae but the degree of cellular specialization is lesser

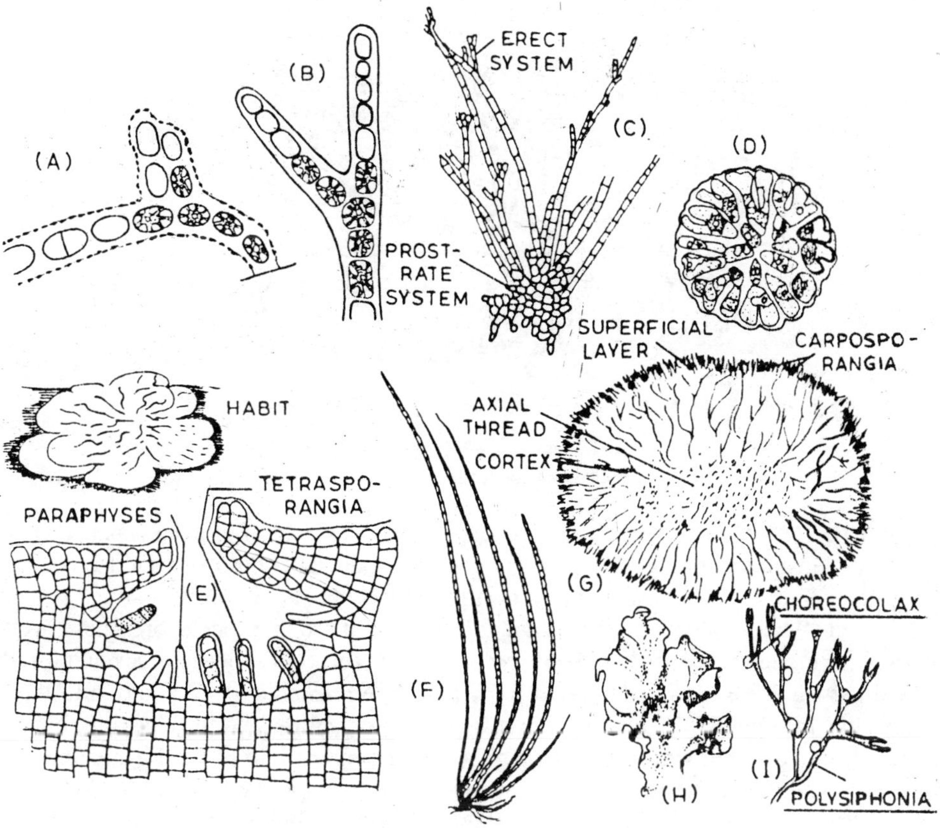

Fig. 8.4. Members of Rhodophyceae showing range of vegetative structure. (A) *Asterocystis* sp. (B) *Goniotrichum* sp. (C) *Acrochaetium attenuatum.* (D) *Erythrocladia subintegra.* (E) Habit and section of tetrasporangial conceptacle of *Hildenbrandia* sp. (F) Habit of *Lemanea nodosa.* (G) T.s. of mature thallus of *Nemalion multifidum.* (H) Habit of *Porphyra perforata.* (I) *Choreocolax polysiphoniae* (parasite) on *Polysiphonia* (host).

than that found in the members of brown algae. In multiaxial forms like *Polysiphonia,* the axial cell divides vertically to form a central and a number of pericentral cells which produce multicellular cortex after periclinal and anticlinal divisions. These may serve as the initials of lateral branches and eventually cylindrial thallus is produced with differentiation into a central medulla with elongated hyphae, cortex for the storage of food and peripheral layer of cells having chromatophores. In some foliose forms like *Weeksia,* four different cell types are distinct in anatomy: secondary medulla which is formed from the cells of inner cortex, primary axial cell, inner cortical cell and outer cortical cell (Fig. 8.5). Medullary cells in red algae may be composed of either enlarged isodiametric cells or uniformly filamentous cells. In few members medulla consists of a mixture of large and small cells (*Callophyllis*).

REPRODUCTION

Vegetative : Vegetative division of the cell into two daugher cells is the most common type of multiplication in unicellular forms. Rarely in some freshwater Bangiophycidae *(Asterocystis),* the fragmentation of thallus and germination of the fragments have been observed.

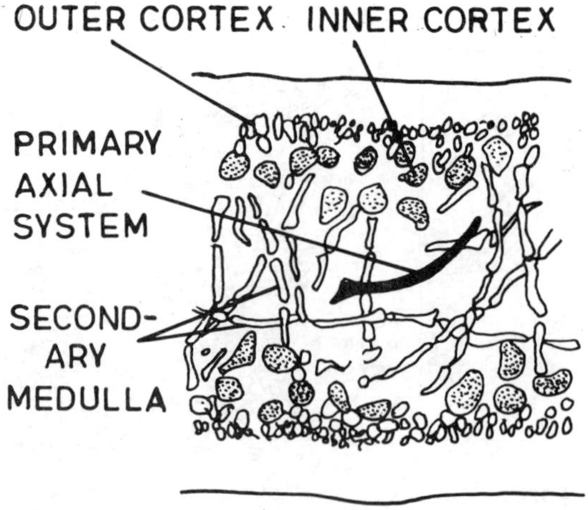

Fig.8.5. Anatomy of *Weeksia fryeana* blade showing four different cell types.

Asexual reproduction : Multicellular forms produce various kinds of asexual spores (Fig. 8.6). **Monospores** are the most common asexual spores which are produced either from undifferentiated cells or from differentiated cells known as **monosporangia**. Monospores are larger structures and are invariably pigmented. When the cell contents divide to produce more than one spore, these are termed as **carpospores**. This mode of reproduction is of common occurrence in Florideae. Asexual reproduction by several-celled propagules is also exhibited in some red algae *(Deucalion)* in addition to their usual carpospore formation. In some Florideophycidae, **tetraspores, polyspores** and **bispores** are found. These are formed after meiotic division within the sporangium and germinate into haploid gametophytic plant. Meiosis does not occur in **parasporangia** (more than one spore per sporangium). In *Seirospora* of the Ceramiaceae, a series of spors or **seirospores** (Fig. 8.6 B) is found. In a freshwater red alga (*Hildenbrandia rivularis*), formation of multicellular **gemmae** has been observed.

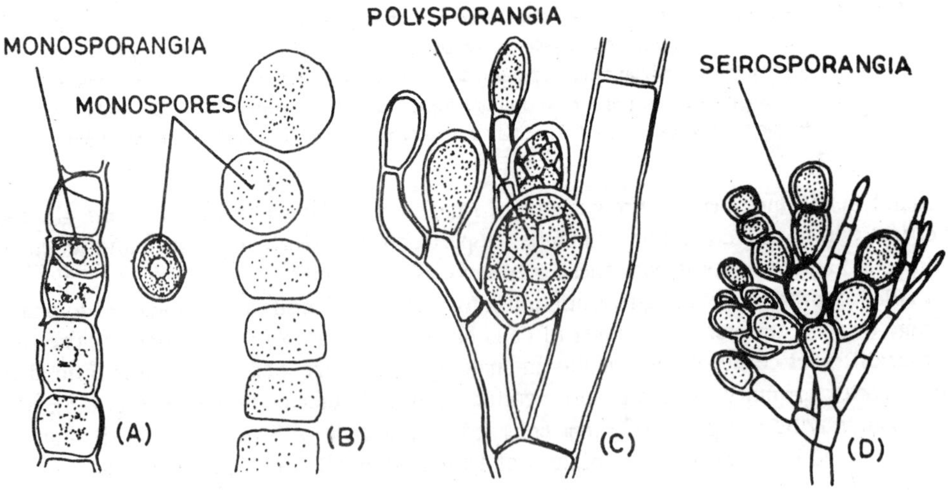

Fig. 8.6. Methods of spore formation in red algae. (A) *Erythrotrichia* sp. showing monosporangia and monospore. Smaller cell formed after unequal cell division forms monosporangium. (B) Formation of monospore in *Goniotrichum* sp. by direct transformation of vegetative cells. (C) Polysporangia of *Tiffaniella* sp. (D) Formation of seirosporangia in *Seirospora* sp.

Sexual reproduction : Sexual reproduction in red algae is invariably oogamous which involves non-motile male gamete **spermatia**) and the specialized female organ (**carpogonia**). Sexual reproduction has not been observed in most of the Bangiophycideae except some forms like *Bangia, Porphyra* etc.

Spermatia of the Rhodophyceae are spherical or oblong cells produced singly in **spermatangia** (Fig.8.7 A). The mature spermatium is uninucleate and wall-less but surrounded by mucilage and may or may not contain functional chloroplast. The spermatium is released by the gelatinization of the spermatangial wall near the apex. Carpogonium consists of dilated basal portion and a usually narrow gelatinous elongate tip, the **trichogyne** which receives the spermatium (Fig. 8.7 B). In Bangiophycidae, carpogonium is sessile but in Florideophycidae it terminates into a short often branched 3 to 4 celled lateral (**carpogonial branch**). At the base of carpongonial branch, the **supporting cell** is present. The female gamete lies at the base of carpogonium.

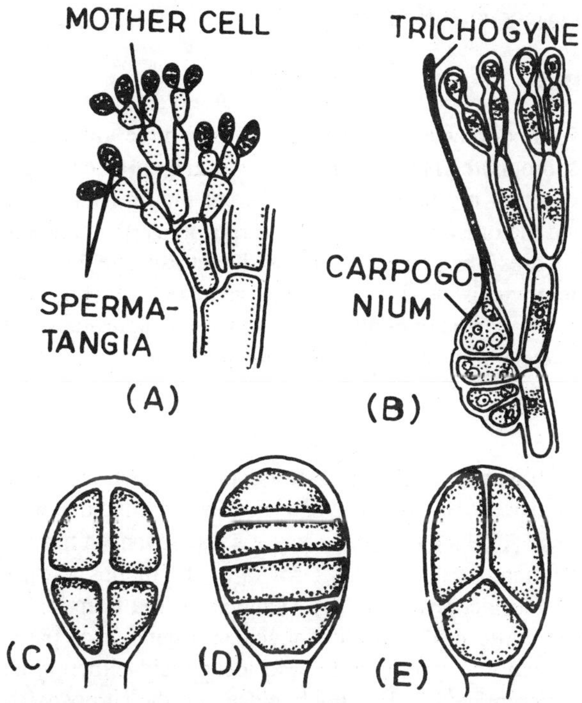

Fig. 8.7. (A) A spermatangial branch of *Acrochaetium* sp. showing spermatangia and spermatangial mother cell. (B) A carpogonial branch of *Liagora* sp. showing carpogonium and trichogyne. (C - E). Pattern of tetrasporangial division. (C) Cruciate. (D) Zonate. (E) Tetrahedral.

The spermatium is usually carried passively by water currents to the trichogyne of the carpogonium. At the point of contact the wall of the spermatium and the trichogyne dissolve and the male nucleus moves into the carpogonium. Fusion of the male and female nuclei takes place in the basal portion of carpogonium. Zygote germinates directly to produce carpospores after meiotic division in Bangiophycidae while in Florideophycidae, the germination of zygote is an indirect process. The fertilized carpogonium produces **gonimoblast filaments, carposporangia** and **carpospores** and this phase is **carposporophyte.** This carposporophytic phase or the "fruiting" stage is also termed as **cystocarp**. Nuclei of the carposporophyte are derived ultimately from the zygote nucleus and in most Florideophycidae (except order Nemalionales) neclear division is mitotic so that the carposporophyte is diploid. The carposporophyte reproduces by the production of carpospores. Diploid carpospores in most of the species do not germinate into gametophyte and they develop into **tetrasporophyte** stage. This is morphologically similar to gametophyte but instead of producing spermatia and carpogonia it produces tetraspores after undergoing reduction division. The four non-motile tetraspores are formed inside tetrasporangia. The pattern of division (Fig. 8.7 C-E) in the tetrasporangium (cruciate, zonate and tetrahedral) is useful in taxonomic relationship. The tetraspores complete the life cycle after forming the gametophyte by germination.

The post-fertilization stages vary from one order to other and this constitute the basis of classification. Although the nuclei in carposporophyte are derived from the zygote, the gonimoblast filaments do not germinate always from the zygote itself. In the advanced orders of Florideophycidae the zygote nucleus migrates to a specialised vegetative cell (**auxiliary cell**). The auxiliary cell may be located near or away from the corpogonium and through a well-defined tube the connection between the two is achieved. According to Papenfuss (1966) there are two types of auxiliary cells, **nutritive** and **generative.** Nutritive auxiliary cells provide nutrients for the developing carposporophyte whereas generative auxiliary cells form gonimoblast filaments.

LIFE HISTORY PATTERNS

In the life cycle stages of Rhodophyceae, three phases are quite distinct i.e., a gametophyte, a carposporophyte and a tetrasporophyte. Thus the life cycle is **trigenic** or **triphasic**. In Nemalionales, there is an alternation between two morphologically different phases, both haploid, one reproducing sexually (the gametophyte) and the other reproducing by carpospores (the carposporophyte). In advanced Florideophycidae, two of the phases (the gametophyte and the tetrasporophyte) are independent and free-living whereas the carposporophyte is dependent on the gametophyte (Fig. 1.5). Gametophytic and tetrasporophytic plants may be similar (**isomorphic**) or dissimilar in appearance (**heteromorphic**). Many variations to the basic triphasic pattern have also been noticed.

CLASSIFICATION

The division Rhodophyta comprises a single class Rhodophyceae which has been divided into two subclasses i.e., the Bangiophycidae or Bangioideae and the Florideophycidae or Florideae. This division is based on the following features: (1) Bangiophycidae possess less advanced forms while Florideophycidae have elaborate pseudoparenchymatous thallus. (2) Growth of the thallus is diffuse in Bangiophycidae whereas it is usually apical in Florideophycidae. (3) Members of the Bangiophycidae normally have a single axile stellate chloroplast while in the Florideophycidae a number of parietal chloroplast is most common. (4) Pit-connections are more common in the Florideophycidae than in Bangiophycidae. (5) Sex organs are simple and little differentiated in Bangiophycidae and the carpospores are produced directly after the division of zygote. In Florideophycidae they are formed indirectly from the zygote.

The subclass Bangiophycidae has single order Bangiales and the Florideophycidae is usually divided into six orders which are distinguished usually on the morphological features and the post-fertilization stages. There is total absence of freeliving diploid stage in the first order i.e., Nemalionales while in the remaining five orders two diploid stages (the carposporophyte and the tetrasporophyte) are present which alternate with haploid gametophytic stage. In the second order Gelidiales auxiliary cells are absent while these are present in the remaining four orders. Out of these four, in Cryptonemiales and Gigartinales, auxiliary cells are formed before fertilization whereas in the Rhodymeniales and Ceramiales they are differentiated after fertilization. In Cryptonemiales the carpogonial branches are special accessory branches whereas in Gigartinales some vegetative branches serve as carpogonial branches. In Rhodymeniales auxiliary cell develops from the outer cell of a two-celled small branch while in Ceramiales, it arises from the supporting cell.

CYSTOCARP

The cystocarp is the fruiting body of the gametophyte in Rhodophyceae. It comprises the mass of carposporangia, the sterile cells of the gonimoblast filaments and the cell or cells subtending them. The cystocarp (carposporophyte) may be borne freely exposed or it may be protected by surrounding vegetative tissue known as **pericarp**. The cystocarp in Nemalionales is haploid while in the rest of the orders it is diploid.

Two views have been put forward about the nature of cystocarp:

(i) It is an integral part of the thallus producing it;

(ii) It is an asexual spore producing generation which develops normally after fertilization and is parasitic upon the thallus bearing the sex organs.

The cystocarp is formed in all Rhodophyceae except Bangiales. Normally the carpogonium is developed terminally on a carpogonial filament which arises from a supporting cell. In primitive members of the class, the zygote formed as a result of fertilization remains in carpogonium and divides meiotically in Nemalionales and mitotically in other orders to form haploid or diploid gonimoblast filaments. In higher forms the zygote or one of its descendants migrates to another cell, called **auxiliary cell,** through a tublar connection called **ooblast** and there it divides and forms **gonimoblast filaments.** The cells of these gonimablast filaments produce **carpospores.** The development of these structures associated with the cystocarp undergoes specialization from Nemalionales to Ceramiales (Fig. 8.8) and thus the cystocarp of Rhodophyceae can be arranged in progressive series as follows.

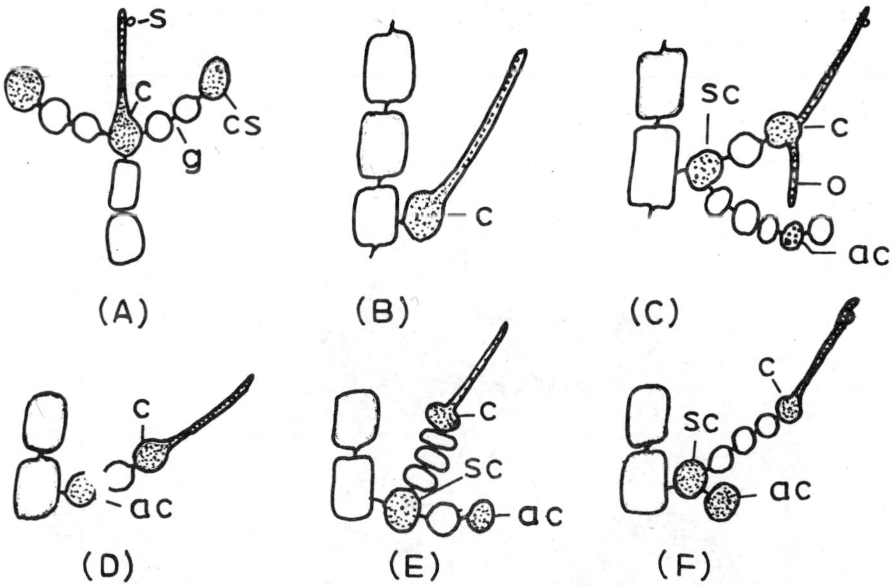

Fig. 8.8. Simple diagrams to show the main features of sexual reproduction in the higher orders of red algae. (A) Gonimoblast arises from carpogonium or hypogenous cell (Acrochaetiales, Lemaneales, Nemaliales). (B) Gelidiales showing single-celled carpogonial branch. (C) Formation of auxiliary cell before fertilization on a special branch (Cryptonemiales). (D) Gigartinales showing vegetative cell behaving as an auxiliary cell. (E) Formation of auxiliary cell as terminal cell on a 2 - 3 celled filament (Rhodymeniales). (F) Formation of auxiliary cell after fertilization on 4 - celled carpogonial branch (Ceramiales). (s - spermatium, c - carpogonium, cs - carposporangium, g - gonimoblast filament, sc- supporting cell, o - ooblast or connecting filament, ac- auxiliary cell). (After Lee, 1980).

1. **Nemalionales**: In this order the simplest type of carpospore formation is seen in *Batrachospermum*, *Acrochaetium* and *Helminthocladia*. Here after fertilization number of haploid cells are cut off from the carpogonium. Each cell contains nuclei resulting from zygotic meiosis (Fig. 8.18). These cells give rise to branched gonimoblast threads. The terminal cell of these filaments produce haploid carpospores.

A little more specialization is observed in *Nemalion* where the daughter nuclei resulting from zygotic meiosis form two cells. The upper cell gives rise to branched gonimoblast threads while the nuclei of the lower or **hypogynous cell** disintegrates (Fig. 8.9). This is followed by the fusion of hypogynous cell with carpogonial filament to form a large **placental cell.** This placental cell furnishes food for the developing carposporophte.

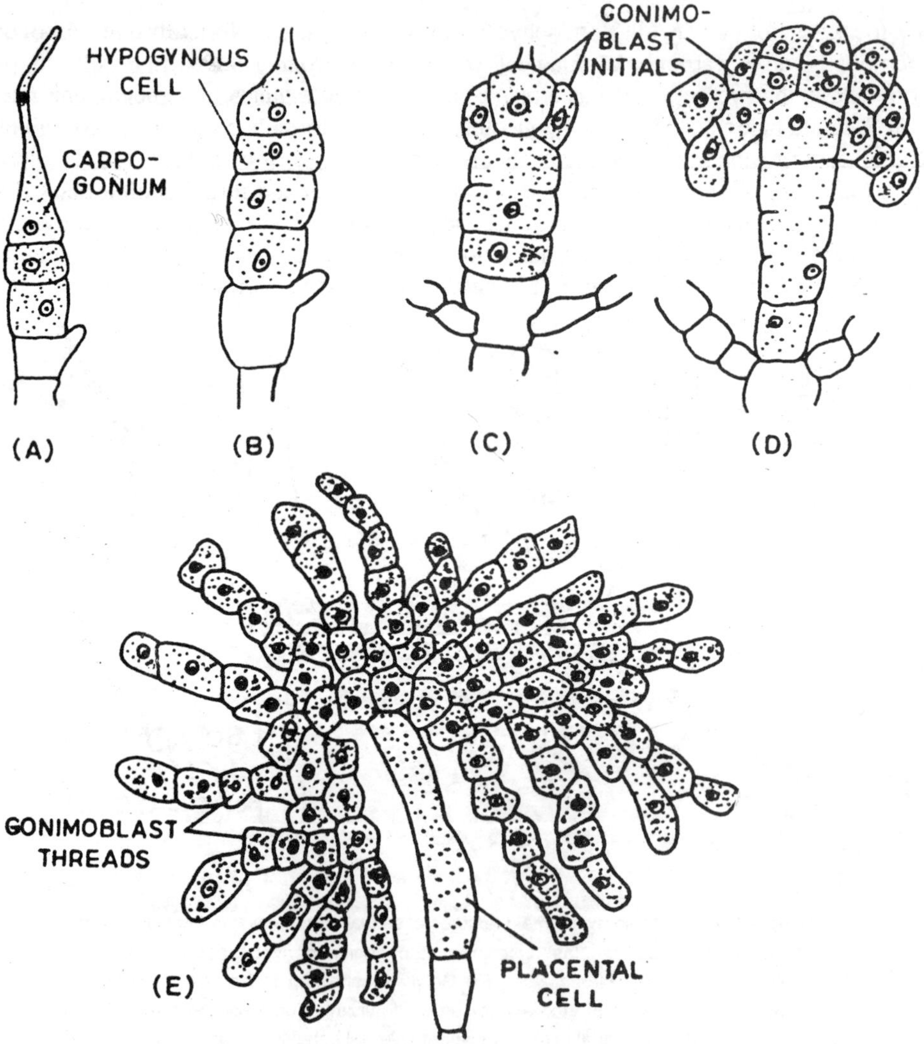

Fig. 8.9. (A - D) Early stages in the development of carposporophyte in *Nemalion multifidum*. (A) Carpogonial branch. (B) Division of fertilized carpogonium into upper cell and lower hypogynous cell. (C) Formation of gonimoblast initials from upper cell. (D) Later stage of development of gonimoblast. (E) Mature gonimoblast.

Members of Chaetangiaceae exhibit greatest specialization among Nemalionales. The formation of cystocarp in this family can be illustrated by *Scinaia* where three-celled carpogonial branch arise, a little below the growing point. Out of these three cells, the first cell forms the carpogonium and trichogyne, the second cell produces four large hypogynous cells with dense contents and the third cell divides to form the sterile covering around the cystocarp (Fig. 8.10).

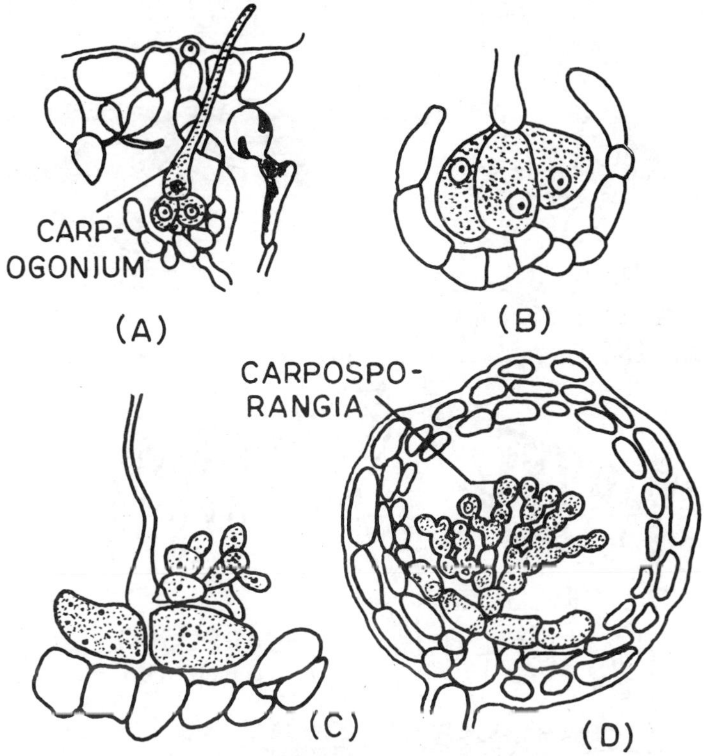

Fig.8.10. Development of cystocarp in *Scinaia furcellata*. (A) Section of the thallus showing carpogonium. (B) Carpogonial branch showing diploid nucleus. (C) Early stage in the development of carposporophyte. (D) Young cystocarp.

2. **Gelidiales** : In the members of this order like *Gelidium*, the carpogonia are borne upon the lowermost cells of vegetative filament growing out from the central axis. Each carpogonium after fertilization gives rise to many celled gonimoblast thread which grows longitudinally along the fertile axis in between certain sterile filaments. The sterile filaments serve as nurse tissue for the growing carposporophyte. Eventually gonimoblast threads form numerous one-celled carposporangia. This is followed by the upward arching of the overlying vegetative tissue and differentiation of the **ostiole** (Fig. 8.11). Thus the cystocarp in this order is an aggregation of carposporophytes formed from several carpogonia.
3. **Cryptonemiales** : In the members of this order the auxiliary cells are borne on special filament of the gametophyte. **Auxiliary cell filaments** may be borne remote from or upon the supporting cells of the carpogonial filament. In the members of some families like Endocladiaceae, the carpogonial filament lacks nutritive cells while in members of Dumontiaceae (*Dudresnaya*) some cells of the carpogonial filament function as nurse cells. After fertilization, the zygote or one of its descendents migrates to the auxiliary cell through ooblast, developed from the base of

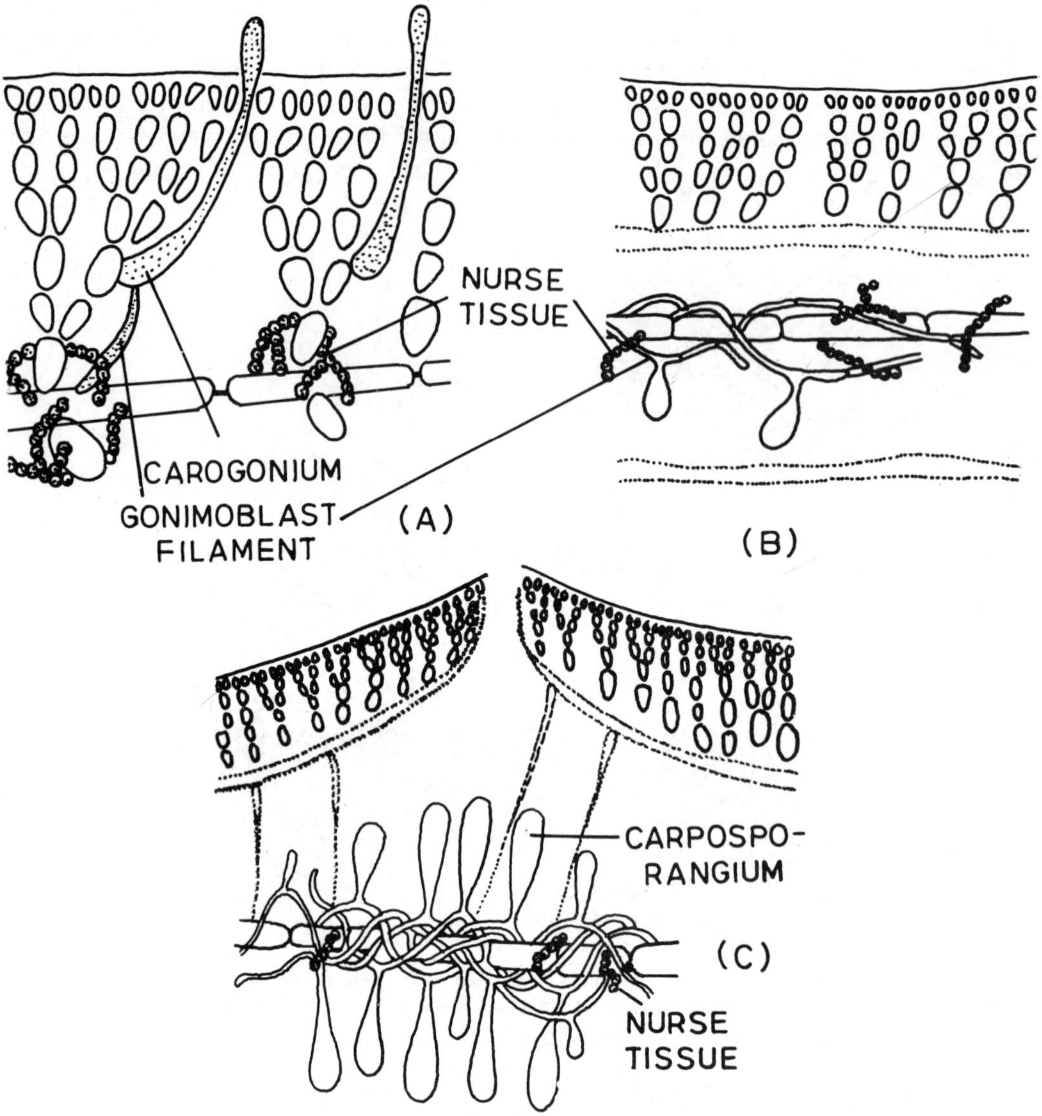

Fig.8.11. Diagrammatic longitudinal section through thalli of *Gelidium cartilagineum* showing developing cystocarp. (A) Carpogonium showing production of gonimoblast filament. (B) Carposporophyte with young carposporangia. (C) Carposporophyte with mature carposporangia.

carpogonium (Fig. 8.12 A). This auxiliary cell may again give rise to secondary ooblast and to another auxiliary cell filament through which the daughter nuclei of zygote moves to the other auxiliary cell. Similarly, tertiary ooblasts are also formed. This is followed by the development of gonimoblast filaments from the auxiliary cell. All gonimoblast filaments produce carpospores. Thus several carposporophytes are formed from single gametic union.

The members of Corallinaceae (*Epilithon*) exhibit greatest specialization amongst Cryptonemiales. Here, the **procarps** (carpogonial branch and its associated auxiliary cells) are borne in **nemathecia** (sori) which lie in a conceptacle. Only the carpogonial filaments at the

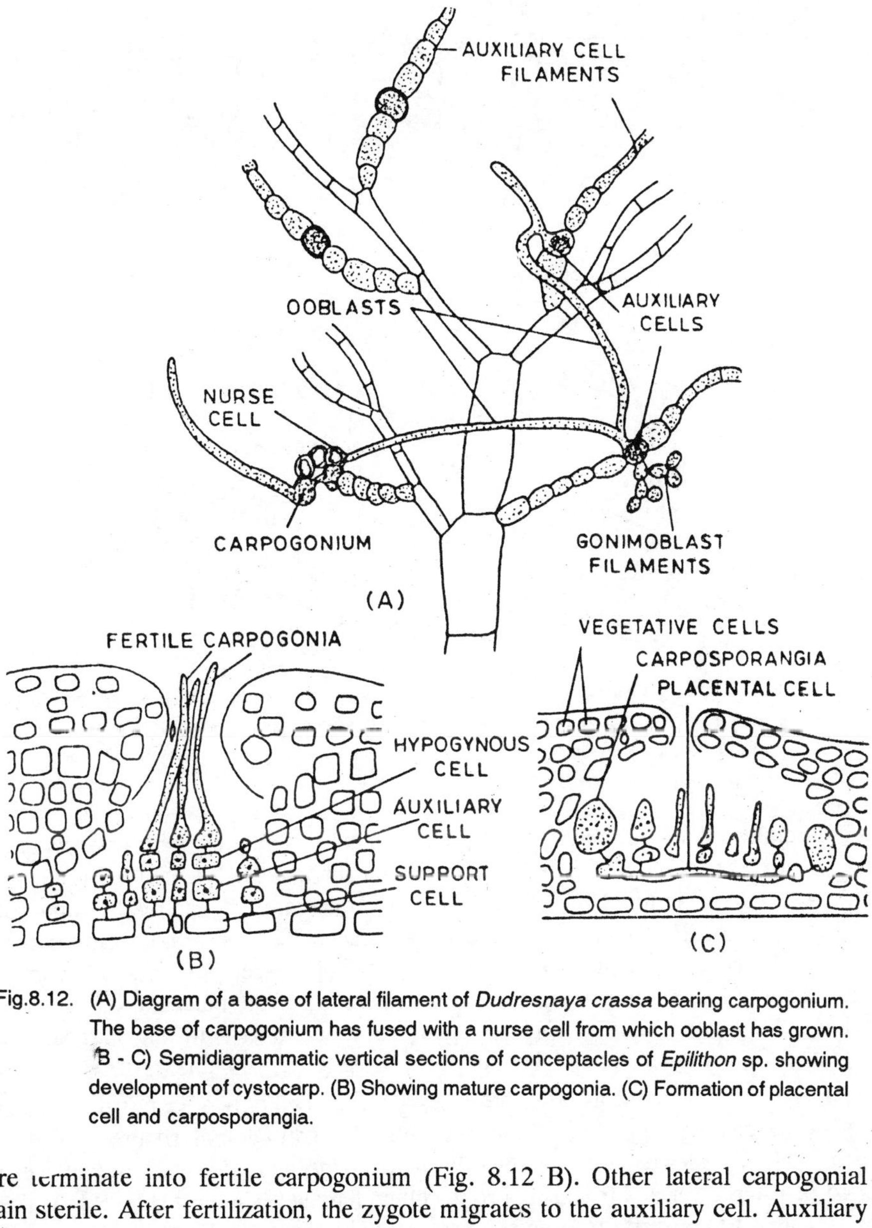

Fig.8.12. (A) Diagram of a base of lateral filament of *Dudresnaya crassa* bearing carpogonium. The base of carpogonium has fused with a nurse cell from which ooblast has grown. (B - C) Semidiagrammatic vertical sections of conceptacles of *Epilithon* sp. showing development of cystocarp. (B) Showing mature carpogonia. (C) Formation of placental cell and carposporangia.

centre terminate into fertile carpogonium (Fig. 8.12 B). Other lateral carpogonial filaments remain sterile. After fertilization, the zygote migrates to the auxiliary cell. Auxiliary cells fuse laterally and give rise to a large disc-schaped placental cell (Fig. 8. 12 C). The gonimoblast filaments arise from the periphery of this placental cell and form carposporangia terminally in basipetal succession.

4. **Gigartinales**: In Gigartinales, the auxiliary cell is a vegetative cell of the gametophyte. In some genera it is the supporting cell of a carpogonial filament (Fig. 8.13 A), while in others it lies in vegetative filaments remote from the carpogonial filament. Some members of this order like *Agardhiella, Opuntiella* etc. are without procarp while others have procarp as in *Gigartina.* After fertilization, the zygote migrates to the auxiliary cell and forms gonimoblast filaments. The gonimoblast filaments draw their nutrition from surrounding vegetative tissue. The carposporophyte formed get surrounded by vegetative tissue which has an opening called **ostiole** (Fig. 8.13 D).

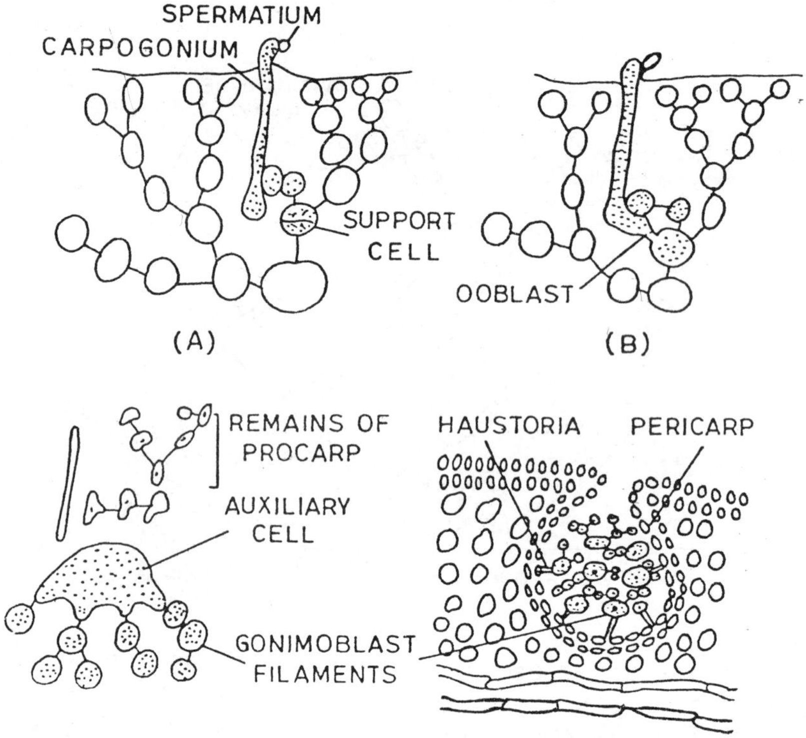

Fig.8.13. Development of cystocarp in procarpial *Gigartina.* (A) Showing support cell which behaves as an auxiliary cell. (B -C) Development of carposporophyte. (D) Formation of pericarp which surrounds carposporophyte.

5. **Rhodymeniales** : Although the auxiliary cell does not differentiate before fertilization, the small immature auxiliary cell which is indistinguishable from a vegetative cell is formed before fusion. Auxiliary cell is the terminal cell of a 2 to 3-celled filament produced by the supporting cell (Fig. 8.14 A). After fertilization, a diploid daughter nucleus of the zygote moves to each of the auxiliary cell and forms gonimoblast initials at its outer face. Cells cut off from these initials develop directly into carposporangia. Meanwhile multinucleate placental cell is formed which is nutritive in function. Simultaneously vegetative tissue lateral to the carposporophyte grow upward and form closed pericarp without an ostiole (Fig. 8.14 C).
6. **Ceramiales** : Members of this order produces auxiliary cell only after fertilization. The auxiliary cell is borne directly upon the supporting cell of a carpogonial filament. The formation of cystocarp in this order can be illustrated by genera *Polysiphonia* of Rhodomelaceae. Here 4 to 5-celled **procarp** is formed from supporting cell on female trichoblast (Fig. 8.15). The supporting cell also cuts off two sterile filaments, one towards its base and the other towards its side. After fertilization, auxiliary cell is cut off from the supporting cell and zygote nucleus migrates to it by means of ooblast. This is followed by the formation of gonimoblast filaments from the upper side of the auxiliary cell. Large placental cell is formed by the fusion of auxiliary cell, supporting cell and the cells of sterile filaments. Meanwhile pericarp also develops from two pericentrals of trichoblast adjacent to the supporting cell.

The members of other families differ only in very minute details. No pericarp is formed in Ceramiaceae.

Thus on the basis of above descriptions, the development of cystocarp can be categorized into three

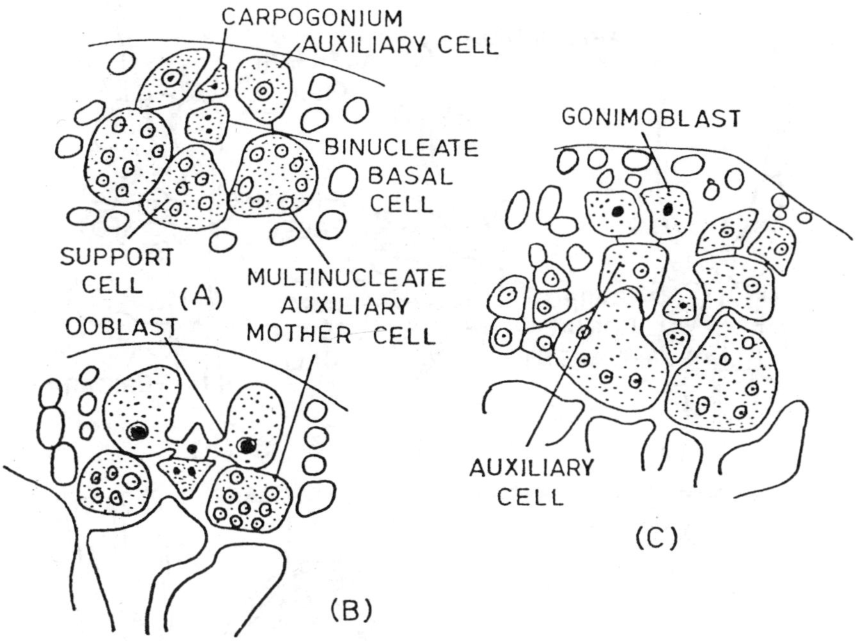

Fig.8.14. Development of cystocarp in *Gastroclonium ovale.* (A) Procarp before fertilization. (B) Procarp after fertilization. (C) Formation of gonimoblast.

distinct series. The first series can be represented by the order Nemalionales, Where the meiotic division occurs in carpogonium. The second line is seen in Gelidiales where the zygote nucleus does not divide meiotically and no auxiliary cells are formed. The third line of evolution can be traced in the order Gigartinales and others where the zygote does not divide meiotically and the diploid phase culminates into the carposporangium and gives rise to tetraspores. These orders are further characterized by the development of the nurse tissue, primary and secondary auxiliary cells and oablasts. The development of cystocarp is so characteristic of each order that it forms basis of classification of subclass Florideophycidae to different orders and families.

PHYLOGENY AND INTERRELATIONSHIPS

Fossil evidences suggest that the Rhodophyta is probably next algal group to have evolved after Cyanophyta during the Precambrian era. Both groups resemble each other in many morphological and biochemical features. Although the Bangiophycidae are less elaborate than the Florideophycidae but it is believed that they represent divergent lines of evolution from a common ancestral stock. Some of the membes of Bangiophycidae lack sexual reproduction and resemble Cyanophyta in cell structure also. The most important resemblances between Rhodophyta and Cyanophyta are: (1) presence of biliprotein pigments, (2) total lack of flagella, (3) similarity between rhodophycean and cyanophycean starch, (4) thylakoids are arranged singly and are widely separated, (5) presence of free trehalose and galactose and linoleic acid, 6) xylans as chief component in cell wall, (7) sulphated galactose, uronic acid, glucose and xylose as main components in the mucilage of Cyanophyta and Bangiales, Nemalionales of Rhodophyta, (8) presence of pit-connections in many Florideophycidae and Stigonematales of Cyanophyta etc. Inspite of these similarities some fundamental points of variations are as follows: (1) Rhodophyta is eucaryotic while Cyanophyta is procaryotic. (2) Difference in spectral properties between biloproteins of both the groups, (3) presence of chlorophyll **d** in Rhodophyta, (4) complex sexual reproduction and life cycle in Rhodophyta, (5) absence of myxoxanthin, myxoxanthophyll, oscillaxanthin in Rhodophyta, (6) absence

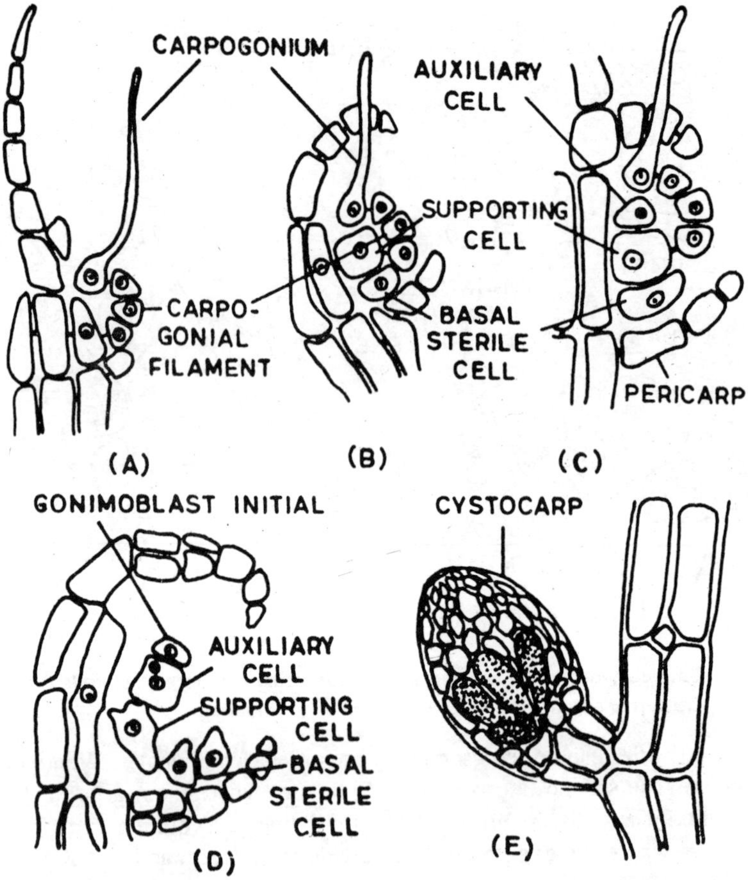

Fig.8.15. Cystocarp formation in *Polysiphonia* sp. (A) Mature carpogonial filament. (B) After formation of sterile filament initial. (C) Formation of the auxiliary cell after fertilization. (D) Procarp after formation of first gonimoblast initial. (E) Mature cystocarp.

of cyanophycin granules and other characteristic structures of Cyanophyta in red algae etc.

On account of above similarities and differences it is widely believed that Cyanophyta and Rhodophyta might have evolved from the same common ancestor.

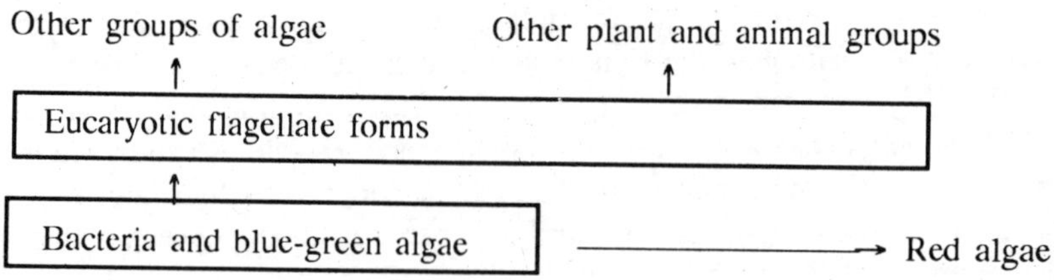

Few workers consider some relationship with Cryptophyta due to presence of biliproteins. The relationships with Chrysophyta and Chlorophyta do not seem justified as there are large number of basic dissimilarities. The present state of affairs regarding the phylogeny and affinities of the Rhodophyta reveals that it is a distinct taxon of a primitive eucaryotic algae with many similarities with Cyanophyta.

LIFE HISTORIES OF SELECTED MEMBERS

Subclass - **Bangiophycidae (Bangiodeae)**
Order - **Bangiales**
Family - **Erythrotrichiaceae**
Genus - ***Compsopogon***

This algae is mainly confined to freshwaters of tropical and subtropical regions. Occasionally it is found in bunches along the edge of the ponds being attached on submerged aquatic plants.

The thallus consists of branched filaments having diffuse growth with blue or purple green colour. The younger parts of the filament are uniseriate with discoid cells (Fig. 8.16 B). A short distance behind

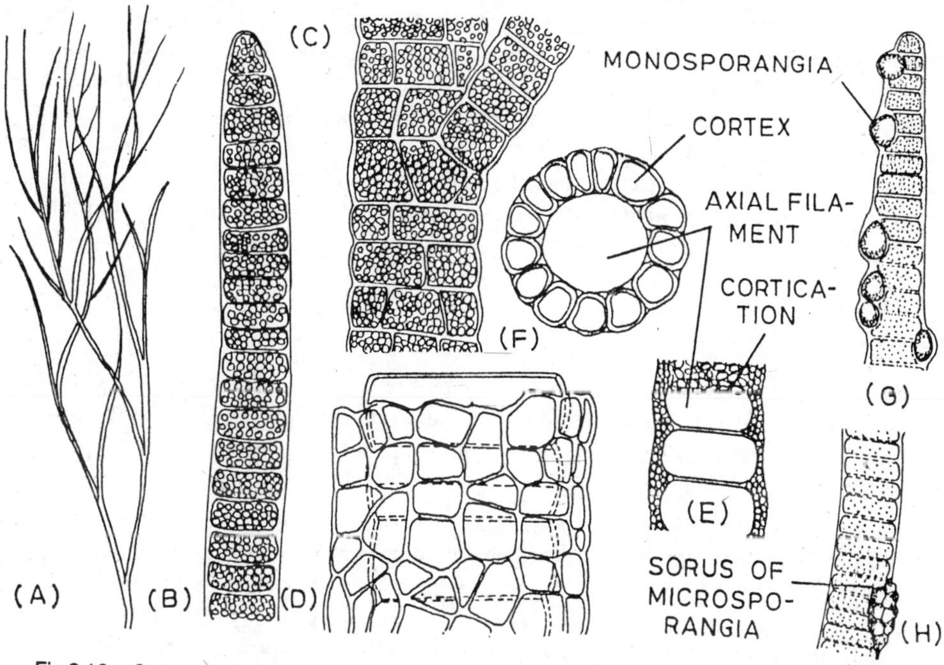

Fig.8.16. *Compsopogon* sp. (A) Habit sketch of a portion of a thallus. (B) Apex of a branch. (C) Surface view of a young branch showing corticated central axis. (D) Older portion of the same branch. (E) Part of a thallus showing cortication. (F) T.s. of a corticated thallus. (G) Apical part of thallus showing monosporangia. (H) Part of thallus showing sori of microsporangia.

the apex, the cells divide vertically and by repeated vertical divisions form peripheral layer of **cortical cells** around a central **axial cell**. The cells at the periphery undergo further vertical divisions forming a small-celled pigmented cortex. The internal cells of the axial row continue to enlarge and it is easy to recognise the cells of this axial row beneath the cortex. This cortical layer is lacking in the basal part of the thallus. Here the cells are commonly overgrown by septate rhizoids which perform the function of anchorage. Cells are uninucleate and in young condition contain a single parietal chloroplast which later become fragmented into many small, spherical chloroplasts lacking pyrenoids. Some of the chromatophores lie immediately near the nucleus but the majority are towards the periphery of the protoplast. Pit connections are absent. Branches arise from the thallus prior to the production of cortex. Secondary branches may arise from cortical cells also.

Sexual reproduction is absent in this alga and asexual reproduction is accomplished with the help of special spores (**monospores**). These asexual spores are non-flagellate having a large nucleus within dense cytoplasm. They are produced either in sori of **microsporangia** or inside solitary **macrosporangia** (Fig. 8.16 G). These sporangia are formed from cortical cells or from uncorticated cells. The macrosporangia develop in specific areas of the filament which is initiated by the formation of a septum. The septum separates the cell into two daughter cells and the prominent one develops into macrosporangium. A single, spherical macrospore is liberated through a minute pore and the vacant place is filled up by the thallus growth. The microsporangial sori are almost hemispherical and contain 15 to 25 microsporangia (Fig. 8.16 H). Initially, sporangia are spherical but soon assume a polyhedral outline as a result of mutual pressure exerted by excessive cell growth. They may develop from any cell of the filament. The spherical spores are formed singly in each sporangium after contraction of the cytoplasmic content. These spores are liberated with a jerk at maturity and secrete a wall.

During germination, the spores become flat on the substratum and divide to form 3-4 celled germling. The central cell of this young germling divides longitudinally to form the filament initial and the apical cell. Subsequent divisions in the apical cell produce discoidal cells which ultimately divide to form the axial and cortical cells. The development of the prostrate system is restricted after the formation of erect filaments and its distal ends produce the rhizoids. The erect filament is uniseriate in young condition and become multiseriate in latter stages involving lateral expansion of the thallus. Thus three regions are distinct in the filament - the growing region, the region differentiated into cortical and axial cells and the rhizoidal region.

Family - Bangiaceae

Genus - *Porphyra*

The blades of *Porphyra* are commonly found on rocky shores from polar to tropical seas. Most species lare present in the intertidal zone but some inhabit the sublittoral region. The base of the blade is attached to the substratum by numerous rhizoidal cells and the length of the blade may reach upto 75 cm. Some species of *Porphyra* are epiphytic.

The thallus arises from a holdfast and on the basis of cell layers and number of chloroplasts, the genus may be subdivided into three types: species with monostromatic blades having a single chloroplast in each cell, species with monostromatic blades having two chloroplasts in each cell and species with distromatic blades having a single chloroplast in each cell (Fig. 8.17). The shape of the blade, the distribution and the pattern of division of reproductive cells are important taxonomic features.

The thallus produces three types of spores-**monospores, α-spores** and **ß-spores**. By the release of a vegetative cell from the thallus, the monospores are formed. They germinate in **bipolar** manner forming a rhizoid at one end and the thallus at the other. α-spores or carpospores are formed by the successive divisions of vegetative cells into 16 pigmented cells. ß-spores or spermatia are developed in same fashion but here 16 to 128 smaller colourless cells are produced. By the gelatinization of the cell walls of the parent thallus, both types of spores are released. Most investigators believe that ß-spores are functionless and degenerate after release. A filamentous outgrowth is formed after the germination of α-spores. This ultimately develops into *Conchocelis* stage. Under long day conditions the *Conchocelis* stage differentiates into monospores which regerminate into *Conchocelis* stage. *Conchocelis* stage forms **conchosporangia** under short day conditions. Each cell of conchosporangia produces a **conchospore** which is liberated from the sporangia under low temperature conditions. After their liberation, conchospores germinate in a bipolar manner and form the thallus (Fig. 8.17). A functional **phytochrome** system is operative as there is true photoperiodic response. The *Conchocelis* stage persists during the period when blade phase is absent. In different species, *Conchocelis* stages can be differentiated with regard to colour, mode of spore liberation and response to environmental factors.

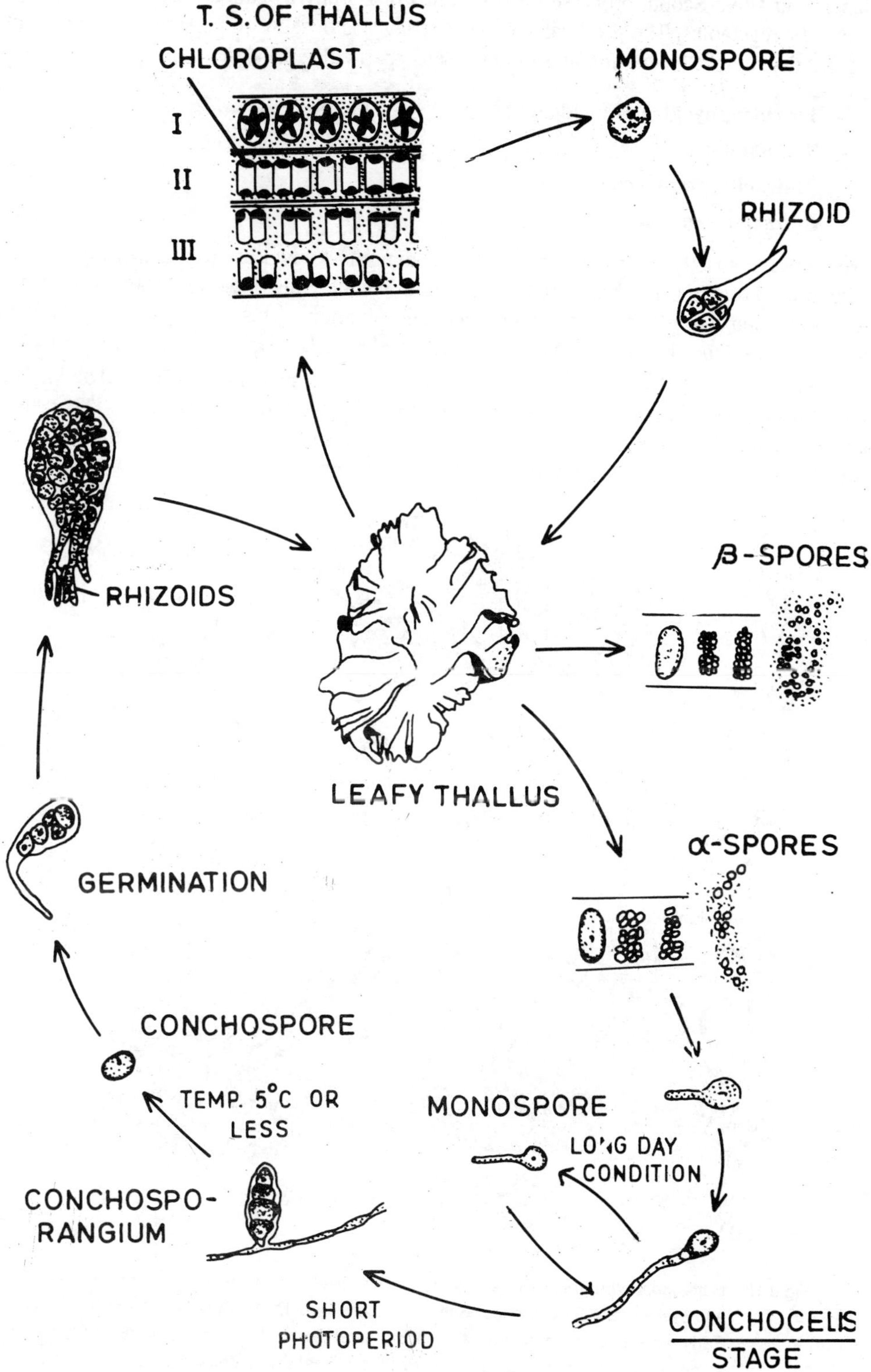

Fig.8.17. Life cycle of *Porphyra* sp.

In Japan and Nova Scotia, processed leafy thallus of *Porphyra* called nori and laver respectively are consumed as vegetables. The food value of nori or laver is because of their high protein content (25 to 30% of the dry weight), vitamins and mineral salts, particularly iodine.

Subclass - Florideophycidae (Florideae)

Order - Nemalionales

Family - Batrachospermaceae

Genus - *Batrachospermum*

Batrachospermum, a strictly freshwater member is commonly known as frog-spawn alga. Most of the species inhabit cold running streams, ponds and lakes. It is widely distributed in tropical, subtropical and temperate regions and the members prefer well-aerated situations. Most of the species are annuals but a few are perennials. The thallus appears in varying shades of blue-green, olive, violet or red.

Plants have soft, gelatinous branched axes with a beaded appearance (Fig. 8.18 A). Numerous branches of limited and unlimited growth arise in whorls from the primary axis. The primary axis develops from an apical cell which cuts off a series of segments at its base. Each segment produces four lateral projections which are cut off as **pericentral cells**. The axial cells do not divide again but continue

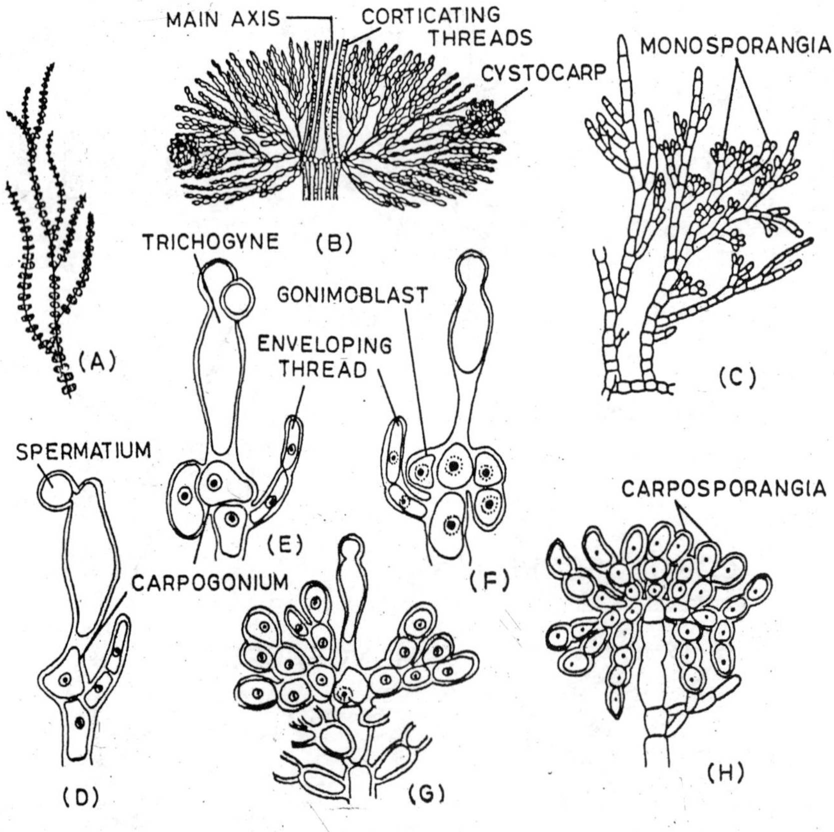

Fig.8.18. *Batrachospermum* sp. (A) Habit of the thallus. (B) Part of the thallus showing branches and cystocarp. (C) Adult shoot with monosporangia. (D - H) Cystocarp formation. (D) Early stage showing attachment of spermatia at the tip of carpogonium. (E - F) Later stages of development. (G - H) Mature gonimoblast with carposporangia.

to increase in length and breadth farther away from the apex. Each pericentral cell forms lateral branches of limited growth. This arrangement of branches differentiates the main axis into nodes and internodes (Fig. 8. 18 B). The branches of unlimited growth contain the same features of nodes and internodes as that of main axis and possess longer cells than those of laterals of limited growth. In addition to the laterally projecting filaments, corticating filaments also grow from the lower most cells of the nodal branches which creep down ward and ensheath the enlarged axial cell, giving the central axis a multicellular appearance. Cells are uninucleate and have a number of parietal lobed chromatophores with a single pyrenoid.

Both asexual and sexual reproduction occur in the life cycle of *Batrachospermum* which is **haplobiontic** type. Non-flagellate uninucleate asexual **monospores** are formed singly within **monosporangia** (Fig. 8.18 C) which develop during Chantransia stage. Sexual reproduction is oogamous. The plants may be monoecious or dioecious and bear spermatangia as male and carpogonia as female reproductive structures.

Spermatangia or antheridia are borne in clusters at the tip of lateral branches. These are single-celled spherical structures which form a single non-motile spherical male gamete (**spermatium**). Through a norrow apical slit of antheridial wall, spermatium comes out in water and is finally carried away to the **carpogonium**, the female sex organ. The carpogonium or female sex organs are borne terminally on a 3-4 celled carpogonial branch. The carpogonium has a basal dilated portion which contains the female nucleus and a prolonged neck, the **trichogyne** (Fig. 8. 18 D). The wall of trichogyne is dissolved when spermatia come in contact with it and the contents of spermatium move down to the base of carpogonium where both male and female nuclei fuse. Trichogyne is separated from the carpogonuum by a mucilage plug. Now the trichogyne either disappears immediately or persists for sometime.

The diploid nucleus formed after gametic union divides meiotically and **gonimoblast initials** are developed from the carpogonium in form of lateral protusion (Fig. 8.18 F). Each gonimoblast cell divides mitotically and form branched threads of **gonimoblast filaments**. The terminal cell of each filaments enlarges and develops into **carposporangium.** Single non-motile carpospore is formed from each carposporangium. The gonimoblast filaments in association with carposporangia and carpospore consititute **carposporophyte** which is parasitic upon the female gametophyte. Along with these developments some enveloping threads are developed from the base of carpogonium which envelope the carposporophyte. The entire structure is termed as fruiting body or cystocarp. The carpospore, after coming out from carposporangium germinates into a heterotrichous filament which is morphologically dissimilar from the main thallus. This microscopic, filamentous system bears monosporangia and is called *Chantrantia* stage or *Pseudochantransia* or **juvenile stage**. The new filament of *Batrachospermum* is formed as lateral outgrowth from this stage.

In the life cycle of *Batrachospermum*, two gametophytic phases (free-living gametophyte and parasitic carposporophyte) alternate with one sporophytic phase. *Chantrantia* stage is an early gametophytic phase. Thus the alga reveals **trigenic** or **trimorphic** or **haplobiontic** life cycle with zygote as only sporophytic phase.

Family - Helminthocladiaceae

Genus - *Liagora*

Plants of *Liagora* inhabit the sublittoral region and are usually found with an external deposition of lime. The genus is mostly confined to warm water such as the Red Sea and the Mediterranean. Thallus may be barely to heavily impregnated with calcium carbonate. Some species do not have calcification. Plants are usually with highly branched tufts, soft and flexible in some species and somewhat rigid in heavily calcified forms. Plants may attain heights of about 20 cm. Branching may be alternate, dichotomous, irregular or at right angles to one another giving a test-tube brush effect (Fig. 8.19). Internally thallus shows multiaxial construction with central axis containing large cells intertwined with slender filaments.

Members are either dioecious or monoecious. In the upper portion of the plant many carpogonial branches are found. The trichogyne of the carpogonium is very long and extends beyond the periphery of the thallus. After fertilization the carpogonium divides transversely into lower **stalk cell** and upper **gonimoblast initial** which forms branched gonimoblast filaments. Filamentous **tetrasporophyte** is produced after the germination of carpospores. The tetrasporophyte bears **tetrasporangia** which produces four tetraspores after meiotic division. The tetraspores germinate to form a second filamentous protonemal phase that produces the gametophyte as buds from the filaments.

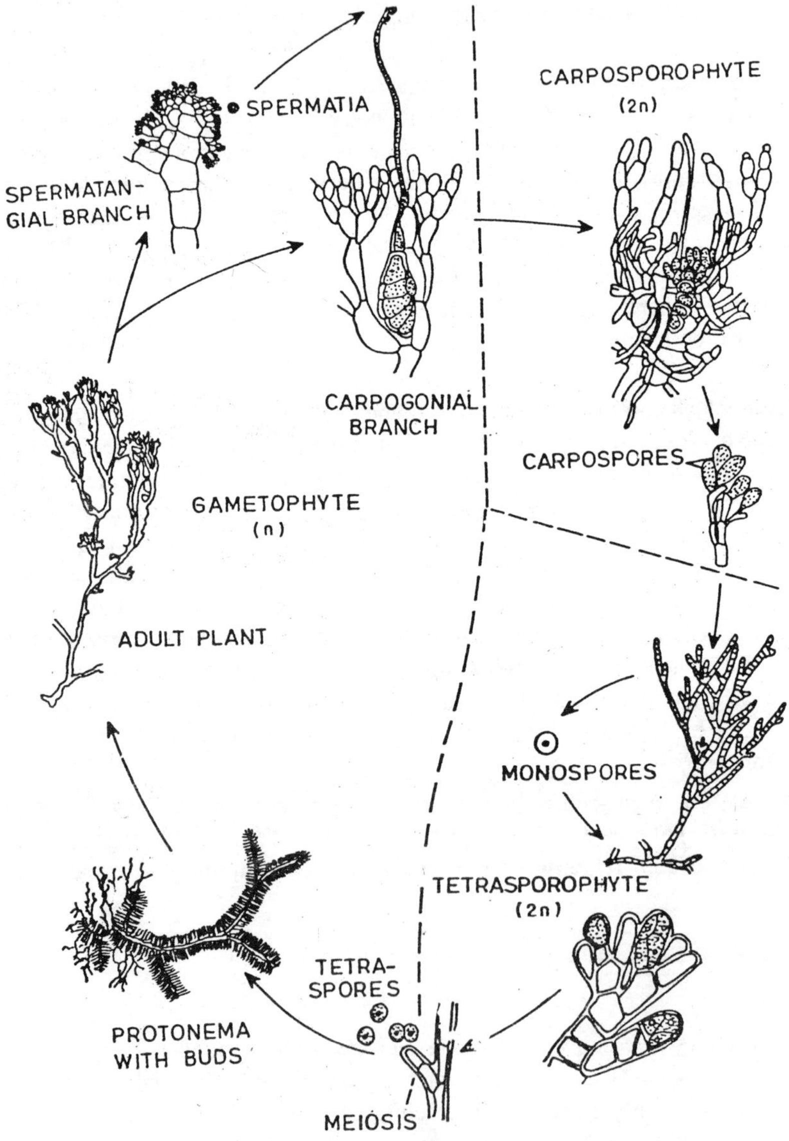

Fig.8.19. Life cycle of *Liagora* sp.

Magne (1964) observed that the zygotic nucleus underwent mitotic divisions rather than meiotic division. Coute (1971), however, found evidence of meiotic division during the formation of spore tetrad. Thus the life cycle is actually **haplobiontic** without free-living tetrasporophyte.

Order - Ceramiales

Family - Ceramiaceae

Genus - *Ceramium*

The species of *Ceramium* usually prefer tidal zone. Some species are frequent in the deeper waters along the Mediterranean coast. A few members are abundant in North Atlantic and Pacific coast. Depending on their depth of occurrence, the plant may appear as red, reddish-brown or yellowish green.

The central or axial cells of the filament are large and show uniaxial construction. Bands of small corticating cells develop at the juncture of these axial cells giving a banded appearance (Fig. 8.20 A). The degree of nodal cortication varies from species to species. In some species, the corticating cells are restricted to the nodes while in others the axes become completely corticated by the upward and downward growth of the corticating filaments. In some species spines are produced. The tips of most

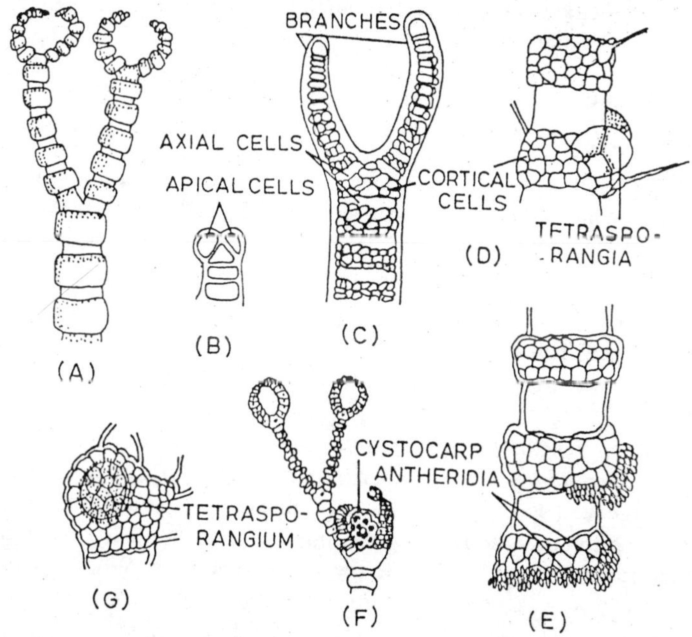

Fig.8.20. *Ceramium* sp. (A) Habit of the apex of a branch. (B) Branch system showing apical forking. (C) A branch showing dichotomy. (D) Node showing cortical cells and tetrasporangia. (E) Part of the thallus showing antheridia. (F) Dichotomous branching of young threads bearing cystocarp. (G) Thallus showing tetrasporangium.

species are forcipate or pincer-like giving the impression of dichotomous branching, although tru dichotomy is absent. The branching is **pseudodichotomous** or **evection** type originating from dome shaped apical cell (Fig. 8.20 B,C). This is due to the rotation of both lateral branch and the parent axi after the formation of lateral branch. In young condition the thallus is attached to the substratum by cushion-like structure which develops into rhizoids during maturity.

Plants of *Ceramium* may be either dioecious or monoecious. **Spermatangia** are borne on th cortical bands which occur in clusters on the upperside of lateral branches (Fig. 8.20 E). Eac antheridium or spermatangium is unicellular and spherical to oblong which forms single, uninucleate and vacuolated **spermatium**. Through a narrow apical opening on the antheridial wall, the spermatium is liberated.

One or two **carpogonial branches** are produced from a first-formed pericentral cell at a node. The **carpogonium** is differentiated into swollen basal portion carrying female nucleus and a narrow elongated **trichogyne**. An **auxiliary cell** is formed after fertilization which gets connected with the base of carpogonium. The descendent of diploid nucleus comes in the auxiliary cell, from where **gonimoblast threads** originate. Subsequently **carposporangia** are produced which get surrounded by a loose envelope of sterile branches. Thus the entire structure develops into a **cystocarp**. A diploid free-living **tetrasporophytic plant** is produced from the germination of **carpospore**. This tetrasprophytic plant morphologically resembles with gametophyte and bears **tetrasporangia**. Tetrasporangia develop in clusters from the cells of the cortical bands (Fig. 8.20 G). Four tetraspores are developed in each tetrasporangium after meiotic division which germinate into gametophytic plant.

The life cycle is typical **diplobiontic** type as there are two diploidic phases (carposporophyte and tetrasporophyte) and a haploid gametophytic phase. These three phases alternate successively with morphologically similar gametophyte and tetrasporophyte. Some species are known to possess **parasporangia** and **paraspores** in cultural conditions.

Family - Rhodomelaceae

Genus - *Polysiphonia*

The genus *Polysiphonia* is of wide occurrence and is distributed throughout the seas of the world. Most species prefer quiet water and appear brownish-red to purple-red. Some occur either as lithophytes or as epiphytes. In India, it is found in west coast. Plants show bushy appearance and are either annual or perennial.

The species of *Polysiphonia* are freely branched with **polysiphonous construction** (Fig. 8. 21 A, B). The dome-shaped apical cell cuts off segments proximally which produce lateral branches before dividing longitudinally into **central** and **pericentral cells**. Although the pericentral cells of *Ceramium* are smaller than the central (axial) cells, in *Polysiphonia* the pericentral cells are of same length as the axial cell (Fig. 8.21 C), so that the primary cortication of the axial filament is carried out by a smaller number of cells than in *Ceramium*. The number of pericentral cell varies from four to twenty four and their number and arrangement is of taxonomic importance. The pericentral cells are connected to the central cell by **primary pit connections** and get connected to each other peripherally as well as to cells immediately above and below them by **secondary pit connections**. When growing on solid substrate, some of the polysiphonous axes creep over the substratum to which they are firmly attached by thickwalled, lobed rhizoids. The lateral branches are of two kinds, the ordinary branches and the **trichoblasts**. Similar to the main axis, the ordinary branches are polysiphonous with limited growth. The **trichoblasts** are uniseriate, usually colourless and bear sex organs. In each cell one nucleus and many discoid plastids are embedded inside dense cytoplasm.

Most of the species of *Polysiphonia* are dioecious. The male plants bear **antheridial sori** on a trichoblast consisting of two-celled stalk and the fertile region, **spermatangia** (Fig. 8.21 D). Spermatangial branches are dichotomously branched and only one branch closer to the main axis becomes fertile while the other remains sterile. Fertile region becomes polysiphonous and the pericentrals divide copiously to form a compact layer of mother cells, each of which gives rise to two to three antheridia.

The **carpogonium** is developed terminally on a four-celled lateral (carpogonial branch) located on the fertile female trichoblasts. At the base of the carpogonial branch, supporting cell is situated. The carpogonium has a long neck, the **trichogyne** and the swollen base. After the spermatium fertilizes, the auxiliary cell is cut off from the supporting cell. The auxiliary cell then fuses with the carpogonium. The

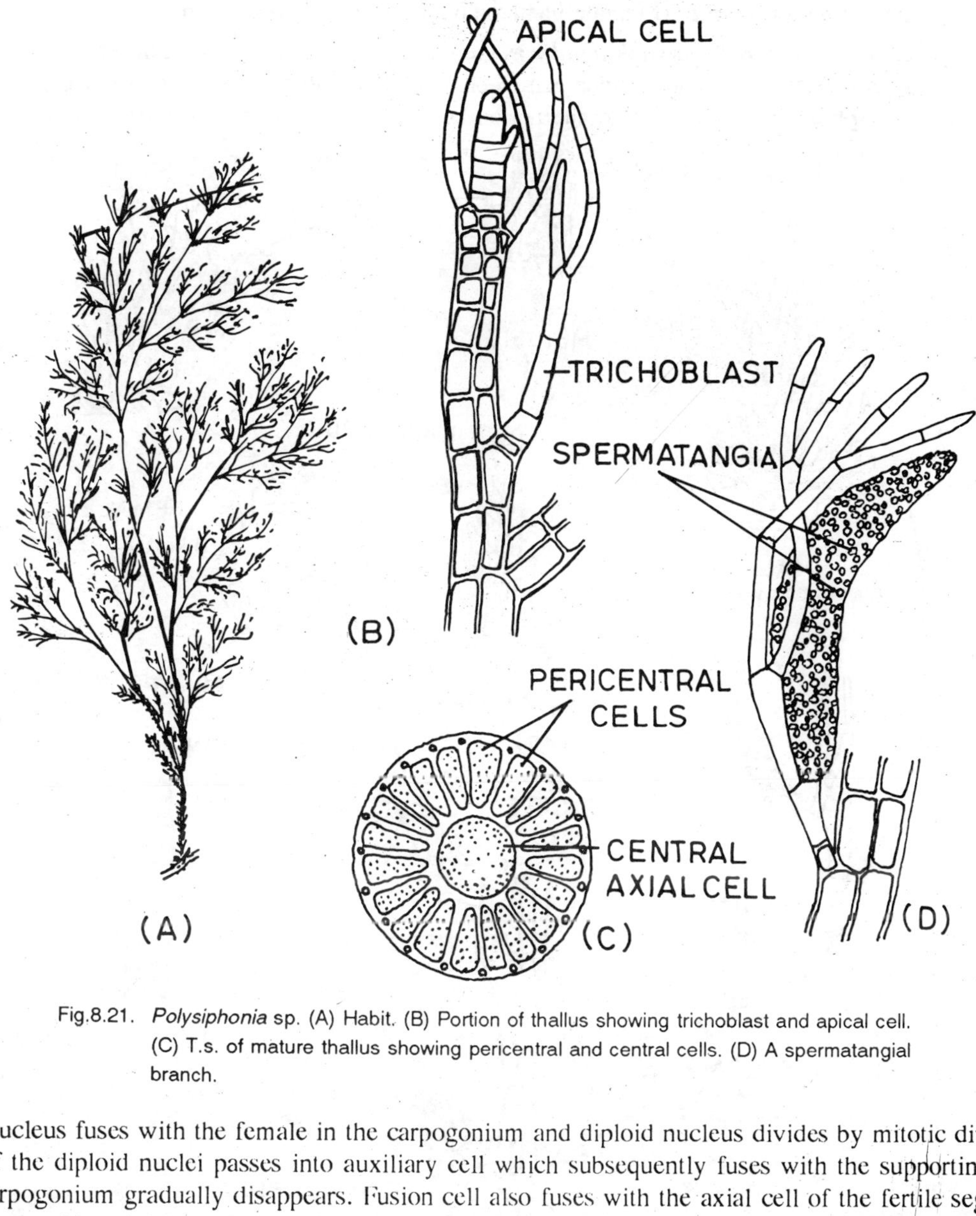

Fig.8.21. *Polysiphonia* sp. (A) Habit. (B) Portion of thallus showing trichoblast and apical cell. (C) T.s. of mature thallus showing pericentral and central cells. (D) A spermatangial branch.

male nucleus fuses with the female in the carpogonium and diploid nucleus divides by mitotic division. One of the diploid nuclei passes into auxiliary cell which subsequently fuses with the supporting cell. The carpogonium gradually disappears. Fusion cell also fuses with the axial cell of the fertile segment. The **gonimoblast initials** are cut off from the fusion cell and develop into a number of **gonimoblast filaments.** The terminal cells of these filaments develop into pear-shaped **carposporangia** and into **carpospores** subsequently. Large placental cell is formed with the gradual fusion of the supporting cell, the auxiliary cell and the cells of the sterile filament. Along with these developments lateral sterile cells closer to carpogonium grow out to form sterile filaments. The sterile filaments form one or two layered walls **(pericarp)** which surrounds carposporangia. This entire fleshy urn-shaped structure is termed as **cystocarp** and there is an opening **(ostiole)** at its apical region.

After their release diploid carpospores germinate into free living **tetrasporophytic** plant which morphologically resemble with gametophytic plant. **Tetrasporangia** are produced on this plant which eventually form tetrahedrally arranged **tetraspores** after meiotic division. Out of four tetraspores, two germinate into male and two into female gametophyte which bear sex-organs.

The life cycle (Fig. 8.22) is **diplobiontic** as there is succession of haploidic dioecious gametophyte, diploidic carposporophyte dependent upon gametophyte and diploid free-living tetrasporophyte. Thus the life cycle is **triphasic** with isomorphic alternation between gametophyte and tetrasporophyte.

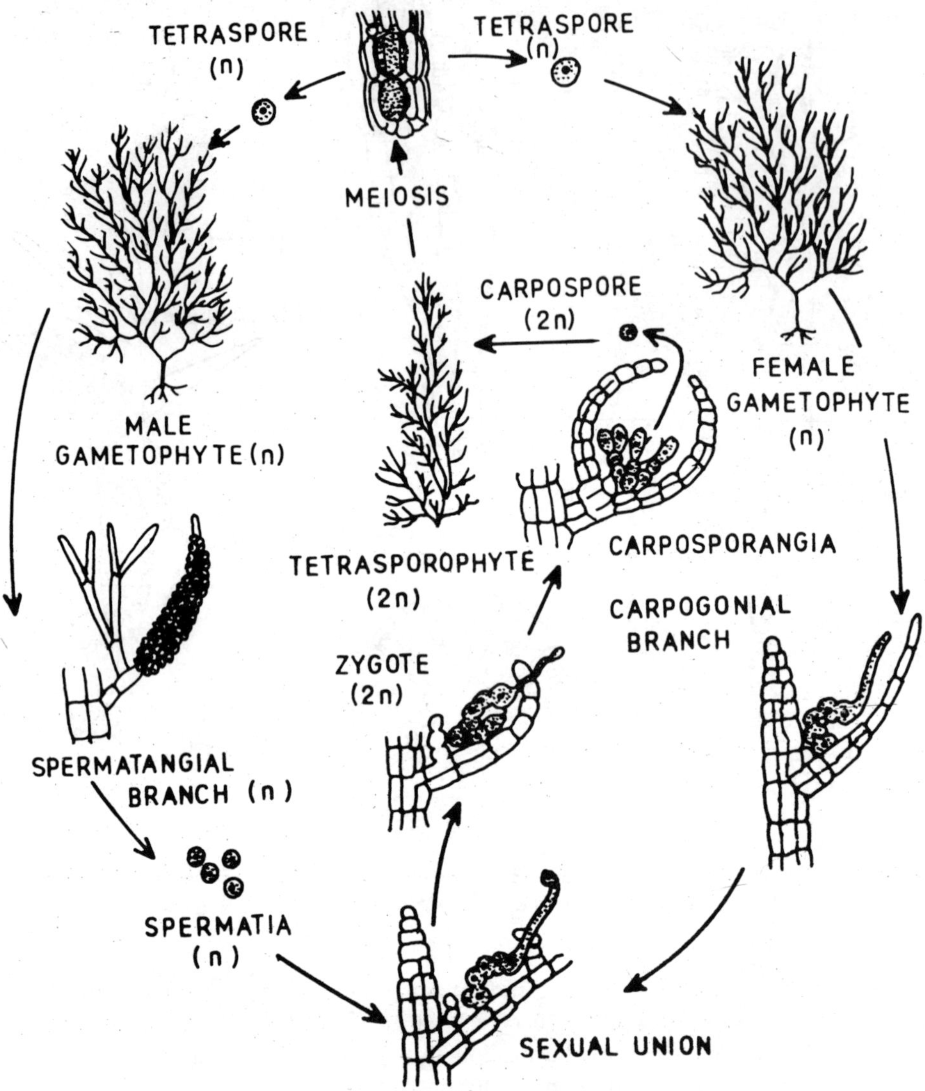

Fig.8.22. Life cycle of *Polysiphonia* sp.

9

DIVISION - EUGLENOPHYTA

Class - Euglenophyceae

GENERAL FEATURES

The members of Euglenophyta (**euglenoids**) are characterised by the following features. (1) Their photosynthetic pigments resemble Chlorophyceae and Charophyceae and they possess chlorophyll **a** and **b**, **ß**- carotene, lutein, neoxanthin, astaxanthin and antheraxanthin. (2) They have one, two or three pantonematic flagella arising from an invagination at the anterior end of the cell. (3) Cells of most of the euglenoids are naked. (4) They have one membrane of chloroplast endoplasmic reticulum. (5) The storage product is **paramylon** or **paramylum** which is found in cytoplasm and not in the chloroplasts. (6) They have mesocaryotic nucleus and (7) They do not have sexual reproduction.

Members exhibit both plant-like as well as some animal-like features. Absence of firm cell wall, occurrence of gullet and holozoic nutrition are their important resemblances with Protozoans.

OCCURRENCE AND HABITAT

Euglenoids are widely distributed and are mainly found in freshwater habitats. Some members are found in brackish and marine waters and also occur on muddy soils. They often predominate and form water blooms in waters contaminated by animal pollution or by decaying organic matter. This may be correlated with facultatively heterotrophic existence of some species. Relatively clean water of rivers, lakes and reservoirs have lesser euglenoids. Some parasitic members are also found in this class.

CELL STRUCTURE

Cell envelope : There is no cell wall in euglenoids but the cells are bounded by plasmalemma and just inside it a proteinaceous **pellicle** or **periplast** is present. It is composed of upto 80% proteins and remainder of lipids and carbohydrates. The pellicle consists of a number of helically wound strips, each of which has a thick edge and a thinner flange end (Fig. 9.1). The thick end of one strip fits under the thin flange end of a second strip, thus appearing as alternate ridges and furrows. The strips of the pellicle continue upto the canal. **Muciferous bodies** and **microtubules** are present under the pellicle strips and

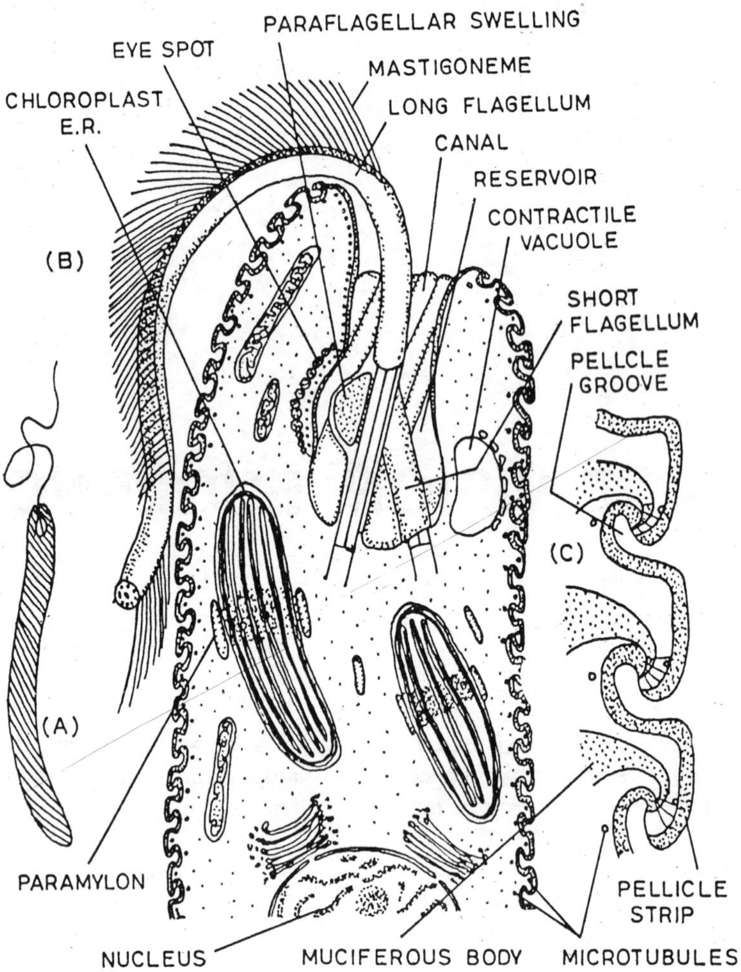

Fig. 9.1. (A) Outline of *Euglena* cell under light microscope. (B) Diagram of the fine structure of the anterior part of *Euglena* cell. (C) A part of pellicle showing structural details. (After Mignot, 1967).

they also follow the same helical pattern. In some members like *Phacus* the pellicle is rigid and has a fixed shape while in others like *Euglena* it is flexible which shows a flowing movement known as **englenoid movement** or **metaboly**. Some species like *Euglena gracilis* changes its shape continuously. Elastic nature of pellicle is due to the presence of fibrous elastic protein. Thus the majority of euglenoid cells are naked except a few like *Trachelomonas* where a special envelope (**lorica**) surrounds the cell. There is well marked space between the envelope and the protoplast (Fig. 9.2). The lorica may be rigid or gelatinous and no cellulose is found.

Canal and Contractile vacuole : Euglenoid cells have an anterior flask-shaped invagination which consists of an enlarged basal **reservoir** and narrow tubular portion, the **canal**. The canal is a rigid structure whereas reservoir changes its shape which gets distorted regularly by the discharge of the contractile vacuole. Because of the presence of microtubules and pellicle covering, rigidity is maintained in canal while these are absent in reservoir which is covered by plasmalemma only.

A **contractile vacuole** is present in the anterior region of the cell near the reservoir and has

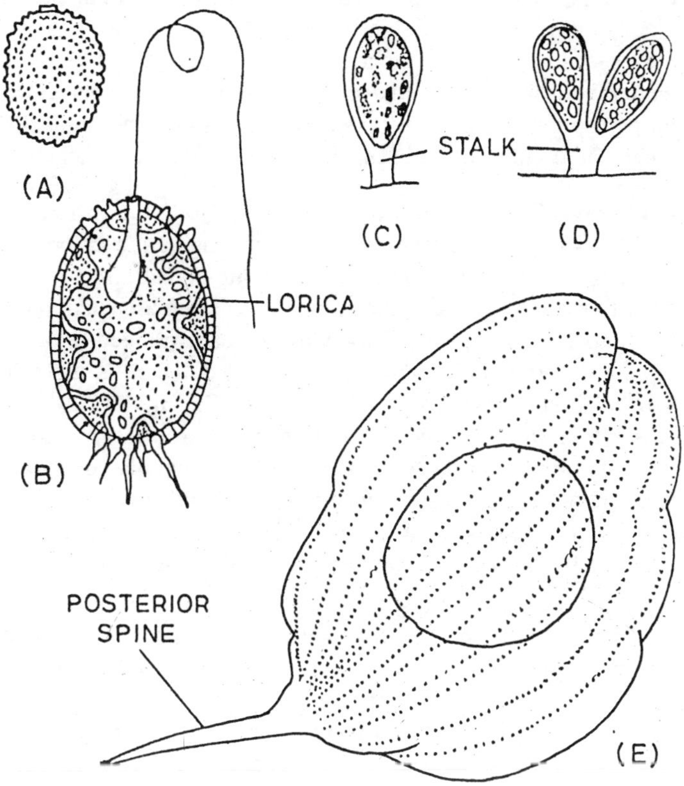

Fig. 9.2. Common members of Euglenophyceae. (A) *Trachelomonas granulosa.* (B) *T. armata* showing lorica. (C-D) *Colacium vesiculosum* showing mucilaginous stalk. (E) *Phacus* sp. showing flattened cell, longitudinal striae and posterior spine.

osmoregulatory function. The vacuole discharges periodically into the reservoir and a new contractile vacuole is developed by fusion of many smaller vacuoles.

Flagella : Euglenoids are flagellate except when they are encysted or in palmelloid phase. The number of flagella is commonly two which are pantonematic. The locomotory flagellum or flagella emerge from the anterior invagination of the cell. Euglenoid flagella have the usual 9+2 arrangement of microtubules and are similar to those of the Dinophyceae in having a flagellar rod that runs the length of the flagellum. The flagella emerge from basal bodies located just beneath the reservoir wall. They possess single helical row of fibrillar hairs. In addition to relatively long hairs (2-3 μm), shorter fibrillar hairs are also found which coat the flagella. Two basic types of flagellar movement have been found in the members. The first group has the flagellum continuously motile from base to apex and in second group the flagellum remains straight in front of the cell having restricted motility at the tip region only. As the emergent flagellum passes from the reservoir to canal, it may be thickened laterally and form **flagellar swelling** or **paraflagellar body** inside the flagellar membrane which is believed to be photoreceptor. In *Euglena* the swelling is on the longer flagellum.

Eyespot and Phototaxis : The prominent eyespot or stigma located in the anterior region is a collection of usually about 20 to 50 orange-red lipid droplets that occur in the cytoplasm independent of the chloroplast. It is mainly composed of ß-carotene, astaxanthin and echinenone and ensheaths the neck of the reservoir on the dorsal side. The stigma lies at about the level of the flagellar swelling. DNA is also found along with carotenoids in the stigma of a strain of *E. gracilis*. All euglenoids with an eyespot and flagellar swelling exhibit **phototaxis**. The members are positively phototactic to light of low intensity and negatively phototactic with respect to bright light and darkness. Two theories have been proposed regarding the role of eyespot in light perception. First view suggests that eyespots receive light stimuli and control the movement and direction of the cell. According to the other view the flagellar swelling acts as the site of light perception and the eyespots function as a shading organelle. Although the stigma is present in all the pigmented cells, it is absent from some colourless members.

Chloroplast and storage products : The chloroplast varies among the different species and genera. It is usually discoid or plate-like with a central pyrenoid. Sometimes the chloroplast may be ribbon-like and arranged in stellate fashion. It is well defined membrane-bound structure having two surrounding membranes of chloroplast envelope and one membrane of chloroplast endoplasmic reticulum which is not continuous with the nuclear membrane. The thylakoids are arranged into bands of three with two thylakoid bands traversing the granular stroma of the pyrenoid. There is a shield of paramylon grains outside the chloroplast which surrounds pyrenoid (Fig. 9.1). **Paramylon**, a ß-1, 3 glucose polymer is present in the colourless cytoplasm. Pyrenoids, present in the chloroplasts may be of various types which are the centres for paramylon formation in some species. Paramylon also develops without showing any relationship to the pyrenoid or plastid. In some members the liquid storage product, the chrysolaminarin acts as storage product which occurs in form of vacuoles in the anterior region of the cell. The chloroplasts contain chlorophyll **a** and **b**, ß-carotene and several xanthophylls.

Euglena cells lose all of their chlorophyll and possess only protochlorophyll after several generations of growth in the dark. These cells have about 30 proplastids with one or two internal membranes and a few ribosomes. As soon as these cells are placed in light, the proplastids begin developing into chloroplasts. The conversion of chloroplasts to proplastids in dividing cells but not in nondividing cells help *Euglena* to adapt its ecological conditions. If the medium is devoid of carbon sources and growth is impossible in darkness, the organisms maintain its plastids which would be advantageous with the re-appearance of light. Bleaching of *Euglena* accompanied by the loss of chloroplasts can also be caused by variety of physical and chemical agents like ultraviolet radiation, exposure to streptomycin, growth at 34°C etc.

Muciferous bodies and extracellular structures : In all species of euglenoid flagellates muciferous bodies occur in helical rows under the pellicle. These are in communication with the exterior of the cell by canals through which mucilage may get extruded. Muciferous bodies contain a water-soluble polysaccharide. Members with large muciferous bodies eject their contents on irritation and produce a copious slime layer around the cell. The stalks and envelopes of *Colacium, Trachelomonas* respectively and of some other genera are the organized products of mucilage secretion (Fig. 9.2). The process of envelope formation involves mineralization of an initial fibrillar envelope which is probably derived as a secretion of the mucilaginous bodies.

Nucleus and nuclear division : The prominent nucleus of the euglenoid cell is readily visible either in center or in posterior region of the cell. The nucleus is of mesocaryotic type with permanently condensed chromosomes during mitosis, persistent nucleolus (endosome) that does not disperse during nuclear division and an intact nuclear envelope (Fig. 9.3). Mitosis begins during early prophase and the nucleus migrates from the center of the cell to an anterior position. Although microtubules appear in the nucleus but they do not attach to the chromosomes. The chromosomes are arranged with their long axes parallel to the long axis of the division. During metaphase, nucleolus starts elongation along the division axis. At anaphase, there is elongation of intact nuclear envelope and the nucleolus divides. The chromatids separate gradually along their long axes and migrate to the poles. Centriole has not been found to be involved. Cytokinesis occurs by longitudinal cleavage of the protoplast.

The chromosome number in euglenoids is usually high and some genera also exhibit polyploidy.

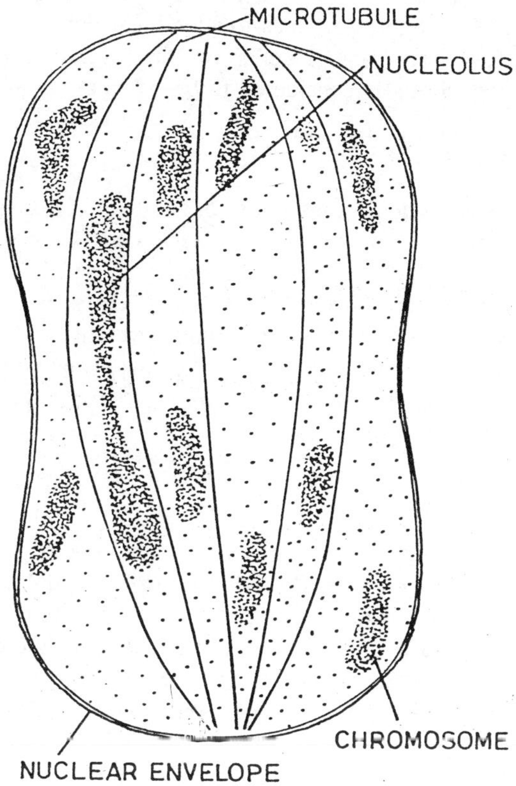

Fig. 9.3. A diagram of an anaphase nucleus of *Euglena* showing intact nuclear envelope, pinching of nucleolus into two and non-attachment of chromosomes to the spindle microtubules.

A close relationship between the flagellar apparatus and the nucleus has been suggested due to the fact that nucleus moves towards the anterior of the cell immediately before division.

CELL DIVISION, GROWTH AND NUTRITION

Cell division may occur in the swimming phase or in the palmelloid condition. It begins with mitosis. The original parent flagellum is either lost or gets transferred to one of the daughter cells, the other acquires a new flagellum. Two daughter canals are formed and there is duplication of pellicle strips also. Division of the organelles is usually equal so that the two daughter cells are identical in size and contents. Cell division in the members having elastic pellicle is accompanied by intense euglenoid movement.

The members of Euglenophyceae show different modes of nutrition. Inspite of having chlorophyll a and b, no euglenoid has yet been demonstrated to be fully photoautotrophic. They are rather photoauxotrophic because they require one or more vitamins. *E. gracilis* has an absolute requirement of vitamin B_{12}. Some green species are able to grow in darkness if organic carbon is made available. These are thus facultatively heterotrophs. Their growth may be stimulated in light by the carbon compounds. Green euglenoids are not phagotrophic but some colourless forms reveal such mode. Thus the euglenoids belong to the acetate flagellates having the ability to grow photosynthetically in light or heterotrophically in the dark. Fixed carbon is used as a source of energy and as building blocks for cell constituents in either case. The substrates that can be used for heterotrophic growth vary from one species to other. Acetate and ethanol are the two most readily utilized substrates.

At comparatively higher temperatures (32-35°C), the strains of *Euglena gracilis* have been found to be bleached. Effect of ultraviolet light also leads to such bleaching. These are unable to turn green again as they lack plastids. The eyespot and flagellar swelling may be retained. The development of colourless euglenoids from green forms is considered to be of great phylogenetic significance.

REPRODUCTION

The usual method of reproduction in euglenoids is by longitudinal cell division. In order to withstand unfavourable environmental conditions some members like *Phacus*, *Euglena* and *Trachelomonas* have the capacity to encyst. With the onset of favourable conditions these thick walled cysts germinate and form motile cells. Palmelloid stage embedded in a gelatinous matrix is occasionally formed in some species of *Euglena*. This stage is, however, temporary and vegetative division does not occur in this stage. In *Colacium* a palmelloid or dendroid colony is the normal vegetative form and vegetative division is restricted to the palmelloid stage. There are few reports of sexual reproduction but these require confirmation.

CLASSIFICATION

Euglenophyceae has been divided into six orders by Leedale (1967). Eutreptiales, Euglenales and Heteronematales are the three important orders. The members of Eutreptiales have two emergent flagella, one directed anteriorly and the other laterally or posteriorly and there is no special organelle for ingestion. In Euglenales, out of two flagella only one emerges from the canal and special ingestion organelle is absent. Members of Heteronematales show two emergent flagella. During swimming the longer flagellum is directed anteriorly and the shorter one is directed posteriorly. Special ingestion organelle is also found.

PHYLOGENY AND INTERRELATIONSHIPS

The euglenoids exhibit both plant as well as animal-like characters. Due to presence of the palmelloid genus *Colacium*, possession of photosynthetic pigment and simplest vegetative form they are obviously close to algae. On the other hand they are classified as Protozoa by zoologists due to presence of gullet, holozoic mode of nutrition and absence of chlorophyll in some species. In *Euglena gracilis* and *Trachelomonas volvocina*, keto-carotenoids have been identified which play a role in the biosynthesis of astaxanthin from ß-carotene in many animal systems. In plants biosynthesis of carotenoids does not involve such keto-carotenoids.

It may, therefore, be concluded that euglenoids are basically plants, some members of which have lost their chlorophyll and thus started animal like ingestion of food. Some biologists are of this view that both plant and animal groups probably have evolved from this primitive stock. Some earlier taxonomists had placed these forms in Protista with bacteria, slime molds and similar other flagellates. As the members are different from other classes of algae in various ways, separate class Euglenophyceae has been assigned for these members. Some affinity with the green algae has been sought due to presence of chlorophyll a and b. However, differences are in food reserve, nuclear division and in other metabolic characteristics. Lysine biosynthesis pathway of euglenoids resemble with many fungi. Stewart and Mattox (1975) are of the assumption that euglenoid chloroplast may have descended from eucaryotic green algal forms in course of evolutionary reduction.

10

DIVISION - DINOPHYTA (PYRRHOPHYTA)

Class - Dinophyceae

GENERAL FEATURES

The division Dinophyta comprises diverse assemblage of biflagellate unicellular organisms (**dinoflagellates**) and are important members of the plankton in both fresh and marine waters. In addition to the flagellate forms, nonmotile forms are also found in diverse adaptations. Members are characterised by following features. (1) Photosynthetic pigments include chlorophyll a and c together with several xanthophylls (peridinin, dinoxanthin and neoperidinin). (2) A typical motile dinoflagellate consists of an **epicone** and **hypocone** divided by the transverse **girdle** or **cingulum**. The epicone and hypocone are usually divided into a number of thecal plates. A longitudinal **sulcus** running perpendicular to the girdle is also present. (3) The longitudinal and transverse flagella arise through the thecal plates in the region where the girdle and sulcus meet. The longitudinal flagellum is acronematic, posteriorly directed and projects out from the cell whereas the transverse flagellum is wave-like and is closely adpressed to the girdle. (4) Most members contain conspicuous membrane - bound **pusules** which have resemblances to contractile vacuoles but show no contractile activity. (5) The cells can be photosynthetic or colourless and heterotrophic. In photosynthetic members chloroplast is surrounded by one membrane of chloroplast E.R. which is not continuons with the outer membrane of the nuclear envelope. (6) The food reserve is starch which is found in the cytoplasm and wall or theca is primarily composed of cellulose. (7) The nucleus is large and dinocaryotic or mesocaryotic type with permanently condensed chromosomes.

OCCURRENCE AND HABITAT

Members inhabit mainly in marine waters but some forms are found in freshwater, estuarine and brackish habitats. Generally the marine forms are less important in the colder polar waters than in warmer waters. Besides most prevalent motile forms, nonmotile forms are represented by coccoid, palmelloid, amoeboid and filamentous members. In addition to photosynthetic forms, saprophytic, parasitic, symbiotic and holozoic patterns are also found in the members of this division.

CELL STRUCTURE

Cell covering : The cell covering of dinoflagellates is called as **amphiesma** and cells may be naked (unarmored) or with thecal covering (armored). The thecal structure consists of an outer plasmalemma and inner single layer of flattened vesicles which usually contain cellulosic plates giving the theca its characteristic structure (Fig. 10.1). Numerous pores are generally present within the amphiesma having a **trichocyst** beneath each pore. The cell surface has a transverse and longitudinal furrow. The former

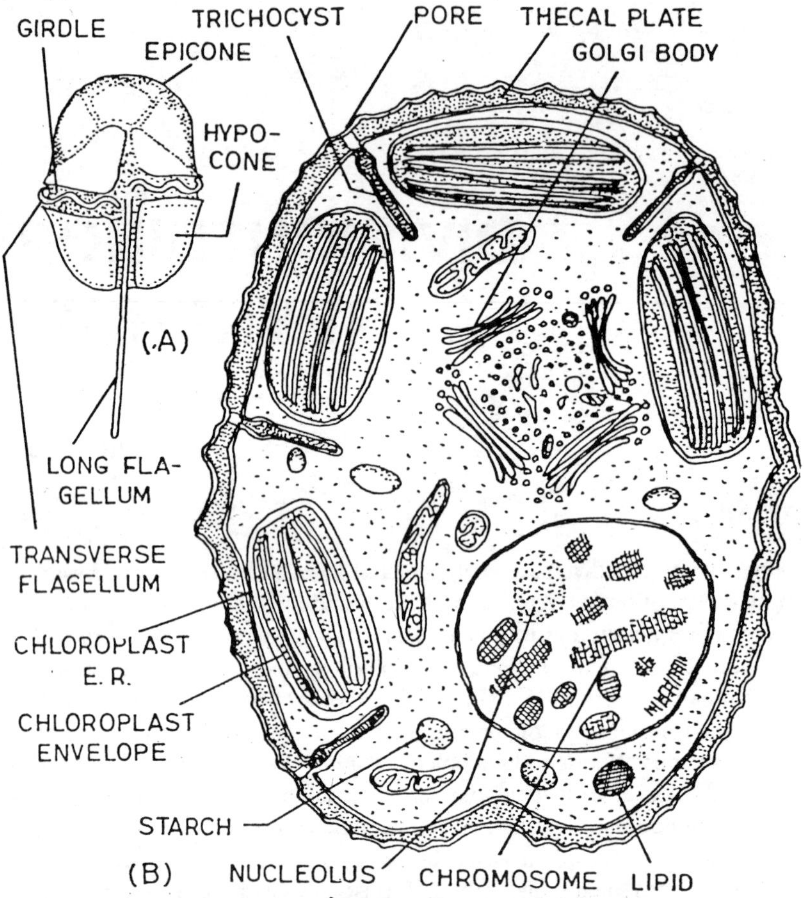

Fig.10.1. (A) A light microscopic and (B) Electron microscopic diagram of *Peridinium* sp. (dinoflagellate) showing structural details. (After Lee, 1980).

runs round the cell helically and the latter runs vertically to connect the two ends of transverse furrow. Additional sculpturing of the surface produces many hexagonal platelets (**thecal plates**) and the detailed structure and arrangement of these constitute criteria for classifying species and genera of dinoflagellates. The **epitheca** or epicone is divided into two complete transverse series of plates. These are **apical** and **precingular plates**. On the dorsal surface of the epitheca, an incomplete seies of plates (**anterior intercalary plates**) are present which varies from one to three in number (Fig. 10.2). The **hypotheca** or hypocone is also divided into **postcingular** and **antapical plates**. In some genera , the hypotheca also has incomplete series of **posterior intercalary plates.**.Although thecal plates are also present in cingulum, sulcus and in the apical pore but these are difficult to analyse.

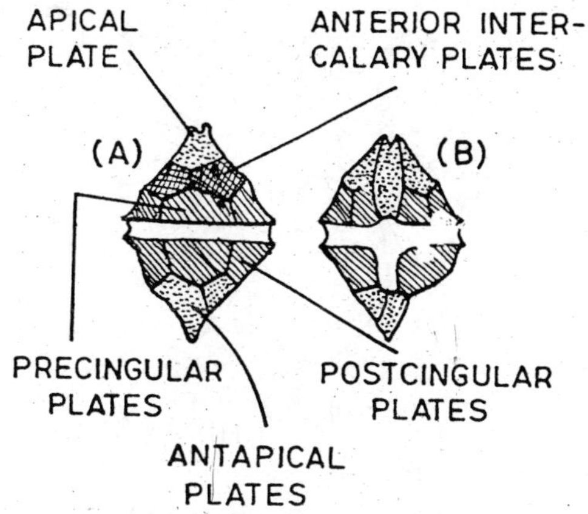

Fig.10.2. Arrangement of the plates in *Peridinium* sp. (A - B) Lateral view from the front and rear.

Although members of the Peridiniales (*Peridinium, Ceratium,* Fig. 10.3 A,B) are flattened dorsiventrally, members of the Dinophysiales (*Dinophysis,* Fig. 10.3 C) tend to be flattened laterally. These can also be recognised by wing like structure which develops from the cingular region as a conspicuous projection. A diverse ornamentation in the thecal plates exists in different orders of armored dinoflagellates. These may bear reticulations, spines, ridges etc.

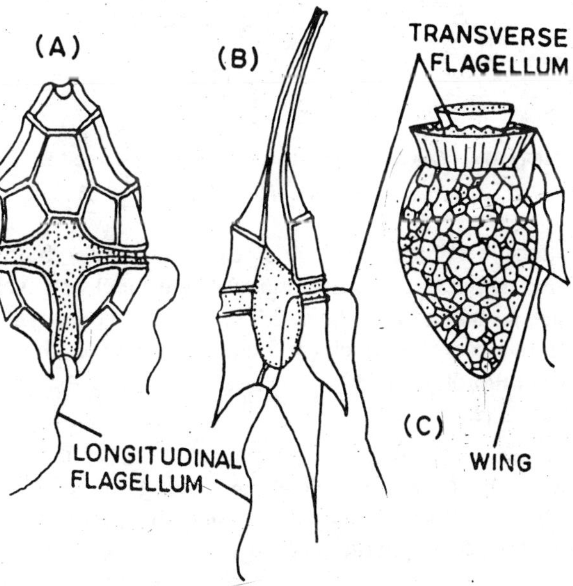

Fig.10.3. Common members of Dinophyta. (A) *Peridinium bipes.* (B) *Ceratium cornutum.* (C) *Dinophysis* sp. showing wing.

Flagella : Flagella in dinoflagellates are of two types - transverse and longitudinal. Longitudinal flagellum projects out from the longitudinal sulcus while the transverse flagellum emerges from the cell at the upper end of transverse furrow (Fig. 10.4). The basal portion of longitudinal flagellum is wide and apical portion is thinner. The two flagella are inserted in the area of the intersection of the girdle and sulcus. The length of transverse flagellum is 2-3 times more than that of longitudinal flagellum and it has helical outline. It consists of an axoneme, flagellar sheath and fibrillar hairs. The wave produced from the flagellum causes a direct forward movement and it also causes the cell to rotate .

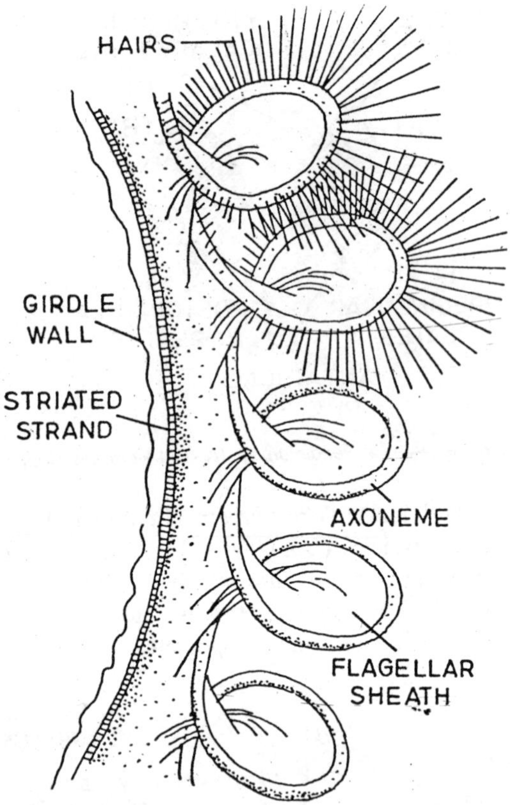

Fig.10.4. Diagram of a part of transverse flagellum of *Peridinium cinctum*. (After Berdach, 1977).

Chloroplasts : The chloroplasts in Dinophyceae are surrounded by the chloroplast envelope and one membrane of chloroplast E.R., which is not in continuation with nuclear envelope. Chlorophyll a and C_2 are present with peridinin as major carotenoid. Most of the chlorophyll a and peridinin remain together in a water-soluble protein complex which are found on the surface of thylakoids and act as light-harvesting pigments like phycobiliproteins of blue-green and red algae. A type of chromatic adaptation has been revealed by dinoflagellates.

Chloroplasts are of variable form. In most of the members many peripherally arranged discoid chloroplasts are common .Some species of *Peridinium* have a single lobed axile chloroplast. Pyrenoids may or may not be present . Some species of many genera like *Gymnodinium* are colourless. The mode of nutrition in these forms may be saprophytic or holozoic.

Pusule : In certain marine and freshwater species, a saclike structure (**pusule**) is present that opens by means of a pore into the flagellar canal. Like that of contractile vacuole pusules have an osmoregulatory function. Some consider these as a possible floatation device. The pusule consists of about 40 vesicles that open by small pores into the flagellar canal (Fig. 10.5). The pusule vesicles are lined by double membrane while the flagellar canal is surrounded by single membrane in continuation with the plasma membrane.

Eyespots : An eyespot is mostly found in naked freshwater species. It is absent from marine species and from those species having cell wall. In the Dinophyceae, eyespots range from collections of lipid

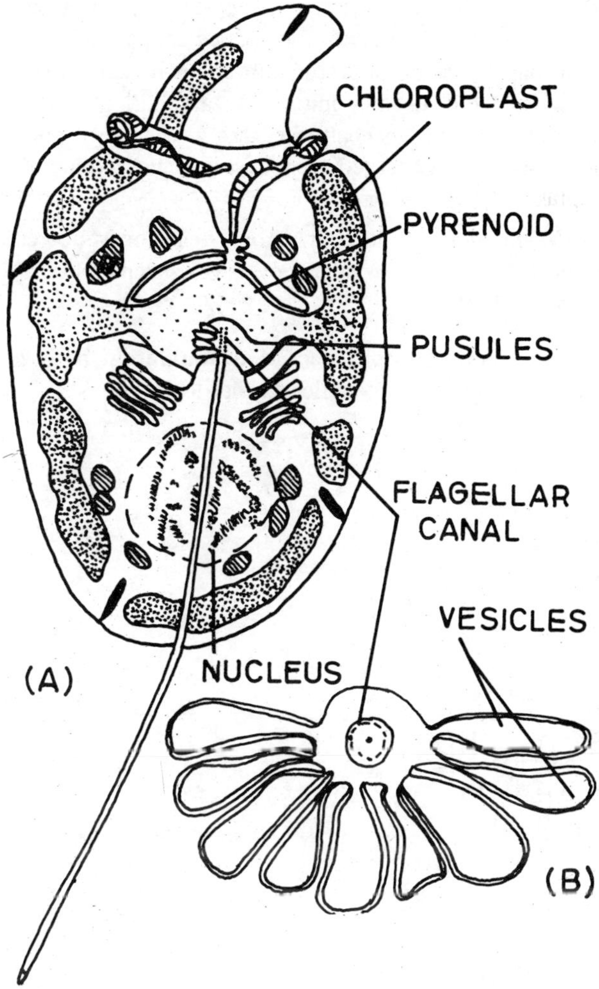

Fig.10.5. (A) Diagram of *Amphidinium carteri* to show the position of pusule. (B) T.s. of a pusule showing the pusule vesicles opening into flagellar canal. (After Dodge and Crawford, 1968).

globules containing carotenoids and lacking membranes to more complex lipid containing globules surrounded by membrane envelopes. Dodge (1984) reported three types of dinoflagellate eyespots:

(i) independent and not membrane-bound,

(ii) independent and surrounded by three membranes and

(iii) situated at the periphery of the cloroplast. All are located behind the sulcus and through a strand of microtubules directional stimuli transmit from the eyespot to the flagellum.

Some marine species possess most complex photosensitive organelle (**ocellus**) in the algae. It consists of a colourless refractive **lens** partially buried in a pigmented **melanosome**.

Projectiles : A number of different projectiles (trichocysts, nematocysts) are found in Dinophyceae. After being irritated these are fired out of the cell resulting in a sudden movement of the cell in the opposite direction.

Both in naked and armored dinoflagellates, **trichocysts** may be present in hundreds per cell. A trichocyst has a membrane-bounded proteinaceous rod-shaped crystalline core (Fig. 10.6) measuring upto 200 µm in length. The outer part of the trichocyst membrane is attached to the plasme membrane . The trichocysts arise in vesicles from dictyosomes initially as a spherical vesicle that eventually becomes spindle-shaped. These are finally located beneath the narrow trichocyst pores that perforate the thecal plates (Fig. 10.1) in armored species. A charged trichocyst is converted to a discharged one in a fraction of second by the rapid uptake of water on irritation.

In only two genera (*Nematodinium* and *Polykrikos*), **nematocysts** or **cnidocysts** are found which are explosive projectiles somewhat similar to nematocysts of coelenterates. The number of nematocysts per cell varies usually from 8 to 10.

Distinctive organelles : In symbiotic Dinophyceae an **accumulation body** is commonly present. This is a large vesicle possessing the digestive organelles similar to the digestive vesicles of other flagellates.

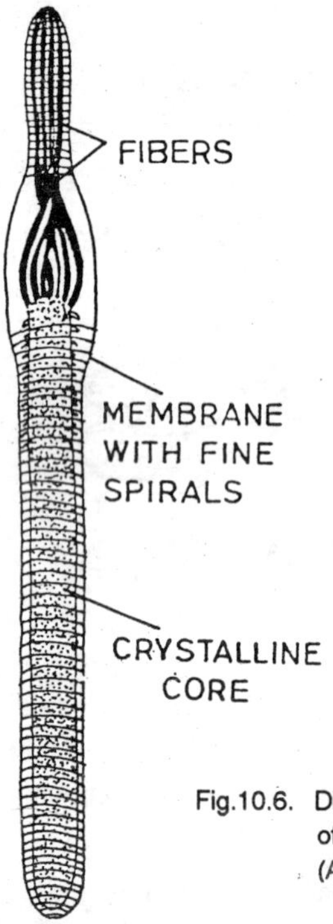

Fig.10.6. Diagram of a longitudinal section through a charged trichocyst of *Gonyaulax polyedra* showing structural details. (After Bouck and Sweeney, 1966).

In some phagotrophic dinoflagellates an extendable structure (**peduncle**) composed of compact row of microtubules emerges at the juncture of cingulum and sulcus. It helps in capturing pray. During active feeding the cytoplasm of the prey flows through the peduncle into a food vacuole in the predator cell The peduncle is withdrawn into the cytoplasm when the cell remains in nonfeeding condition.

Just beneath the cell membrane, vesicles of mucilaginous material (**muciferous bodies**) are present. They facilitate attachment of the cell to sand grains along the seashore and their presence in some species is correlated with **psammophilous** existence of those species.

Nucleus and nuclear division : Due to many unusual properties, dinophycean nucleus is referred to as **dinocaryotic** or **mesocaryotic** which is unlike that of either eucaryotic or procaryotic cells. In the interphase nucleus the chromosomes remain in condensed stage like mataphase and anaphase conditions.

The chemical composition of the chromosomes is also unusual. As compared to eucaryotes , the DNA of dinophycean chromosomes has very little nucleohistone. Dinophyceae appears to be intermediate between procaryotic and eucaryotic organisms in DNA synthesis. In primitive dinoflagellates, DNA synthesis is continuous (procaryotic feature) whereas complex dinoflagellates have discontinuous DNA synthesis (eucaryotic feature). Centrioles are lacking except in some parasitic species.The nucleus is relatively large.

Nuclear division is mitotic and chromosomes divide longitudinally. The nuclear envelope and the nucleous are persistent throughout nuclear division. At metaphase, metaphase plate is not formed as is common with eucaryotes and the chromosomes are still scattered . Centromere and spindle have not been found. The chromosome upon dividing assume a V or Y shape and the apices of these are closely associated with the nuclear envelope through extranuclear microtubules, supposed to be extranuclear spindle.

REPRODUCTION

The most common method of reproduction in dinoflagellates is cell division. Parent cell divides by longitudinal, transverse or oblique bipartition. In armorned dinoflagellates either new theca is formed after splitting of parental theca or before division theca may be shed and thus necessitating the formation of new theca. In naked dinoflagellates the new amphiesma is formed around each product during the course of their separation.

Filamentous members like *Dinothrix* releases zoospores which after division do not separate resulting into filamentous habit. Some species produce one or two aplanospores.

Both isogamous and anisogamous mode of sexual reproduction has been recorded in dinoflagellates. Gametic union results into **planozygote**. In freshwater species, planozygote forms resting cyst where meiotic division takes place usually. In some freshwater species of *Peridinium* a nitrogen-deficient medium was observed to induce sexuality.

Encysted stages: During unfavourable conditions most of the dinoflagellates form resting spores which are morphologically distinct. Some cysts resemble vegetative cells while others are very different being covered with tubular projections . During the process of resting spore formation or encystment, theca is replaced by a thin amorphous wall which gradually thickens. Cytoplasmic inclusions like chloroplasts, golgi bodies and pusules are gradually reduced or disappeared. Fossilized resting spores are called **hystrichosphaerids** or **hystrichospores.** In sexually reproducing members, cysts represent the nonmotile zygote stage. A definite opening (**archeopyle**) is formed when cysts rupture. The morphological variations of cysts provide some phylogenetic relationship.

BIOLUMINESCENCE AND CIRCADIAN RHYTHMS

Bioluminescence is the process of emission of light by a living organism. Many of the marine dinoflagellates reveal bioluminescence by emitting a bluish flash of light for 0.1 second after the cells get agitated. It is of lesser magnitude in day and about 60 times brighter at night. The intensity of luminescence at night is based on the amount of light received during the day. This process in dinoflagellates is associated with a cytoplasmic particle called a **scintillon.** The compound responsible for bioluminescence is **luciferin** which is oxidized by enzyme luciferase resulting in the emission of light.

An **endogenous circadian rhythm** has been observed in many dinophycean members. The amount of light emitted is found to be low during the period of illumination which reaches a peak during the middle of the dark period. This cycle of bioluminescence peak continues for several days.

11

DIVISION - CHRYSOPHYTA

Class - Chrysophyceae (Golden-brown algae)

GENERAL FEATURES

The members of Chrsophyceae is characterised by the following features. (1) The chloroplast are surrounded by two membranes of chloroplast E.R., the outer being continuous with the outer membrane of the nuclear envelope. (2) Photosynthetic pigments include chlorophyll **a** and carotenoids like fucoxanthin and diadinoxanthin. (3) The storage product is oil or chrysolaminarin (leucosin). (4) The cells are basically biflagellate. Flagella are unequal in length and one of them has tripartite hairs. (5) Cellulose wall is absent and members have marked tendency to become silicified. (6) The cells always have one eucaryotic nucleus and sexual reproduction has not been established conclusively. Many members produce statospores (cysts) which are enclosed by a silicified wall having a terminal pore.

The silica-scale bearing Chrysophyceae (Mallomonadaceae and Paraphysomonadaceae) present special problems in identification and in their study because in most cases electron microscopy of their silica structures is necessary for reliable identifications.

OCURRENCE AND HABITAT

Most of the members are freshwater and they constitute important group of phytoplankton. Small ponds and swamps often contain the largest diversity of silica-scaled Chrysophyceae. They inhabit dystrophic, oligotrophic and mesotrophic waters. Many species prefer unpolluted aquatic freshwater habitats and relatively cool water . Saha and Wujek (1990) reported the abundance of some scale-bearing chrysophytes from eutrophic tropical waters. The filamentous and coccoid forms are mostly found in cold springs where they occur as crustaceous or gelatinous growth. Members are very sensitive to the environmental changes and thus they are increasingly used as **environmental indicators** . Only a few species seem to occur exclusively in marine waters.

Though motile unicellular condition is most frequent, different other vegetative forms (motile colonial forms, coccoid, palmelloid, dendroid and filamentous forms like that of Chlorophyta and Xanthophyra are also present showing parallel line of evolution. Most of the members of this class reveal photosynthesis but some colourless forms are also available. The mode of heterotrophic nutrition is both osmotrophic and phagotrophic. Species of *Ochromonas* represent both the heterotrophic nutrition as well as autotrophic mode.

CELL STRUCTURE

Cell covering : Many motile Chrysophyceae have naked protoplast bounded by a periplast. Periplast may be soft or firm. Other species have cell coverings including scales, loricas and cell walls. The scales, spines and bristles of Mallomonadaceae and Paraphysomonadaceae are considered extensively in identification (Asmund and Kristiansen, 1986). In Synuraceae, the silicified scales are deposited outside plasmalemma in a thin pectic envelope. The silicified scales of *Mallomonas* are composed of a dome, shield and bristle (Fig. 11.1). Other types of scales are crownlike *(Synura)* and box-like *(Paraphysomonas).* Like the frustule of diatoms the scales of Chrysophyceae are deposited inside a silica deposition vesicle (SDV) which originates from endoplasmic reticulum (Lee, 1978). The scale morophogenesis in Mallomonadaceae can be divided into three steps: (1) formation of the primary vesicle, (2) formation of the scale - forming vesicle and (3) finishing and transport on the cell surface. The details about these has been studied by Wujek and Kristiansen (1978), Mignot and Brugerolle (1982), Leadbeater (1986) etc. In addition to scales many *Mallomonas* species contain 1-2 types of bristles which sometimes exceed the length of the total cell. Organic scales without any mineralization are found in some Chrysophyceae.

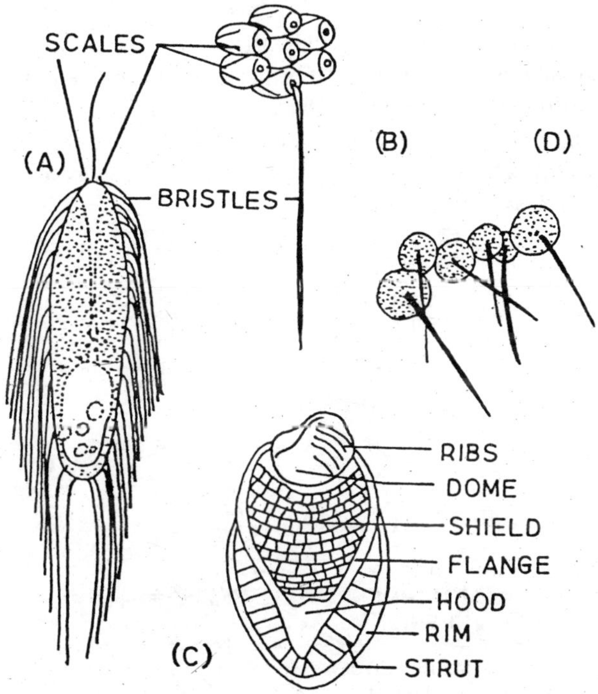

Fig.11.1. (A-C) *Mallomonas zellensis.* (A) Whole cell. (B) Scales. (C) A scale with bristle removed. (After Fott, 1962). (D) Pointed Scales of *Paraphysomonas vesitata.*

In some members **lorica** is found. In these the protoplast is surrounded by a rigid cell envelope and is separated from it by a space (Fig. 11.2). The loricas of *Dinobryon* are composed of interwoven system of microfibrils of cellulose with secondary mineralization. The formation of lorica begins when a small funnel-shaped piece arises from the cell. The walls and loricas may be of cellulosic nature with or without deposition of calcium carbonate or iron compounds. In some species lorica is thin and without mineralization while heavy impregnation of iron compounds is observed is some genus.

Flagella and Eyespot : There are two heterokontic flagella, both arising from a slightly subapical position. One of the two heterodynamic flagella is pleuronematic which bears two rows of hair like

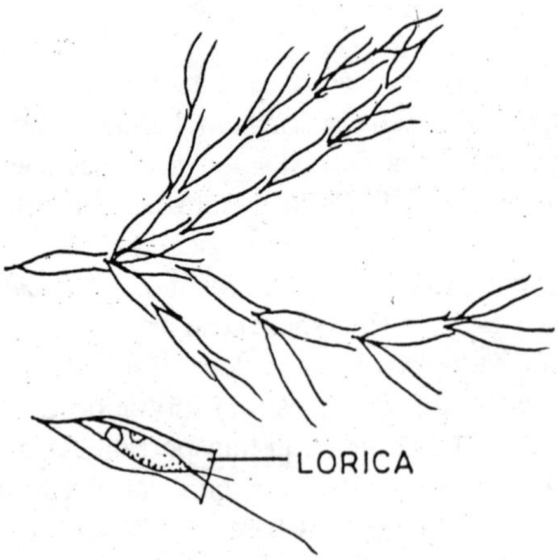

Fig.11.2. *Dinobryon* sp. showing lorica.

appendages called mastigonemes. Other one is whiplash or acronematic without any appendages. In the tinsel flagellum of *Ochromonas,* tripartite and fibrillar hairs have been found (Fig. 11.3). Whiplash flagellum is usually posterior and reduced one. It has a swelling at its base (flagellar swelling, Fig. 11.4). Just beneath this swelling there is a depression inside the chloroplast in which the eyespot is located. The eyespot consists of lipid granules and is found between chloroplast envelope and thylakoids band in the anterior region.

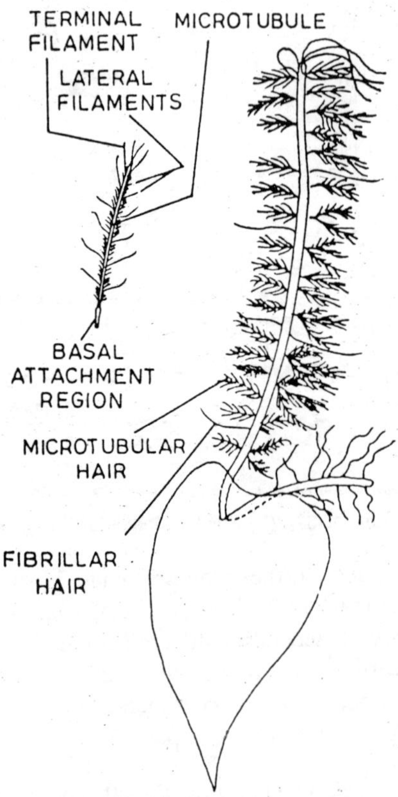

Fig.11.3. *Ochromonas danica* showing flagella and structural details of tripartite hairs. Internal cell contents have not been drawn. (After Bouck, 1971).

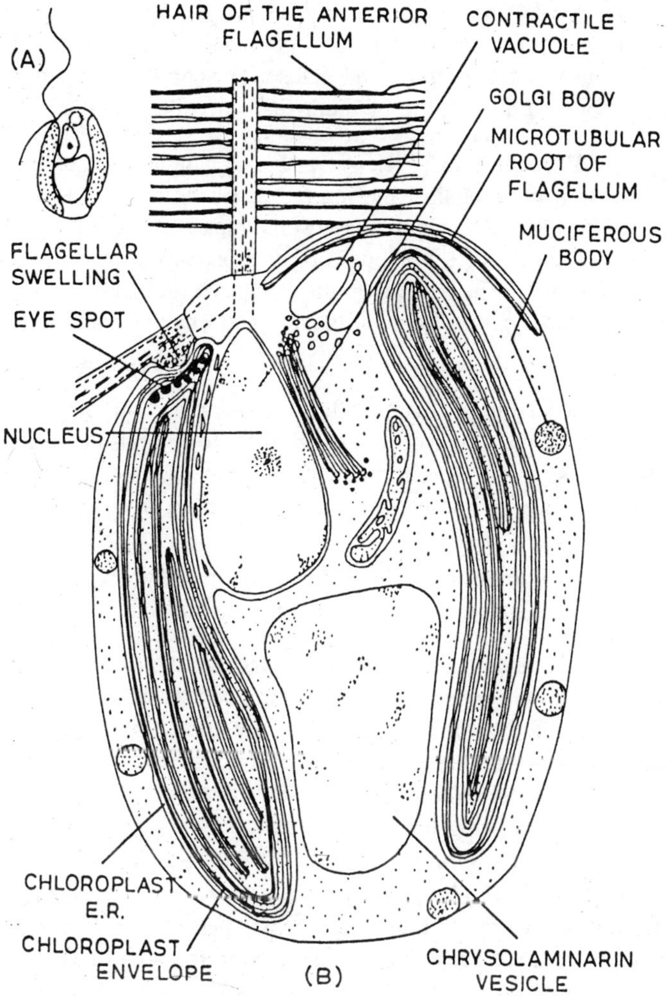

Fig.11.4. Diagram of (A) Light microscopic and (B) Electron microscopic view of the basic organisation of a chrysophycean cell. (After Hibberd, 1976).

The flagella of *Mallomonas* and *Synura* have basal bodies arranged in parallel fashion while the genera of Paraphysomonadaceae have basal bodies arranged perpendicular to each other. Both flagella show usual 9+2 pattern of microtubules with the exception of *Mallomonas caudata* (Wujek, 1978).

Chloroplasts and associated structures: The chloroplasts are parietal and are surrounded by two membranes of chloroplast E.R. The C.E.R. is in continuation with the outer membrane of the nuclear envelope. The number of chloroplast is usually few, often only one or two. The chief carotenoid is fucoxanthin while chlorophyll is represented by chlorophyll **a**, $\mathbf{c_1}$ and $\mathbf{c_2}$. ß-carotene was found to be present. The thylakoids are grouped in a band of three. Lipid droplets (**plastoglobuli**) are distributed randomly in the chloroplast matrix. Chloroplast ribosomes and chloroplast DNA have not been reported in scaled Chrysophyceae while in *Ochromonas* these are known to be present. Species of *Paraphysomonas* contain a leucoplast which lacks any trace of thylakoids.

Pyrenoids are of common occurrence in the chloroplast of the Chrysophyceae. They possess a granular area having difference in appearance with that of stroma. In silica-scaled Chrysophyceae pyrenoids are of immersed or semi-immersed type and are free of thylakoids.

Chrysophyceae contains chrysolaminarin (leucosin) as storage product which is composed of ß - 1,3 D-glucose residues and are found in posterior vesicle. In *Ochromonas* they possess several polysaccharides composed of D-glucose, D-galactose, D-mannose and xylose. The chrysolaminarin vesicle is much larger in organisms grown in the dark on a synthetic medium than in cells grown in the light.

Nucleus, Golgi Body and Mitochondria : The pear-shaped single nucleus in Mallomonadaceae occupies a characteristic position lying against the inner face of the right chloroplast with its tip close to the flagellar basal body (Fig. 11.4). The single dictyosome making up the Golgi body lies against the nucleus towards the anterior end of the cell. The edges of dictyosomes are usually inflated and one edge is directed towards the flagellar base. The nucleus in the Paraphysomonadaceae occupies a more or less central position. The nucleus has a normal double-membraned envelope and a single nucleolus. The mitochondria contains tubular cristae.

Contractile vacuoles and Lipid droplets : Contractile vacuoles are common and usually occur in the anterior region of the cell next to the Golgi apparatus. Similar to the pusule structure of Dinophyceae there is complex system of vesicles associated with contractile vacuole. The position of contractile vacuole is reported to be posterior in Mallomonadaceae, lateral in freshwater Paraphysomonadaceae except *Chrysophaerella.*

Lipid globules are frequent and are found in the protoplasm. In young cells there are usually few lipid bodies and their number and size increase with aging until they fill the protoplasm.

Projectiles or Vesicles : Muciferous bodies and discobolocysts are the two different types of projectiles found in Chrysophyceae. The **muciferous bodies** are bounded by a single membrane and possess granular material. The contents of the vesicle after discharge form fibrillar network outside the cell. These vesicles are usually located peripherally and size of these also varies considerably. In some members it is extremely large and pyramid-shaped. By the discharge of contents of these muciferous bodies scales and colonies are presumably held in position.

The **discobolocysts** are present in the outer layer of the cytoplasm and consists of a single membrane-bounded vesicle. The discharge of the discobolocyst is explosive. As the discharge occurs, the cell jerks violently. After the discharge of both these projectiles, the protoplast recovers without any deleterious effects. The origin of these projectiles takes place in the area of Golgi bodies. Unidentified vesicles containing microtubules and a granular ring near the basal bodies are reported by Wujek (1978).

Cytoplasmic processes: Different types of cytoplasmic processes have been found in the members of Chrysophyceae. **Dendricule,** the most simple process is relatively short membrane-bounded containing no ogranelles. In **tentacles** and **rhizopodia,** usually three mocrotubules run along the length of the process. These are membrane-bounded with number varying between 10 and 15 per cell. Organelles have not been found inside these processes also, although sometimes electron - dense granules may be seen.In *Pedinella,* a tail-like process **(peduncle)** is found which trails passively during swimming behind the organism. Large number of microtubules have been found in the central core of the peduncle.

REPRODUCTION

The most common method of reproduction is by longitudinal division of the cell. Cell division has not been studied at the fine-structural level in case of silica-scaled Chrysophyceae. Cell division may occur when the cell is motile but usually cell loses flagella before the initiation of division. In some forms the daughter cells fail to separate and extensive palmelloid masses can be formed.

The characteristic feature in both the silica-scale chrysomonads and other members of Chrysophyceae is the formation of silicious **statospore (cysts** or **resting spore** or **stomatocyst**).These are variable in shape, mostly spherical, ovate or ellipsoidal in outline. Their outer surface may be smooth or ornamented with arms, spines or warts. These ornamentation is of taxonomic importance. At the apical end of statospores a pore is present with collar that is closed by a plug (Fig. 11.5).

The formation of statospore is endogenous. It is known that the statospores are produdced within a cytoplasmic silica deposition vesicle, apparently formed by coalescence of Golgi vesicles. In nature, statospores are generally formed after a period of rapid growth. Statospore formation can be induced by lowering the pH and by nutrient deficiency under laboratory conditions. Unlike loricas which are cellulosic, the wall of statospores is always silicious. Thus in loricate Chrysophyceae, two processess of calcification and silicification occur at different phases. There is the dissolution of plug or it is separated from the spore wall during germination of statospore. The propoplast then moves out of the cyst by amoeboid motion. One or more motile spores are produced by the formation of flagella. Due to their silicious composition these are best preserved microfossils and are of great importance in assessing cultural eutrophication and lake acidification.

Asexual reproduction is also accomplished by the production of zoospores or autospores besides cell division and fragmentation.

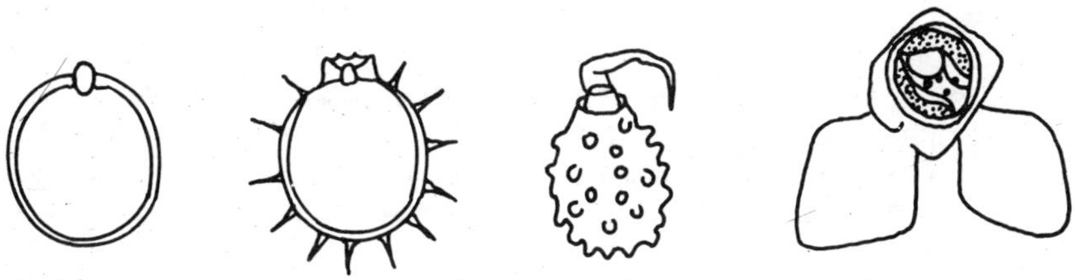

Fig.11.5. (A-C) Different types of statospores found in Chrysophyceae. (D) Sexually formed statospore of *Kephyriopsis* enclosed in an envelope. The loricas of gametes are still attached.

In some genera sexual reproduction has been observed and is of three kinds. In **typical isogamy** motile gametes are produced and fused like isogamous process. In **hologamy**, protoplasts of two vegetative cells fuse and in **autogamy** two nuclei fuse within the cysts which is not preceded by fusion of two protoplasts.

The loricate cells attach at their openings and the protoplasts migrate from the loricas which fuse and form the zygote (Fig. 11.6). The empty loricas normally remain attached to the zygote. In *Dinobryon cylindricum* an erotactin is released from the loricate female strain which attracts nonloricate male strain. Quadriflagellate planozygote is formed by anisogamy which loses its flagella and comes to the mouth of lorica by amoeboid movement. Ultimately a statospore is formed. Both the asexual statospores and the sexual statasporc are morphologically similar. Meiosis takes place at the time of germination of zygote.

Class - Prymnesiophyceae

GENERAL FEATURES

Members of the class Prymnesiophyceae (earlier name Haptophyceae) are characterised by the following features. (1) Motile cells possess a **haptonema** between two smooth flagella (Fig. 11.7). (2) Photosynthetic pigment includes chlorophyll **a**, $\mathbf{c_1}$ and $\mathbf{c_2}$ with fucoxanthin as major carotenoid. (3) Chloroplast E.R. is two-layered which is continuous with the outer membrane of the nuclear envelope. (4) Chrysolaminarin acts as storage product and remains in vesicles in the posterior part of the cell. (5) A large Golgi apparatus is found in the anterior region. (6) The cells are usually covered with scales.

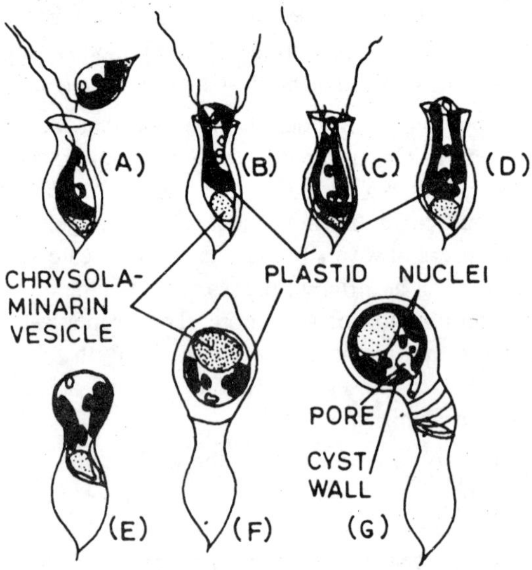

Fig.11.6. Sexual fusion in *Dinobryon cylindricum.* (A) Initial gametic contact. (B) Plasmogamy. (C) Binucleate planozygote. (D) Encystment and loss of flagella in planozygote. (E) Planozygote moving out of the lorica. (F) Immature statospore prior to silica deposition. (G) Mature hypnozygote with silicious cyst wall and a single plugged pore. (After Sandgren, 1981).

OCCURRENCE AND HABITAT

Although there are some freshwater forms but the members are primarily marine. They are important group of nannoplankton and constitutes about 45% of total phytoplankton in South Atlantic. Their number decreases in polar waters. Many members including coccolithophores are known to produce blooms. Gallois (1976) observed that blooms of coccolithophores in the past might have been responsible for the origin of the oil in the North Sea.

Most of the species are photosynthetic but either saprophytic or phagotrophic mode of heterotrophic growth is also possible. This allows the organisms to survive when they sink below the euphotic zone.

CELL STRUCTURE

Cell covering: Most of the members of Prymnesiophyceae have elliptical scales as cell covering. These scales which are embedded in a mucilaginous substance cover both motile and nonmotile stages. The composition of the scales is organic (cellulosic in *Hymenomonas*). In **coccolithophores** which constitute a large group within this class, a layer of calcified coccoliths is deposited outside the scales. Silicious deposition has not been found. The structure of the scales vary considerably. It may constitute either the whole structure of the scale or just the base plate on which the remaining structure is based. The elliptical organic scales have radiating ridges extending to the edge (Fig. 11.8 A). Cells of *Chrysochromulina* may be covered by scales of different morphology in which simple plate like scales are present underlying the more elabroate scales. In *Chrysochromulina pringsheimii* a complex cell covering with four different

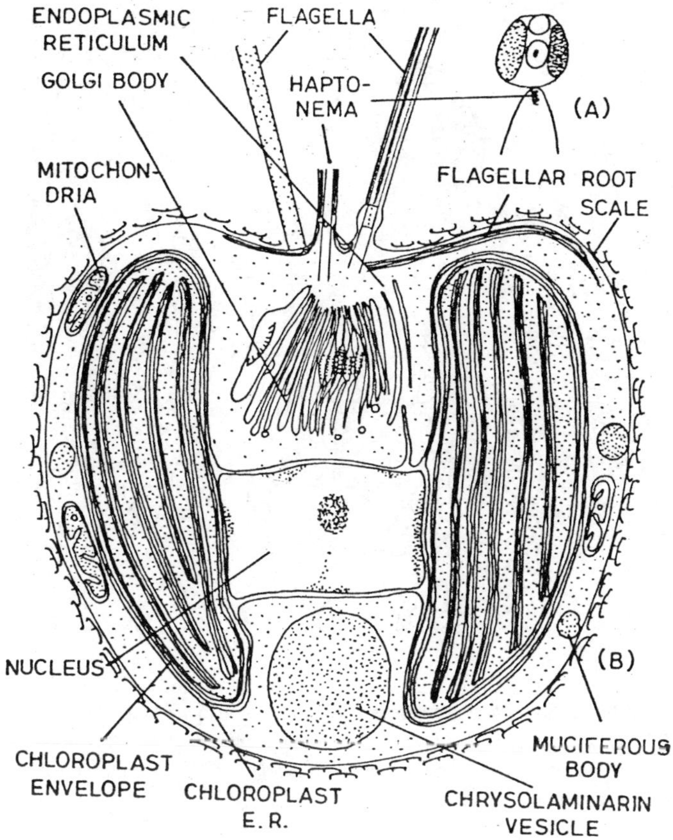

Fig.11.7. Diagram of (A) Light microscopic and (B) Electron microscopic view of a cell of *Chrysochromulina* sp.

types of scales are found (Fig. 11.8 B). At both the end of the cell, long-spined scales are present while rest of body has small-spined scales. Underneath this, there is a layer of plate-scales. Near the base of the flagella, a number of small plate-scales are present. In filamentous genera like *Gloeothamnion* (Fig.11.8 C), the mucilaginous investment thickens with age and gets lamellated. Besides usual circular scales covering the cell body, smaller oval scales may be present on the haptonemal surface (Leadbeater, 1970).

Golgi apparatus has been found to be involved in the production of organic scales (Fig. 11.7). The single polarised Golgi body is placed always between the bases of the flagella and the nucleus. A scale, after its formation inside Golgi vesicle, is closely enveloped by the vesicle membrane but this contact is lost immediately before liberation to the outside of the cell. Diurnal rhythm has been observed in the formation of scales.

The calcified scales of Prymnesiophyceae are called as **coccoliths**. The individual scales are produced within vesicles of Golgi body and collectively form a rigid outer skeleton (**coccosphere**) around the cell (Fig.11.8 D). These are basically organic scales having calcium carbonate deposition on one surface in a characteristic pattern. There are many complex forms of coccoliths. The number of coccoliths borne on a cell depends on its size and age. The coccolithophorids (members with coccoliths) are commonly found in tropical water due to saturation or supersaturation of warm water with calcium carbonate. By lowering the pH of the culture medium, the photochemical process is apparently directly associated with coccolith production because there is sharp fall in coccolith production when light is turned off.

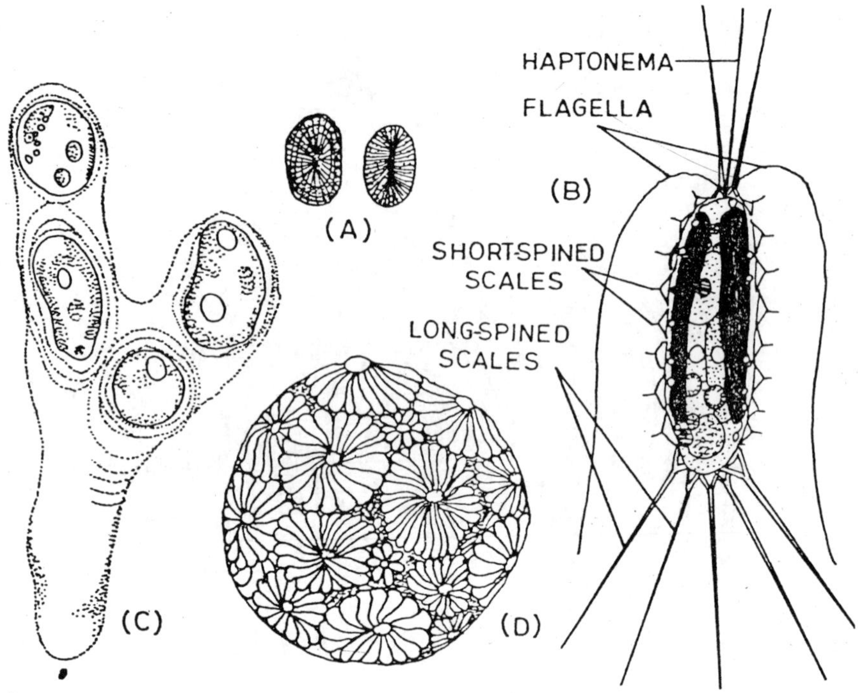

Fig.11.8. (A-B) *Chrysochromulina pringsheimii.* (A) Upper and lower face of small plate scales. (B) Cell showing fully extended haptonema, flagella and different types of scales. (C) *Gloeothamnion* sp. showing lamellated mucilaginous investment. (D) *Calcidiscus* sp. showing coccosphere comprising of coccoliths.

Flagella : Most of the members have two smooth flagella of approximately equal length (Fig. 11.7) with homodynamic or heterodynamic motion. In order Pavlovales, one flagellum is longer than the other and is covered with delicate hairs. Contrary to many other members of Chromophyta there is no flagellar swelling associated with an eyespot (except *Paviova).* Movement of the flagella is usually rapid. The cell swims in one direction only for a short distance after which the flagella move off in the opposite direction.

Haptonema : Near and in addition to flagella, a filamentous appendage (haptonema) arises which is thinner and with different structure and properties. The fine structure of haptonema is unlike that of flagellum (Fig. 11.9). A haptonema in cross section consists of an outer sheath of three concentric membranes and an inner circle of six or seven microtubules. It may vary from a highly coiled appendage to an elongated structure. In some genera (*Prymnesium*) it is stiff and short while *Hymenomonas roseola* shows bulbous haptonema. Leadbeater (1971) observed that the haptonema is capable of autonomous movement and can control the direction of coiling. The haptonema is commonly covered with small body scales.

Chloroplasts : In each cell two elongate discoid chloroplasts are present which are surrounded by four membranes (Fig 11.7). The two membranes of chloroplast envelope are encircled by two additional membrane of chloroplast E.R. The thylakoids are grouped into bands of three. There is no girdle band of thylakoids. In the centre of the chloroplast, a pyrenoid is commonly found.

Other cytoplasmic inclusions : In the cytoplasm two types of membrane-bounded vesicles (lipid containing and storage product containing) are found. Chrysolaminarin (leucosin) usually acts as storage

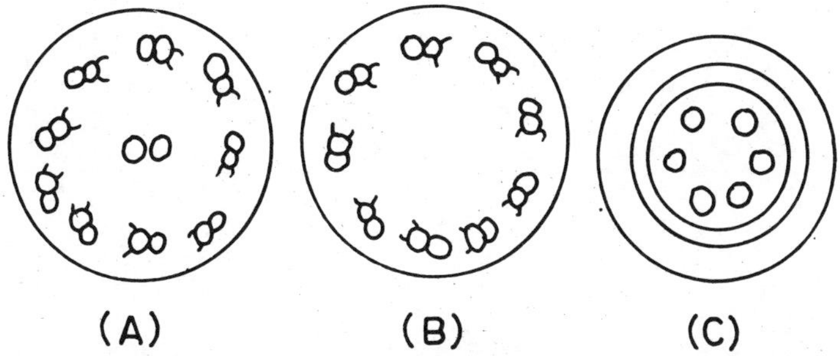

Fig.11.9. Anatomy of flagella and haptonema. (A) Typical eucaryotic flagellum sho^^ing 9+2 pattern. (B) Flagellum of the sperm of centric diatom showing 9+0 pattern. (C) Haptonema showing three concentric layers and 6 microtubules.

product, although ß-1,3 linked glucan similar to paramylon of Euglenophyceae has been seen in *Pavola mesolychnon.*

In *Chrysochromulina* and some other members **muciferous bodies** have been seen under the plasma membrane (Fig. 11.7). Like other classes these bodies consist of single-membrane with granular contents.

Pseudopodia are commonly found in the members of this class. Cells can also extrude trailing fialments from their surfaces. These are known as **filopodia** and may be straight or branched. Many members are phagocytic and have food vesicles in the cytoplasm in which they digest other small algae and bacteria.

REPRODUCTION AND LIFE CYCLE

Usual mode of asexual reproduction in Prymnesiophyceae is by binary cell division. Some members exhibit **pleiomorphism** (diverse morphologies) and transition from one expression to another.

The life cycle of *Hymenomonas* is biphasic with an alternation between *Hymenomonas* stage and *Gloeothamnion* stage. The diploid *Hymenomonas* phase bears coccoliths and multiplies by mitosis. After sometime it undergoes meiosis and produces four haploid products which germinate into a pseudofilamentous *Gloeothamnion* stage which lacks coccoliths. In addition to normal vegetative cell division, two types of motile spores are formed in this stage (Fig. 11.10). von Stosch (1967) reported that the one type of motile cell was sexual and the zygote develops into *Hymenomonas* stage while the other swarmer was asexual germinating into *Gloeothamnion* stage.

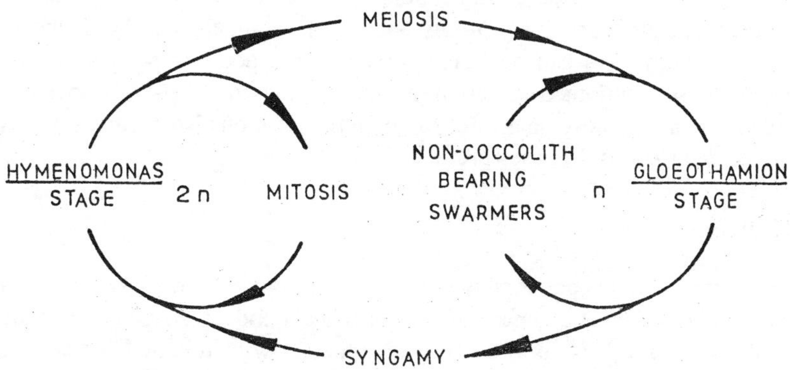

Fig.11.10. Life cycle pattern of *Hymenomonas carterae.* (Based on von Stosch and Leadbeater).

12

DIVISION CRYPTOPHYTA

Class - Cryptophyceae

GENERAL FEATURES

The class Cryptophyceae is a relatively small group which contains biflagellate members (**cryptomonads**). The members are recognised by the following features. (1) The cells are asymmetric and are flattened dorsiventrally with a periplast boundary. (2) The photosynthetic pigments include biloproteins which are different than those of blue-greens and red algae. (3) In most of the members, a tubular gullet is present which passes from anterior to posterior region of the cell. (4) Two equal flagella are located at the anterior gullet which bear hairs. (5) Sexual reproduction has not been reported. (6) A few members lack chloroplasts and are heterotrophic but most of the species have a bilobed chloroplast with a central pyrenoid (Fig. 12.1).

OCCURRENCE AND HABITAT

Most of the species of Cryptophyceae are unicellular and are found in different habitats. In freshwater systems, these constitute a frequent group of phytoplankton. Members are also found in marine habitat. Some species are euryhaline and can be found in salt marsh pools as well as in the open water of estuaries. The example of symbiotic association *(Mesodinium rubrum)* has also been seen in this class. Although most members are photosyntheitc but saprophytic nutrition also occurs. Besides free-swimming condition, palmelloid and coccoid phases are also found.

CELL STRUCTURE

Cell shape : Most of the members are dorsiventrally flattened but the group shows different modification from this normal morphology. Some genera (*Hemiselmis*) have kidney-shaped cells with laterally placed flagella (Fig. 12.2 A). *Hemiselmis* is broadly hemispherical with rounded posterior end and pointed anterior end. Some members are ovoid.

Cell covering : Cells are bounded by a proteinaceous **periplast** which consists of a trilaminate plasmalemma (Fig. 12.1) and has a hexagonal surface pattern. There is a coating of fine firils outside

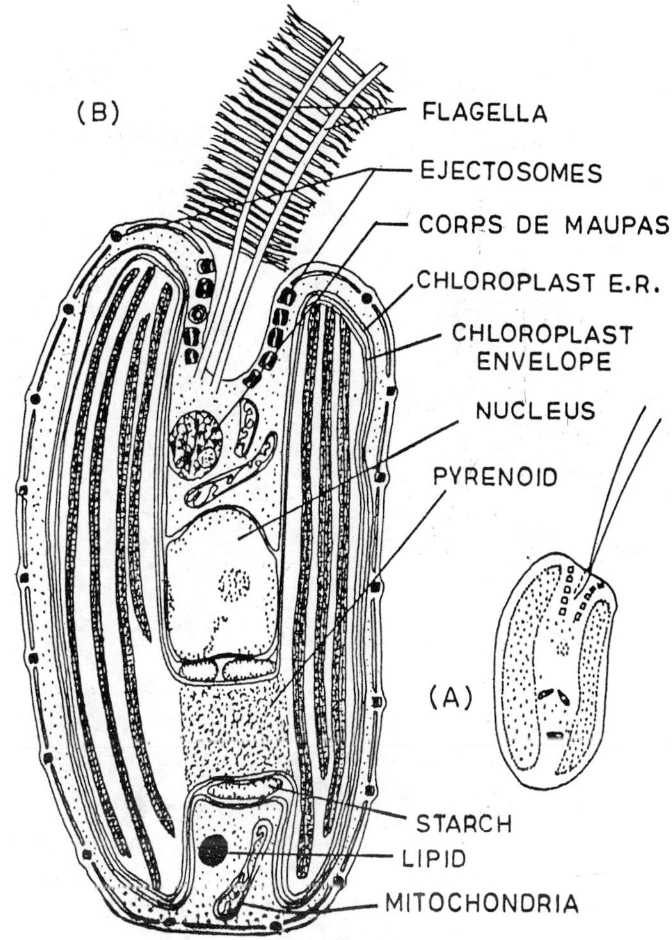

Fig.12.1. Diagram of (A) Light microscopic and (B) Electron microscopic view of *Cryptomonas* sp. (Cryptophyceae). (After Lee, 1980)

:he plasmalemma and a layer of dense material is found inside plasmalemma which is interrupted by ejectosomes (Gantt, 1971). Santore (1977) has observed some additional plates external to the plasmalemma.

Flagella : Two equal (sometimes slightly unequal), homodynamic or heterodynamic flagella are normally located in the anterior gullet or furrow. Each flagellum is about of same length as the body of the cell. Like the flagella of Chrysophyceae and Phaeophyceae, microtubular hairs are present along the surface of the flagella. The hairs on longer flagellum are approximately 2.5 µm long and are arranged in two opposite rows whereas in shorter flagellum 1 µm long hairs are arranged in one row in *Cryptomonas* sp. (Fig. 12.2 B). The presence of two pleuronematic flagella is a distinctive feature of this class.

Chloroplast and associated structures : In most cryptomonad cells only one chloroplast is present which may be bilobed and a single pyrenoid is present in the central region. Few species have several discoid chloroplasts. The chloroplast is surrounded by two inner membranes of chloroplast envelope and two outer membranes of chloroplast E.R. The Cryptophyceae is the only class where thylakoids are grouped in pairs and they are loosely arranged having no connections between each other. This character

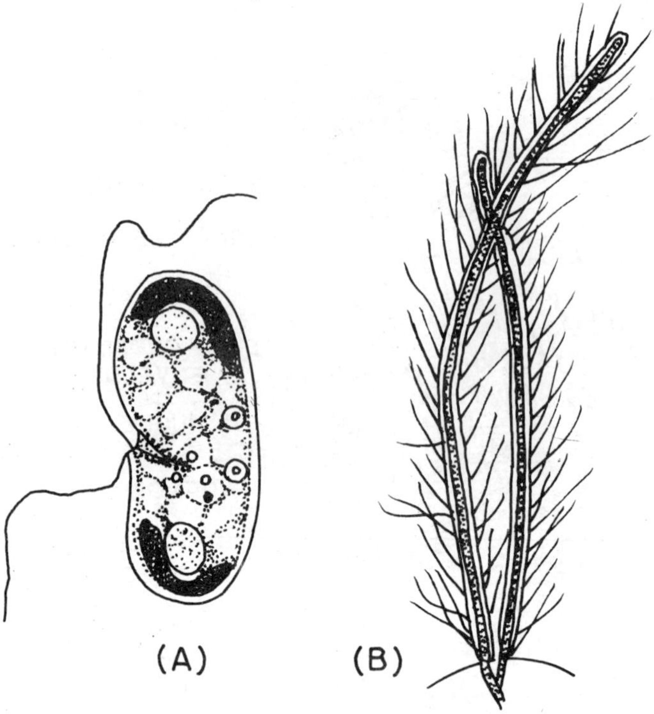

Fig.12.2. (A) *Hemiselmis* sp. showing kidney-shaped morphology and lateral flagella. (B) Flagella of *Cryptomonas* sp. showing two opposite rows of hairs on longer flagellum and only one row of hairs on shorter flagellum.

is considered to be an intermediate feature between the singly occurring thylakoids (Rhodophyceae) and the prevalent bands of three thylakoids found in many algal classes. According to Taylor and Lee (1971) some variations in the typical paired arrangement of thylakoids also occur.

Members of Cryptophyceae may be blue, red, brown, green or olive-yellow. Some colourless forms are also found. Phycobiliproteins are present in the intrathylakoid space (inside the thylakoids) rather than towards stroma. There are three spectral types of phycoerythrin and phycocyanin each, which are different from those of Cyanophyceae and Rhodophyceae. These pigments do not occur as discrete phycobilisome-type aggregations. Photosynthetic pigments include chlorophyll a, c_2 with α -carotene as major carotenoid and diatoxanthin as major xanthophyll. Under different light conditions the amount of pigment varies. In low light the thylakoids are thicker and there is a higher concentration of phycoerythrin. With the increase in light intensity the amounts of phycocyanin and chlorophyll a decrease whereas the chlorophyll c_2 contents remain unchanged (Faust and Gantt, 1973).

In nonphotosynthetic Cryptophyceae there is the absence of chlorophyll. In the colourless species (*Chilomonas paramecium*), a leucoplast is found which contains starch granules and is surrounded by a chloroplast E.R. (Sepsenwol, 1973).

Starch, the reserve product is present in form of 2, 4 or many small granules within the space between the chloroplast envelope and the chloroplast endoplasmic reticulum (periplastidial compartment). This is the only class of algae which has storage product in this area. This compartment also has visicles, ribosomes and distinctive structure known as nucleomorph. Adjacent to the flagellar groove, the contractile vacuole is present. Mitochondria may be unbranched or branched. It is prominent and with finger-like cristae.

Eyespot : Some of the species of Cryptophyceae possess eyespots. They are usually situated beneath the chloroplast membrane. The eyespot consists of a single layer of lipid containing granules and occupy usually a median position within the cell. They remain close to nucleus and in some members exhibit positive phototaxis.

Ejectosomes : The projectiles of Cryptophyceae are called **ejectosomes** which appear as small, refractive dots under light microscope. These form an evident lining on the sides of gullet or groove and are closely related to the R-bodies of the kappa particles of the ciliates. The other type of smaller ejectosomes (about 10 times smaller than larger one) are also found peripherally in some members like *Cryptomonas* which are analogous to the trichocyst of Dinophyceae. The smaller ejectosomes are arranged in a hexagonal pattern. Both types of ejectosomes are similar in structure. Intact ejectosomes are made up of two unequal-sized bodies enclosed with a single membrane (Fig. 12.3). They are arranged in such a fashion that smaller body remains near the surface. After being irritated ejectosomes get discharged. The discharged ejectosomes constitute a long tubular structure with a short portion at an angle to the long portion. The organisms move in opposite direction with the discharge of the ejectosome. The ejectosomes lining the gullet are released into it. The Golgi body is found to be associated with the formation of ejectosomes.

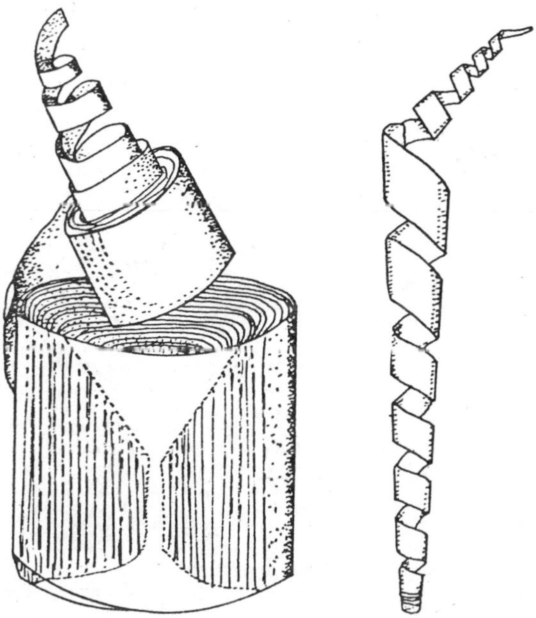

(A) (B)

Fig.12.3. (A) General organisation of an ejectosome showing the two subparts. (B) A model of an ejectosome being fired to the outside of the cell. (After Hovasse et al., 1967).

In the anterior portion of the cell a large vesicle like structure is found. It is known as **Corps de Maupas** (Fig. 12.1) and it probably helps in disposal of unwanted protoplasmic structures by digestion i.e., lysosomal in function (Lucas, 1970).

Nucleus : The cells are uninucleate and the nucleus is situated between the chloroplast lobes. The nucleus is eucaryotic and appears to have well-defined membrane and a prominent nucleolus. The chromosomes are nonmoniliform and the chromatin is granular. Gillott and Gibbs (1980) described the **nucleomorph**

in which nucleic acid has been identified tentatively. It is thought to represent possibly the vestigeal nucleus of a primitive endosymbiont.

REPRODUCTION

Asexual reproduction occurs by longitudinal division of the cell when still motile. In most species of *Cryptomonas* the dividing cells come to rest and become embedded in mucilage. The apical depression is usually halved during division. Some species produce thick-walled cysts. In those members where palmelloid condition is the normal vegetative phase, reproduction is affected by the production of biflagellate swarmers. The coccoid algae *Tetragonidium* also produces zoospores. There is no conclusive report of sexual reproduction.

13

GENERAL TOPICS

(I) DISTRIBUTION OF ALGAE

Although Algae are ubiquitous in distribution but they constitute an important component in freshwater and marine environments. Some algae also live in terrestrial habitats. Thus the distribution of majority of algae may be grouped into three different categories i.e., soil, freshwater and marine algae.

SOIL ALGAE

The algal members which are found on and below the surface of soil are known as soil algae (**edapophytes**). It may be divided into two groups -1. **sapophytes** or epiterrancan and 2. **cryptophytes** or subterranean. Sapophytes are exposed directly to the atmosphere while cryptophytes are found below the surface of the soil. The study of soil algae is not a recent development but since last century many unicellular and colonial forms have been recorded as soil algae. These members belong to diatoms, green, red and blue-green algae, Euglenophyceae, Xanthophyceae and occasionally from other classes. Important among them are:

Chlorophyceae : *Chlorococcum, Ulothrix, Rhizoclonium, Fritschiella, Oedocladium, Cosmarium, Closterium, Zygogonium, Sirocladium, Protosiphon.*

Cyanophyceae : *Chroococcus, Gloeocapsa, Oscillatoria, Lyngbya, Microcoleus, Symploca, Aulosira, Schizothrix, Calothrix, Cylindrospermum, Nostoc, Scytonema.*

Xanthophyceae : *Vaucheria, Botrydium*

Bacillariophyceae : *Navicula, Nitzschia, Caloneis.*

Characteristics : Tropical soils favour the growth of soil algae as compared to temperate soils. Soils having acidic pH (3.7-5) are supposed to be favourable for green algae whereas alkaline soils favour the growth of blue-green algae and diatoms. These soil microflora have some kinds of specialized structures which protect them from the changed environmental conditions. Many soil algae have specialized cell wall, mucilaginous sheaths and viscous protoplasm which enable them to overcome high temperature, prolonged and severe droughts, high salinity and other unfavourable conditions. Algal growth is poor in intense light while profuse growth is found in shaded areas. Reproduction is completed within the limited period particularly when the water is available.

Due to thick coloured mucilaginous sheath, blue-green algae can survive wide variety of soil conditions. In cultivated or manured soils, algal flora is denser than that of uncultivated soils. The moist soils harbour a luxurient population of algae. Soils having higher nitrate and phosphate contents lead to abundance of algae mainly diatoms. Algae may be represented in form of fragments, asexual or resting spores when soil is dry and conditions are adverse. The spores of *Nostoc muscorum* can be viable even after 70 years. The members of Xanthophyceae and Bacillariophyceae are not very common on soils because they can not sustain severe drought.

Cryptophytes (Subterranean algae) : As these algae are found below the surface of soil, they do not get light and hence remain in dormant conditions. They are found in the resting stages and do not impart active vegetative growth. All species occurring in lower layers of soil are also found in surface layers. The maximum depth to which algae have been recorded in soils is 2 meters. Kamat and Patel (1973) reported algae from one meter depth below the surface of the soil in the cultivated fields. With the increase in depth the number of taxa decreases. The common subterranean algae are *Chroococcus, Lyngbya, Phormidium, Nostoc, Scytonema, Chlorella, Oedocladium, Fritschiella, Cylindrocystis, Caloneis, Pinnularia, Navicula, Trachelomonas, Euglena* etc.

During the studies of Nagpur paddyfield soils, Kamat and Patel (1973) reported 30 algal genera belonging to Chlorophyceae, Cyanophyceae, Xanthophyceae, Euglenophyceae and Bacillariophyceae from surface to 30 cms depth, 12 genera of green and blue-green algae between 30 to 50 cms depth, 11 genera between 50 to 70 cms depth, 13 genera between 70 to 90 cms depth and 9 genera between 90 to 100 cms depth. The occurrence of algae in the lower soil layers depends on 1) cultivation, 2) water seepage, 3) animals and 4) self-motility. Some algae like *Euglena* may move down by their own mobility.

Factors affecting the soil algae : A number of factors have been found to be associated with the growth and diversity of soil algae.

Moisture : It is one of the important factor for the growth of soil algae which is required to complete the life cycle.

Temperature : Many blue-green algal forms can tolerate temperature between 60-90°C and due to this they are common in tropical soils. Diatoms can survive very low temperatures.

Light : Although many algae can withstand bright sunlight but their growth is maximum in less bright light.

pH and Salinity : pH has an important role in determining the algal groups. Many green algae can be found in a wide range of pH while blue-green algae prefer neutral or alkaline pH. Blue-greens and diatoms do not prefer acidic soils.

Soil texture and chemical composition : The chemical properties and the texture of soil are important factors which decide the type and growth of algal flora.

Importance : Some of the soil algae like *Vaucheria, Scytonema, Aulosira* etc. are known as good soil binders. Nitrogen-fixing blue-greens like *Nostoc, Aulosira* etc. increase the fertility of soil by increasing nitrogen content. Under water-logged conditions, the blue-green algae may be used for reclamation of the usar soils. The soil algae are the pioneer plants on barren areas or virgin soils. After death and decomposition of algae, organic matter is added to the soil and after some years the soils become suitable for cultivation. Some of the soil algae are known to produce different chemicals which enhance the growth of other plants. Under rice field conditions, algal growth has been estimated to produce approximately 2-3 tons of dry matter per hectare.

ALGAE AND THE AQUATIC ENVIRONMENT

Within an aquatic environment an alga grows as either benthos, periphyton or phytoplankton.

Benthos : Benthos are those organisms that grow on the bottom of a body of water. If the alga is attached to the surface of a rock, it is known as **lithophytic**. Many species of blue-green algae (*Chroococcus, Gloeocapsa, Scytonema, Rivularia, Calothrix, Stigonema* etc.), green algae like *Chlorococcum* grow on exposed rocks. If the alga bores into or lives inside a rock, it is **endolithic**. Some species of Chlorococcales, *Trentepohlia* grow within the small crevices of the rocks. Algae which live on the surface of mud or sand are **epipelic**.

Smaller benthic algae may be associated loosely with submerged substrates but larger species are usually attached firmly. The entire thallus may be appressed to the substrate but in many species the thallus is attached by a special base called the **holdfast**. The benthic flora may be influenced by many factors like topography, physico-chemical characters of water, population dynamics.

Periphyton : Periphyton are organisms attached to submerged vegetation. The periphyton is termed as **epiphytic** if it is attached to the surface of an aquatic plant. The epiphytes may be on leaves (**epiphyllophytes**) or on barks (**epiphloeophytes**). Periphytons living inside the other plant are known as **endophytes**. If an alga is attached to the surface of an animal, it is **epizoic** or **epizoophytes**. **Endozoic** or **endozoophytic** algae are those which live inside an animal.

Phytoplankton : Phytoplankton are those plants which are free-floating and move with the water current. **Net** or **macroplankton** are those phytoplankton which are large enough and trapped in a fine plankton net. If it is too small (75 μm or less) to be trapped in a net, it is known as **nannoplankton**. Certain minute phytoplankton (upto 2 μm) found in aquatic systems are called **picoplankton**. On the basis of local environmental conditions, planktons may be **limnoplankton** (found in lakes), **rheoplankton** or **potamoplankton** (found in running water), **heleoplankton** (found in ponds), **haliplankton** (found in salt water) and **hypalmyroplankton** (found in brackish water). On the basis of life history, planktons may be **holoplankton** (free-floating throughout their life) and **metoplankton** (free-floating only at certain stages of the life cycle).

These phytoplanktons occur both in fresh and marine waters. The floating mass of algae are sometimes referred to as **tychoplanktons** and those that are found in flowing waters as **fluviatile communities**. The floating community at the water-air interface is called the **neuston**. Debris mingled with the plankton is termed as **pseudoplankton**. The **euplankton** (true plankton) is mainly composed of species of Chlorophyta, Cyanophyta, Bacillariophyta and Chrysophyta (Fig. 13.1). The nannoplankton mainly belong to flagellates, coccoid Chorophyta and centric diatoms. Although morphologically the phytoplankton found in fresh water and marine habitat are similar but a very few species are common to the two habitats. Desmids are absent in marine habitats and the blue-green algae are frequently represented by the bloom-forming *Trichodesmium*. Conversely genera such as *Coscinodiscus, Chaetoceros, Rhizosolenia* are almost confined to the marine plankton. Common dinoflagellates (*Peridinium* and *Ceratium*) are found in both but the number of species is more in marine habitat than that in freshwater.

The spatial distribution of phytoplankton is affected by the shape and size of the lake basin, position of inflows, degree of stratification etc. The growth and distribution of marine algae are dependent upon many physico-chemical and biological factors like temperature, light, pressure, turbulance, pH, nutrient conditions etc.

FRESHWATER ENVIRONMENT

Waterbodies can vertically be divided into **euphotic** (photic) and **aphotic** zone (Fig.13.2). In euphotic zone sufficient light penetrates and thus photosynthesis is greater than respiration. The zone where photosynthesis is equal to respiration over a 24 hour period is called as **compensation depth**. This separates euphotic zone from aphotic zore.

Ponds and lakes are two basic types of lentic system and rivers and springs are the common examples of lotic communities. Three different zones are distinct in ponds and lakes.

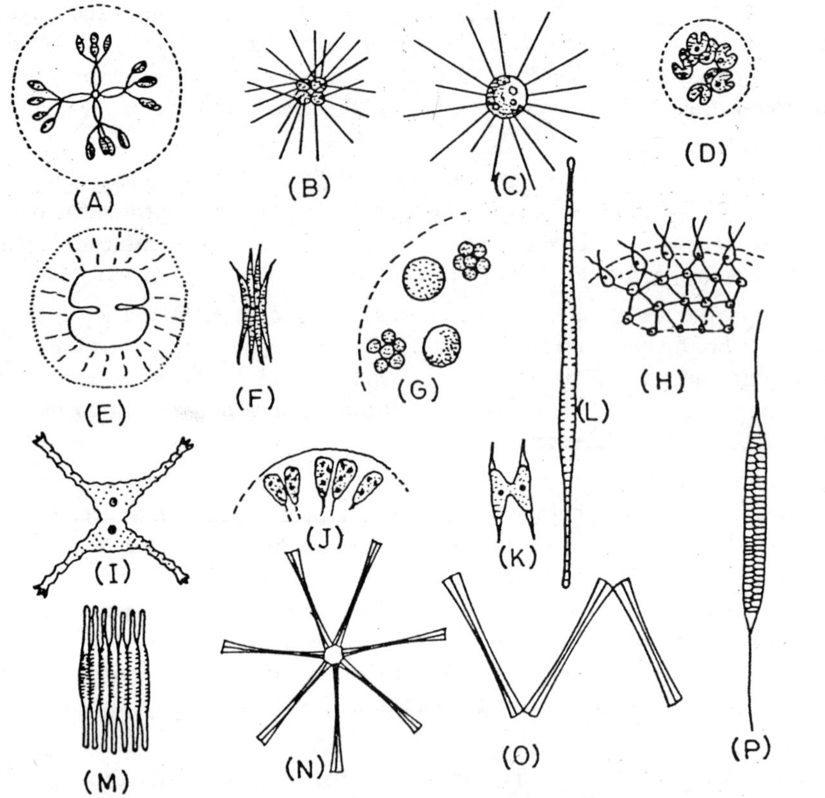

Fig.13.1. Common freshwater planktonic algae. (A) *Dictyosphaerium*. (B) *Micractinium*. (C) *Golenkinia*. (D) *Kirchneriella*. (E) *Cosmarium*. (F) *Ankistrodesmus*. (G) *Sphaerocystis*. (H) *Volvox*. (I) *Staurastrum*. (J) *Gomphosphaeria*. (K) *Arthrodesmus*. (L) *Synedra*. (M) *Fragilaria*. (N) *Asterionella*. (O) *Diatoma*. (P) *Rhizosolenia*.

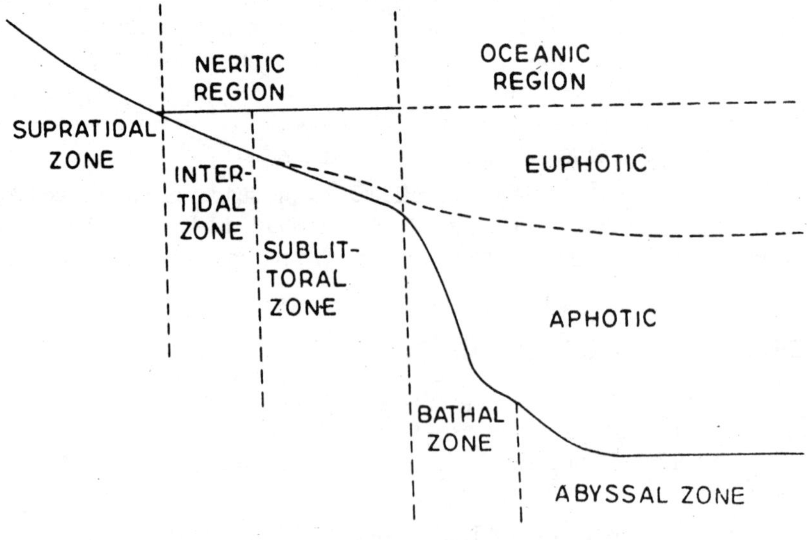

Fig.13.2. Different regions and zones of an ocean.

(1) The **supralittoral zone** is above the edge of standing water which is exposed to waves and winds. Very few algal forms are found in this zone because of these adverse conditions. (2) The **littoral zone** extends from the waters edge to the depth of about 6 m. Large amount of periphytic algae mainly the diatoms and desmids are found in this region. In terms of biomass, this is the most productive area of the lake. (3) The **sublittoral zone** is the zone upto compensation depth. (4) The **profundal zone** extends beneath the compensation depth. It contains heterotrophic algal cells and resting spores of photosynthetic algae.

When the surface water is warmed by radiation, lakes reveal **thermal stratification.** The surface water is warmer and lighter than the colder and denser water beneath it. Upper zone of warmer water is called the **epilimnion** and the lower zone of denser water is **hypolimnion. Thermocline** is the layer of water which has a sharp temperature drop within a few meters depth and this separates the epilimnion from hypolimnion.

Majority of the algae found in lotic systems (springs and rivers) are able to attach themselves to a stable bottom substrate. In swiftly moving streams, phytoplankton population remains little while in slowly moving rivers, a considerable amount of phytoplankton is present. Some green algae like *Cladophora, Stigeoclonium*, blue-green algae and attached diatoms are the common constituent in smaller streams and rivers.

MARINE ENVIRONMENT

In the ocean, two basic regions are **neritic** and **oceanic**. Neritic region extends from the high-tide mark to a depth of 200 m and the oceanic region extends upto bottom from a depth of 200 m. The **splash zone** or **supralittoral zone** is above the high-tide level and receives splash water from the waves. Mostly the blue-green algae form a blackish scum on the rock in this zone. Neritic zone is divided into intertidal and sublittoral zone.**Intertidal** zone occupies the region between the low- and high-tide marks of the spring (greatest) tides. If the intertidal zone is sandy then few algae will grow while rocky intertidal zone leads to the growth of large number of lithophytic marine algae. This region is well-lighted and oxygenated. The **sublittoral** zone extends from the lowest low-tide level to a depth of 200 m. Large brown and red algae are common in this zone. Bathal and abyssal zones are the two distinguishable regions in oceanic region. **Bathal** zone includes the continental slope (200 - 1000 m) and the **abyssal** zone covers the deep ocean bottom (Fig. 13.2). Only few algae are present in this because these are in aphotic zones. Thermal stratification also occurs in the ocean which result in the areas of nutrient deficiencies and low phytoplankton growth.

(II) ALGAL ASSOCIATIONS

Besides most common free-living existence, there is a broad range of association between algae and other organisms. Some algae are epiphytic (attached on other plants), epizoic (attached on animals), endophytic (growing inside other animals) and even parasitic in plants and animals. Keeping the syllabus in view here a brief account of symbiotic and parasitic algae is given.

Symbiotic Algae

A German botanist de Bary in 1879 coined the term **'symbiosis'** to describe the mutual relationship between algae and fungi in lichen. There are two schools of thought - one accepts the original meaning of de Bary i.e., living together without any reference to dependence while other group believes on dependency. The association of algae with the other partner may be grouped into two: 1) **Extracellular** association between *Anabaena* and *Azolla*, 2) **Endosymbiotic** association between *Chlorella* and *Paramecium*. The common examples of association of algae with different groups are mentioned here.

Symbiosis between Algae and Algae : The filament of *Calothrix* has been found within *Enteromorpha. Richelia intracellularis*, a blue-green alga is sometimes found inside the frustule of a diatom (*Rhizosolenia*).

In this association photosynthesis is exhibited by *Richelia* while *Rhizosolenia* has little developed chromatophore. *Coleochaete* (Chlorophyceae) grows endophytically in the cell walls of *Nitella.*

Symbiosis between Algae and Fungi : Lichens are the best known case of symbiotic association between algae and fungi. The algal partner in lichens is called **phycobiont** while the fungal partner is called **mycobiont.** Here algal partners are capable of free and independent growth. The common examples of algal genera associated in lichens are *Gloeocapsa, Calothrix, Dichothrix, Nostoc, Scytonema, Stigonema, Trebouxia, Stichococcus* etc. Among all these, *Trebouxia* is one of the most common algal partner in lichens. The algal partners may or may not modify their morphology after this association. Algal symbionts of *Nostoc* and *Scytonema* maintain their morphology in the lichen *Collema* and *Ephebe* while *Dichothrix* (Rivulariaceae) looks like *Nostoc* in the lichen *Placynthium nigrum.* Normally there is one phycobiont in the lichens but it may be two, as in the lichen *Lobaria* two phycobionts (*Coccomyxa* and *Nostoc*) are found.

Electron microscopy has revealed that many mycobionts produce thin walled haustoria which penetrate the protoplasm. The fungal hyphae and the filaments of *Nostoc* are freely mixed in the mucilaginous lichen *Collema.* The fungal partners receive the benefits in form of carbohydrates, nitrogenous products, vitamins etc. while the alga is protected by fungi from the environmental extremes.

Symbiosis between Algae and Bryophytes : Many species of *Nostoc* and *Anabaena* are found to be in symbiotic association with bryophytes like *Anthoceros, Notothylas, Blasia* etc. Mucilage cavities of these bryophytes possess above mentioned nitrogen-fixing blue-greens. As the mucilage cavities have opening through a narrow pore, the blue-greens may enter through this.

Symbiosis between Algae and Pteridophytes : *Anabaena* establishes a symbiotic relationship with Azolla, the fern. Alga is found in the cavities on the dorsal lobe. The growth of *Azolla* has been found to be more in presence of *Anabaena.*

Symbiosis between Algae and Gymnosperms : Among the gymnosperms, cycads are the only examples which possess symbiotic algae (Fig. 13.3 A). A distinct algal zone is found in the cortex of *Cycas* root.

Symbiosis between Algae and Angiosperms : *Nostoc punctiforme* has been found as symbionts in the root nodules of *Trifolium alexandrium.* Wart-like swellings of *Nostoc punctiforme* is found on the leaf bases of the huge leaves of *Gunnera macrophylla* (Fig. 13.3 B).

Symbiosis between Algae and Animals : The algae growing on or in animals are often called *Zooxanthellae* or *Zoochlorellae. Zoochlorellae* is typically found in freshwaters and includes green algae like *Platymonas* in addition to *Chlorella.* The *Zooxanthellae* are almost exclusively marine organisms and include dinoflagellates and cryptophytes. *Zoochlorellae* and *Zooxanthellae* can be found in Protozoa, sponges, flatworms, coelenterates, molluscs, echinoderms, ascidians, annelids and jelly fish. Symbiotic blue-greens are found in *Gloeochaetc wittrockiana, Glaucocystic nostochinearum, Amoeba* etc. In the endodermal digestive systems of *Paramecium bursari, Chlorella* cells upto 18 in number has been observed. *Basicladia* (Chlorophyceae) is found on the carapase of turtles. *Characium* (Chlorophyceae) is found in the antennae of mosquito larva. Taylor (1975) found the members of Dinophyceae as most dominant symbiont among marine invertebrates. Association between blue-green algae and sponges has also been recorded. Sponges grow better when alga is present. The majority of associations with animals are found in the tropical marine environment where the water is nutrient poor.

Parasitic Algae

There are several algae that are parasitic upon the leaves and twigs of land plants. These members mainly belong to Chlorophyceae and are found upon a wide variety of related and unrelated hosts and may or may not cause injury to the host in which they grow. When the organisms involved (host and parasite) are closely related taxonomically, they are termed as **adelphoparasites** and when are not so related they are known as **alloparasites** (Feldmann and Feldmann, 1958). The parasitic forms belonging to different classes are discussed here.

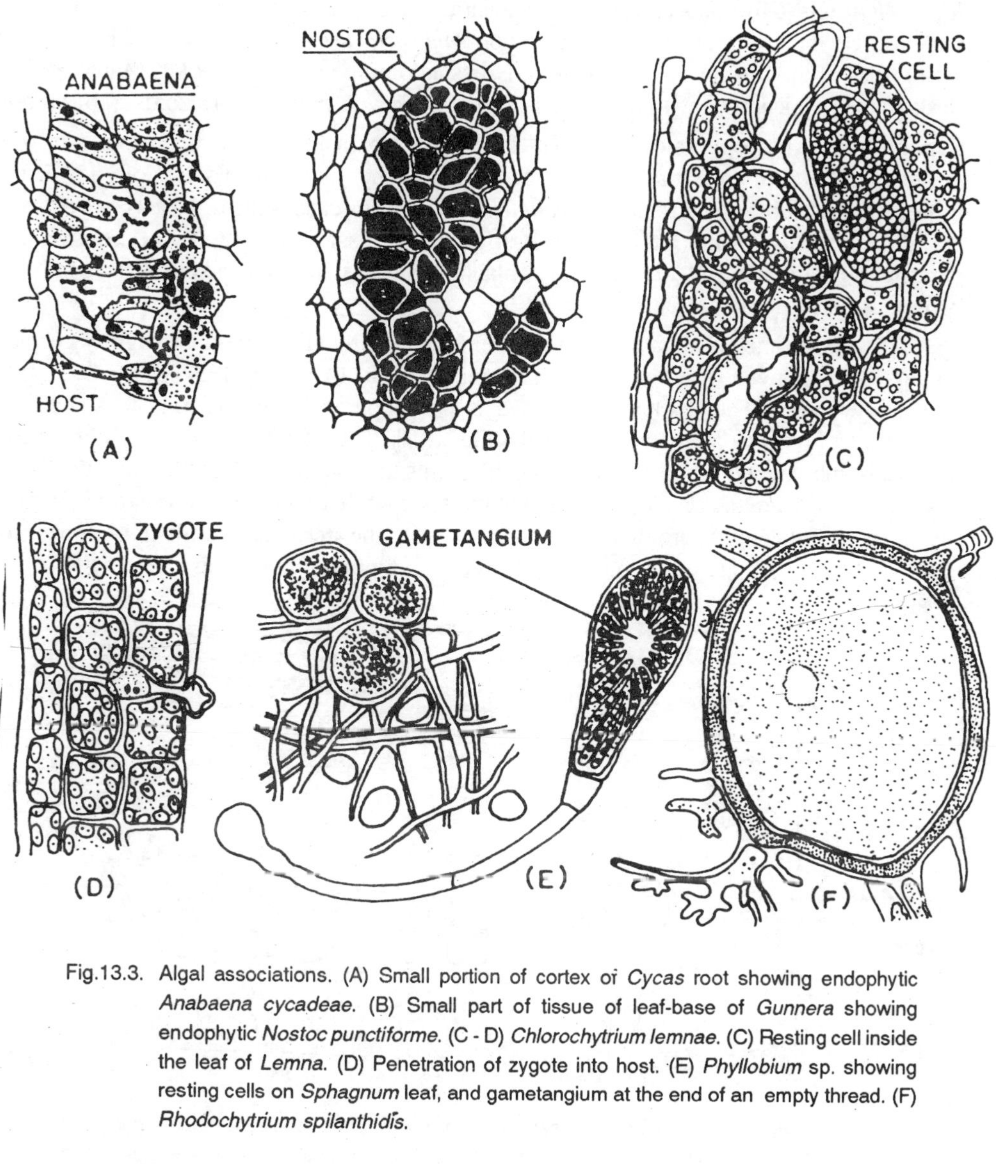

Fig.13.3. Algal associations. (A) Small portion of cortex of *Cycas* root showing endophytic *Anabaena cycadeae.* (B) Small part of tissue of leaf-base of *Gunnera* showing endophytic *Nostoc punctiforme.* (C - D) *Chlorochytrium lemnae.* (C) Resting cell inside the leaf of *Lemna.* (D) Penetration of zygote into host. (E) *Phyllobium* sp. showing resting cells on *Sphagnum* leaf, and gametangium at the end of an empty thread. (F) *Rhodochytrium spilanthidis.*

CHLOROPHYCEAE

(i) *Chlorochytrium* (Chlorococcales) : It grows within the tissues of mosses, *Lemna, Ceratophyllum,* marine algae and lower animals. It is a common endophyte which resembles free-living *Chlorococcum* in morphology (Fig. 13.3 C). The swarmers settle down on aquatic plants and gives rise to tubular projections which enter the host either through stomata or in between two epidermal cells (Fig. 13.3 D). It is an example of space parasite which causes little damage except in the immediate neighbourhood of the endophyte.

(ii) *Phyllobium* (Chlorococcales) : It is an endophyte whose simple or irregularly branched coenocytic tubular cells have globosely inflated tips (Fig. 13.3 E). It is found along the peripheral veins of the leaves of *Lysimachia, Ajuja* and others. *Phyllobium sphagnicolum* seems to be true parasite on the leaves of a bryophyte, *Sphagnum.*

(iii) *Rhodochytrium* (Chlorococcales) : This parasitic unicellular form (Fig. 13.3 F) is found upon a wide variety of hosts, most commonly upon *Ambrosia artemisiifolia.* The parasite attacks all aerial parts of the host but is abundant in tissues adjoining vascular bundles. Mature cells have a flask-shaped appearance with somewhat twisted neck with numerous rhizoids in basal portion.

(iv) *Cephaleuros* (Chaetophorales) : *Cephaleuros* grows parasitically on the leaves and twigs of *Magnolia, Thea, Piper, Rhododendron* and on many other higher plants. It is of wide distribution in tropical and subtropical regions. *Cephaleuros virescens* causes the disease **(red rust of tea)** in tea (*Thea sinensis*) plantation in North East India and Assam. It is highly destructive as a stem parasite. It also infects and damages piper plantation in Easten India and citrus trees in Florida. *C. coffeae* is parasitic on coffee.

The appearance of alga on the host is in form of circular to irregular slightly elevated spot having usual diameter of less than 2 mm but may be upto 1 cm (Fig. 13.4 A). Spots are greenish-grey and velvety. The thallus of the alga is pseudoparenchymatous, one to several cells in thickness and with radiately arranged cells (Fig. 13.4). The upper surface of the algal mass bears numerous unbranched vertically projecting filaments through the cuticle. Some of these filaments bear cluster of gametangia and sporangia at the apex while others remain as sterile hairs. Due to the presence of haematochrome pigment, cells of both the sterile and fertile hairs appear reddish-brown. Chloroplasts are either parietal or discoid and without pyrenoids.

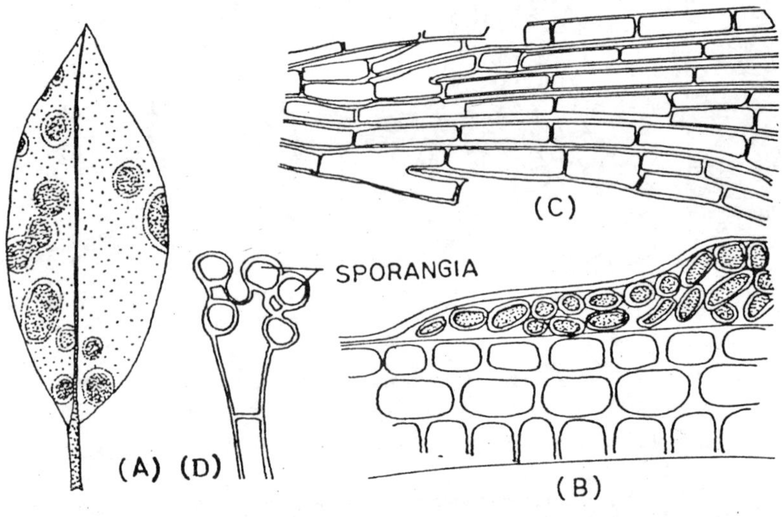

Fig.13.4. *Cephaleuros virescens.* (A) Leaf of *Magnolia* with patches of *Cephaleuros.* (B) Vertical section of the thallus showing pseudoparenchymatous habit. (C) Surface view of the thallus showing radiately arranged cells. (D) Fertile filament with sporangia.

Biflagellate zoospores are produced from the zoosporangia which are broken away and carried by wind action. Zoospore may germinate to form a new thallus if it falls on suitable host.

(v) *Phyllosiphon* (Siphonales) : It grows as an intercellular parasite in stems and leaves of various Araceae like *Arisarum vulgare.* The parasite stops the development of chloroplast by the host. The thallus of *Phyllosiphon* is dichotomously or irregularly branched in which the branches are loosely interwoven with one another (Fig. 13.5 A). The entire coenocyte is densely packed with elliptical chloroplasts which are without pyrenoids.

RHODOPHYCEAE

In red algae the parasitic habit has apparently been adapted more easily when the host and parasite are

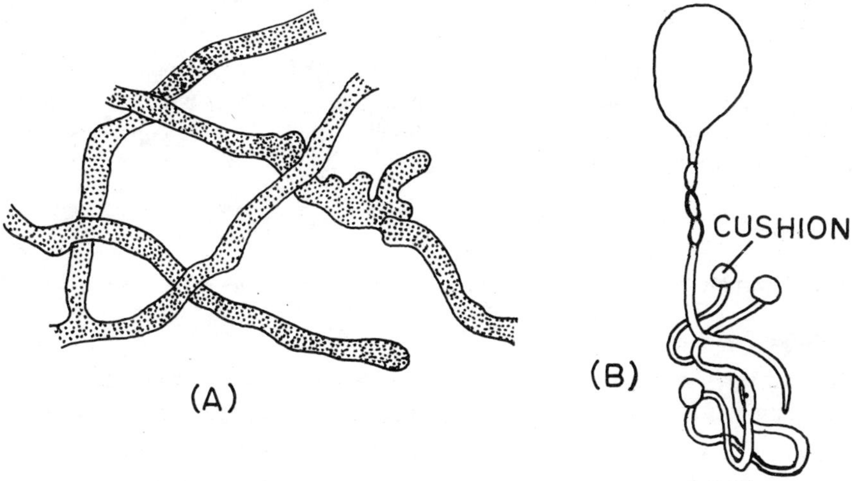

Fig.13.5. (A) Portion of the thallus of *Phyllosiphon.* (B) *Oodinium pauchetii* showing basal cushion for attachment to the host.

closely related (adelphoparasites) than when it is different (alloparasites). It is because of easy penetration of parasite into the host through secondary pit connections and in transferring nutrients.

(i) *Choreocolax* (Cryptonemiales) : *Choreocolax polysiphoniae* is an example of complete parasite on *Polysiphonia fastigata* which itself is epiphytic on *Ascophyllum*. This is a type of alloparasitism as *Choreocolax* belongs to Cryptonemiales and *Polysiphonia* is in Ceramiales. It is enclosed by gelatinous matter and consists of dichotomously branched filaments (Fig. 8.4 I) and a mass of haustorial cells growing inside the host. Secondary pit connections are established between the haustorial filaments of the parasite and the larger cells of the host.

(ii) *Holmsella pachyderma* : This rhodophycean member infects the host *Gracilaria verrucosa* (Rhodophyceae). In this association, the main product of photosynthesis (floridcoside) is transferred from the host to the parasite where it gets accumulated as floridеoside, mannitol and starch (Evans et al., 1973).

(iii) *Colacodasya inconspicua* : It is parasitic on the species of *Heterosiphonia.*

(iv) *Ceratocolax hartzii* : It forms pinkish tufts on *Phyllophora brodiaei.* Sometimes the parasite invades the medulla and kills some of its cells.

(v) *Harveyella mirabilis* : There is complete absence of photosynthetic pigment in this parasitic member which is as big as the cystocarp of the host. It is found on *Rhodomela subfusca.*

(vi) In *Stromatocarpus* and *Choreonema*, the greater degree of parasitism is observed. *Stromatocarpus* is found on *Polysiphonia virgata* in which the entire vegetative system is endophytic and the fructifying fronds alone appear above the surface of the host. The parasitic *Choreonema* is completely enclosed in the tissues of *Corallina.* The presence of parasite becomes evident only when the superficial conceptacles are formed. The genus *Gracilariophila oryzoides* appears like the grain of rice and infects the host *Gracilaria.*

Fan (1961) postulated two theories for the origin of parasitism in the Rhodophyceae. According to him, alloparasites and some of the adelphoparasites probably evolved from epiphytes while some of the adelphoparasites probably evolved from mutated spores. In the first theory parasites evolved along the progression from epiphytes to semiparasites to parasites. The second theory proposes the evolution of parasites from mutated tetraspores or carpospores that germinated on the parent plant. New species, germinated after the mutation of these spores, will be so close to the original that the intrusive tissue of the new species would not be rejected by the host.

DINOPHYCEAE

Some members of marine Dinophyceae like *Oodinium* (Fig. 13.5 B) are ectoparasites while *Blastodinium, Schizodinium* etc. are endoparasites. *Oodinium* is attached to the host with the help of basal cushion. *Blastodinium* appears as sickle-shaped cyst inside the intestines of marine copepods. The parasitic stage takes a naked amoeboid plasmodial form in *Paradinium*.

PHAEOPHYCEAE

Streblonema (Ectocarpales) has been found to be endophytic in *Fucus* (Fig. 7.3 H).

Pathogenic algae

Besides these parasitic forms some pathogenic algae are also reported which can infect certain living organisms including man. Since algae are basically autotrophs, they are poorly adapted as pathogens.

Prototheca, a member of order Chlorococcales (Chlorophyceae) has been reported to infect humans and animals. It is a colourless unicellular algae appearing like *Chlorella* and causes disease **protothecosis** in animals and mans. *Prototheca* occurs in variety of habitats like soil, freshwater and marine conditions. The disease usually involves cutaneous lesions with some subcutaneous infiltration and occasional joint involvement (Ahbel et al., 1980). In animals, dogs and the cattles have been found to be infected by *Prototheca*. Since it is resistant to many antibiotics, the treatment of protothecosis is difficult. In animals most of the reported cases reveal severe systemic infections such as massive invasions of the blood stream that have resulted into the death of the animal within a short time (Migaki et al., 1969).

Beskow (1978) reported the disease **silicosis** in Sweden due to inhalation of diatomaceous earth. Morse et al. (1981) have reported the association of a siphonaceous chlorophyte with tumorous growth in sea fans (*Gorgonia*). *Entocladia endozoica*, a member of Chlorophyceae seems to cause abnormal growths in corals (*Pseudoplexaura*).

(III) ECONOMIC IMPORTANCE OF ALGAE

Early records indicate that both marine and freshwater algae were gathered for more than 4000 years by humans. These were and have been utilised for various purposes. Historically, almost any country with a coastal region has had atleast some interest in their local marine algal flora. The economic aspects of algae can be studied as their benefical and harmful roles.

BENEFICIAL ROLES

1. **Algae as human food** : From ancient times large number of algae have been used as human food. They are often mixed with rice and fish and consumed as salads, soups etc. The edible forms are called **limu** in Hawaii, **tsao** in China and **rimu** in Tihiti. Chapman and Chapman (1980) listed 160 species of seaweeds eaten by humans (25 Chlorophyceae, 54 Phaeophyceae and 81 Rhodophyceae). The extensive use of seaweeds as human food can be seen in Japan, China, Hawaii, Philippines, Malaysia, Korea, Indonesia etc.

A number of dishes named as Kombu, Nori, Doshi, Koshi-Kombu etc. are prepared out of seaweeds in Japan. **Nori** or **amanoria** is prepared from *Porphyra* (Rhodophyceae) after washing and boiling.It is cultivated in large scale in Japan. During October-November, the cultivation ground is prepared by fixing bamboos at the sea shore. Nylon nets or coconut fibres are fixed on these. Besides high protein content (25-30% of its dry weight), different vitamins and mineral salts are also found in good proportions. Every 100 g of *Porphyra tenera* yield 11.4 g water, 35.6 g protein, 0.7 g fat, 44.3 g carbohydrate, 8 g ash and different vitamins. **Kombu** is prepared from different species of brown algae like *Laminaria, Alaria* etc. Kombu is boiled with meat, fish or used with rice. Powdered kombu is used in soups, sauces and curry. Irish Moss

(*Chondrus*) is most widely used in North America and Western Europe. The collection of *Chondrus*(Rhodophyceae) is mainly done in France, Canada, Ireland and U.S.A. A palatable dish **blancmanges** is prepared from this after drying and cooking with milk or seasoning with vanilla or fruits. In Ireland, *Rhodymenia*(Rhodophyceae) is known as **dillish** and is chewed fresh or after drying. It is known as **dulse** in Scotland and **sol** in Iceland. Dulse contains half as much vitamin C as oranges. *Ulva lactuca* (Chlorophyceae) is used as salads and soups in Scotland and Ireland. Products of some algae like gelatinous carbohydrates (carragheen) from *Gigartina stellate* and *Chondrus crispus* are used as puddings and as stabiliser in ice creams. *Durvillaea antarctica* is used in large quantities by indigenous population of *Chile. Caulerpa racemosa* is cultivated in Philippines as a source of food. The value of seaweeds lies more in the way that they stimulate apetite, provide necessary vitamins and trace elements.

Some of the freshwater algae are also used as human food. *Chlorella,* a future hope of food supplement, is grown in large scale in many countries. This has been used as flour for making crackers and biscuits on experimental baisis. *Chlorella* yields about 30% protein, 15% lipids, 30% carbohydrates and 5% ash. Under normal conditions it contains about 50% of dry weight of proteins containing 17 amino acids. A large number of vitamins have been reported in *Chlorella* by Tamiya (1960). Although the efficient digestibility of cell wall of *Chlorella* is a problem but its vitamin B content equals to that of lemon juice. *Chlorella ellipsoidea* is used successfully with tea in Japan. A terrestrial form of *Nostoc* is used in China and Brazil. *Nostoc commune* is consumed as food and in called **yuyucho**. *Scenedesmus* (Chlorophyceae) and *Spirulina* (Cyanophyceae) are also used as food and are grown on a mass scale. *Spirulina* has a protein content in excess of 60%. Thus the different marine and freshwater algae used as food are rich in proteins, fats, carbohydrates, vitamins, inorganic substances like iodine. Due to lack of suitable enzymes in the alimentary canal some of the carbohydrates of seaweed are not digested by human beings.

2. **Algae as fodder** : Seaweeds mainly the members of brown algae are used as feed for domestic animals in different parts of the world like Scotland, France, Norway, Great Britain, America, New Zealand etc. Some of them have established industries for production of feed for cattle, poultry and piggery. *Sargassum, Fucus* and *Laminaria* are consumed by cattles in Scotland. *Pelvetia canaliculata* is used as food for young live-stock. In Finland, *Laminaria* and *Alaria* are used for cattles. For fattening the pigs, *Pelvetia* and *Ascophyllum* are used in Scotland. *Rhodymenia* is consumed by cattle in France. In China and Hongkong, dried *Sargassum* is used as fodder.

The use of dried seaweeds like *Pelvetia* as cattle food has enhanced ten percent milk yield. The milk does not have any taste of algae after the seaweed diet. The high mineral and vitamin G content of kelp meal has increased its use in various poultry and other animal rations. Egg-yolks of fowls fed on chopped seaweeds have increased iodine and carotene content. *Macrocystis* sp. (Phaeophyceae) being rich in vitamin A and E is also used. Seaweeds appear to be good source of cobalt which is about ten times more than that of grasses. Among various components the most important digestible component appears to be laminarin while much of the nitrogenous material is undigested.

3. **Algae and nitrogen fixation** : Conversion of atmospheric nitrogen into nitrogenous compound i.e., nitrogen fixation is one of the important attribute of blue-green algae. A large number of members belonging to order Chroococcales, Nostocales and Stigonematales have been found to perform this function. Singh (1961) stated that average amount of nitrogen fixed over the whole world is approximately 700 mg/m^2/year and the greatest part of this comes from photosynthetic nitrogen-fixing algae. Majority of nitrogen-fixing algae are heterocystous and the details about the structure and function of heterocyst can be seen in chapter III. Burris (1966) has given the working hypothesis for N_2 fixation in which for the production of ATP, the pyruvate and acetyl phosphate are involved. Both ATP and reduced ferredoxin operate the mechanism step by step from N - N to NH_3 -NH_3. (For details please see "Algae in Biotechnology in this chapter).

The recent finding has established that nitrogen fixation is controlled by genes (**nif**). Each complex has one operator gene and a few structural (nif) genes being regulated by a common regulator gene.

4. **Algal role in fisheries** : Various algae have successfully been utilized in fish culture programmes. It has been found from the successful culture of the Siamese fish, *Tilapia mossambica* that this is voracious feeder of filamentous algae. This fish has been introduced in different parts of India. For the culture of *T. mossambica*, a culture of *Scenedesmus* is often exclusively used as a daily dose of fish meal.

The members of planktonic and periphytic algae serve as primary food for fishes and other aquatic animals. *Zebrasoma flavescens*, the surgeon fish feeds on red alga (*Amansia glomerate*) and brown alga (*Ectocarpus*). Species of *Oedogonium, Microspora, Ulothrix, Spirogyra, Cladophora, Pithophora*, diatoms etc. serve as fish food in freshwater systems. *Lyngbya aestuarii* (blue-green algae) serves as baby food in fish culture programmes of Philippine Islands. After three months these fishes are transferred to the ponds containing *Cladophora* and *Chaetophora*. After few months they are placed to the pond containing *Enteromorpha*. It has been estimated that about a hundred kg of algae would produce approximately one kg of fish.

As oxygen is vital for the fish life and the higher concentration of carbon dioxide is lethal, the role of photosynthetic algae is very important in aquaculture. The removal of carbon dioxide and the release of oxygen both are achieved by algae during photosynthetic process. Fishes like *Catla catla, Lobeo fimbriatus, Chanos chanos* and *Etroplus suratensis* are mainly phytoplankton feeders. Planktonic and filamentous blue-greens, green algae and diatoms are often found in the guts of fishes. Reserve food materials of these algae are utilized by the fishes.

5. **Algae in soil fertility and land reclamation** : The species of *Nostoc, Scytonema, Aulosira, Lyngbya, Microcoleus, Aphanothece, Cylindrospermum, Anabaena* etc. grow extensively on moist soil surfaces. Many of these are nitrogen-fixing algae which can fix atmospheric nitrogen and increase the soil fertility. Due to their mucilaginous sheath they are able to prevent soil erosion by binding the soil particles firmly. They also reduce the evaporation of water because of thick covering on the soil. Metting (1981) made some estimations of primary production by soil algae which ranged from 0.5-1.4 tons of organic matter per hectare in a Russian soil and upto 6 tons of organic matter per million pounds of soil in an American soil.

On the surface of saline usar soils during rainy season, a thick stratum of *Anabaena, Scytonema* and *Nostoc* develop. These algae have capacity to reclaim usar soils through a series of successive algal growth in a water logged condition. After a year of such reclamation, it is possible to grow rice upto 2000 lb per acre. Due to the presence of these algae, water holding capacity is increased by 40% and pH is reduced from 9.7 to 7.6. Rapid growth of *Hormidium, Zygogonium* (Chlorophyceae), *Chroococcus, Phormidium* (Cyanophyceae) on burnt heath soil reduce the danger of soil erosion. The immense growth of *Nostoc commune* has been shown to increase the nitrogen content by 30-38% in saline usar soils by adding organic matters. After the volcanic eruption of Krakatau island in 1833, blue-green algae like *Plectonema, Dalmetella* and *Hyella* were the first to appear.

6. **Algae as fertiliser** : Long before the recognition of their potash content, seaweeds were used as fertilisers in the United States. In some regions of Britain and along the coast of north-west France rockweeds are used as manure extensively. These algae increase the water holding capacity besides the addition of their chemical constituent in the soil. Seaweeds are richer in potassium but poorer in nitrogen and phosphorous than the farm manure. In Japan they are used in rice fields whereas in China for ground nuts and sweet potatoes. In India *Turbinaria* is used around palm tree. The seaweeds are also used as compost.

7. **Algae as oxygen donors and link of food Chain** : Algae oxygenate the environment by photosynthesis. They are primary link of many diverse food chains. Aquatic animals depend ultimately for food on algae.

8. **Algae in sewage treatment plants** : In sewage oxidation tanks, presence of algae oxygenates the sewage to a great extent. The released oxygen is utilised by bacteria in rapid decomposition of the sewage. Algae used in sewage ponds mainly belong to Chlorococcales, Volvocales (Chlorophyceae) and Euglenophyceae. The common members are *Chlorella, Scenedesmus, Pediastrum, Microactinium, Euglena, Phacus* etc. Algae of sewage disposal ponds utilise nutrients to breakdown sewage. The water coming out of the

oxidation ponds is relatively clear as the suspended particles get oxidized and settle at the bottom. This water is pumped in next oxidation ponds successively till it becomes clear and potable.

9. **Algae and recreational use** : Some selected algae are grown in recreational areas like lakes and streams along with the fishes. Large patch of *Trentepohlia* giving a reddish colour to land looks beautiful. *Oedocladium* imparts valvety appearance to the soil. In the gardens of central India, parrot-green colour of *Vaucheria* is pleasant to see.

10. **Algae as research material** : *Chlorella* culturing tanks are used in space vehicles because the alga restores the oxygen by its photosynthesis. Algae like *Chlamydomonas, Chlorella, Acetabularia* etc. are very much useful in physiological, cytological and genetical studies. Species of *Nitella* are used in the studies of cytoplasmic movement, ion accumulation etc. *Valonia* and *Halicystis* help in understanding the phenomenon of permeability.

11. **Algae and medicinal use** : Brown algae are used in various goitre medicines due to their high iodine content. A medicine **palcato** is prepared from the members of Laminariales and *Sargassum* which is used against goitre and other glandular troubles. Insect diseases to humans (veromious disease) are treated with extract from *Corallina, Digenia, Codium, Alsidium* and *Durvillea*. Because of swelling property, laminarins are used in child birth for the expansion of cervix. Fucoidin and compounds of laminarin are also used as anticoagulants while carrageenin acts as blood coagulant. Algae are used in the treatment of kidney, bladder and lung diseases in China, Japan and Chile. *Gelidium* is used in stomach disorders and in heat induced illness. *Laminaria* stipes are used as surgical tool in the opening of wound due to its gentle swelling property. Different kinds of ointments, pills and other pharmaceutical products are made from agar. In the preparation of antihelmitic medicines, *Digenia simplex* is used in Japan and China. Mannitol powder is odourless and half sweet to sugar which serves as a supplement of sugar for diabetic patients without any harm. *Acetabularia major* is used in kidney and bladder problems. In Hongkong, *Ulva* is used for glandular troubles.

Antibiotic **chlorellin** is extracted from *Chlorella vulgaris* which inhibits the growth of certain bacteria and a few algae. *Rhodomela larix, Ascophyllum nodosum, Pelvetia* etc. have shown some antibiotic properties. The growth of *Escherichia coli* is found to be reduced by *Nitzschia palea* (diatom). *Microcystis* reveals inhibitory action to *Staphylococcus, Closteridium* and zooplanktons like *Cyclops* and *Daphnia*.

12. **Algae as source of growth promoting substances** : Algae have been found to increase the yield even when the soils are well manured. Shukla (1967) found that the presoaking seed treatments with *Phormidium tenue* (Cyanophyceae) results in profuse tillering, increased height and multiple rice yields. There is also an increase in protein contents of grains of treated plants. *Phormidium* is not a nitrogen-fixing algae.

13. **Algae in uptake of heavy metals and radioactive wastes** : Some reports are available which suggest the role of algae in absorbing radioactive wastes and heavy metals. Species of *Chlorella, Euglena, Spirogyra, Cladophora glomerata, Scenedesmus quadricauda* and *Porphyra lucinata* have been found to absorb the radioactive elements and heavy metals.

14. **Lens paper** : For cleaning optical articles, lens paper is produced from *Spirogyra* in Japan.

15. **Algae in the origin of petroleum and gas** : The plankton of the seas is probably of great importance as a source of organic matter. Organic compounds produced from photosynthesis get accumulated in bottom deposits and in an oxygen-free environment gradually converted into oil and gas. Unlike coal which was laid down in inland swamps, oil and gas were formed from organic matter in marine environments. Natural gas (largely methane) can be produced by certain kinds of anaerobic bacteria and oil formation is also associated with this.

16. **Algae and limestone formation** : Some members of green algae, blue-green algae, red algae and flagellates deposit calcium carbonate on their cell wall or gelatinous sheath both in fresh and salt water.

Most important calcareous algae of the seas are red algae which have a significant role in the formation of coral reefs. Algae also play an important role in formation of beds of limestone rocks.

17. **Role as indicator of pollution** : During recent years the role of algae in assessing the pollutional level of environment has greatly been realised. Presence and dominance of a particular alga or a set of algae indicate a specific trophic condition. For details please see separate heading "algae, pollution and their role as indicators" in this chapter.

18. **Role in industries** : Many commercial products are extracted from algae and their cell walls. Here only a brief account of these are given.

(i) **Alginates (Alginic acid)** : Alginate is derived from cellulose free middle lamella and primary wall of the members of Phaeophyceae like *Macrocystic, Laminaria, Ascophyllum, Lessonia* etc. Alginic acid content varies with the location, seasons, temperature and parts of the plants etc. Its content is approximately 30-40% in brown algae on dry weight. It is similar to cellulose and pectic acids in composition consisting of a long unbranched chain of ß-d- mannuronic acid joined by 1:4 glycosidic linkages. The soluble calcium salt of alginic acid is **algin**.

As algin has remarkable water absorbing capacity it is used in many industries where there is the requirement of thickening, suspending, emulsifying, gel-forming, stabilizing etc. Sodium salts of alginic acid are used in textile industry as they form excellent polishing and dressing material. Alginates are also used in food industry for filling creams, thickening soups, sauces; in cosmetics industry as dispensing agents in ointments, creams, tooth pastes, shampoos; in paint industry for suspension of pigments, stabilization of emulsion; in pharmaceutical industry as emulsifyers and as filters in the manufacture of tablets, pills etc. Alginats are also used as gel in the freezing of fish, antibiotics and in emergency transfusion in the treatment of shocks.

(ii) **Agar (Agar-agar)** : This is dried gel-like non-nitrogenous, gelatinous extract obtained from many red algae. This is one of the best known algal products and used as a solidifying agent in the preparation of microbiological culture media. The word itself is of Malayan origin but the material was first used in China in seventeenth century. Dried agar is insoluble in cold water but soluble in hot water. The important algae used for the extraction of agar are *Gelidium, Gracilaria, Pterocladia, Gigartina, Ceramium, Chondrus, Furcellaria* etc. Gelling property varies with the species but it will set at from 35 to 52°C.

The major component of agar is agarose. Agar from *Gelidium cartilagineum* is composed of a chain of alternating D-galactose and 3:6 anhydro -L- galactose residue with a half ester sulphate one about every tenth unit of the galactose. Uronic acid, pyruvic acid, polysaccharides like agarose and agaropectin are also present in agar-agar.

The material is prepared by washing, sorting and boiling or autoclaving. Today most agar comes from Japan, Korea, Australia, New Zealand and Morocco but many countries like Frace, Italy, Spain, India, Ceylon, Egypt, United States, Mexico and some countries in South America are also producing their own agar.

Besides most important use of agar in the preparation of culture media, it has also been used in food industry, cosmetics, leather, textile industry, pharmaceuticals, dental impression mold, meat packing, for clotting of blood and as emulsifyers, laxatives etc.

(iii) **Carrageenan (Carrageenin)** : It is a carbohydrate mucilage named after Irish village Carrageenin which is extracted from red alga *Chondrus crispus* (**Irish Moss**) and to a lesser extent from *Gigartina*. The compound is a cell wall polysaccharide complex of D - galactose - 3, 6 - anhydro - D - galactose and monoesterified sulphuric acid. Two of its major components (kappa-carrageenin and lamda-carrageenin) have been recognised. These compounds are used like alginates in food, textile, leather, pharmaceutical and brewing industries. This gelatinous carbohydrate is variously used with puddings, consumed with milk, fruit and ice-cream. It is also used as clearing agents in beer preparation.

(iv) **Iodine and other compunds** : Members of brown algae such as *Laminaria digitata, Fucus* spp., *Ecklonia* etc. are known for the extraction of iodine. The maximum percentage of iodine (1.23%) has been obtained in Laminarias of Britsh Coasts.

Seaweeds are also known for the presence of macronutrients useful for human consumption like iron, manganese, zinc, copper etc. Bromine, formic acid, acetone, acetic acid are also extracted from seaweeds. Seaweed ash is also used as source of salt and soda. The word **kelp** was used in England to denote ash of brown algae for the production of potassium and sodium salts while in the usual sense it means big seaweeds. A large number of vitamins besides mannitol, laminarin, fucosan, beta-carotene, fucosterol etc. are found in seaweeds which are of one use or the other.

(v) **Funori and funorin** : From *Gloeopeltis furcata* a sizing agent and glue for textile called funori is prepared in Japan. Algae like *Chondrus, Iridaea, Gratilaupia* are also used for other inferior products. The composition of funori is like agar but there is no sulphate ester group in it.

(vi) **Diatomite** : Diatom wall with high percentage of silicon dioxide are constantly being deposited in freshwater and marine sediments. In some areas, because of bottom contours, currents and other factors large concentration of diatoms were deposited over millions of years. Some of these deposits were elevated and now are seen on the land which are known as **diatomaceous earth** or **kieselguhr**. Large deposits of diatomaceous earth have been found in California (USA), Germany, Britain, France, Japan, Spain, Australia, Nicobar islands (India) etc. In Lompoc, California the largest deposit of diatoms (about 425 m thick and over 30 sq km area) is found. The diatomaceous deposits can be either marine or freshwater and can date from the Jurassic, Cretaceous, Tertiary or quarternary. The marine deposits are of great depth than those formed in freshwater environments.

Diatomite is used in industrial filtration processes, sugar refining and brewing industries. It is also used as an absorbent for nitroglycerine in the manufacture of dynamite. Other uses of diatomaceous earth include polishing, insulation, in paper products, plastic industries and as a filter. It is also used as catalyst in several industries, as a cleaning agent in soap, toothpaste, for insulation of boilers and blast furnaces as a loose jacketing powder, as bricks when the temperature is much higher (1000°F), for reducing danger of explosions in coal mines etc.

HARMFUL ROLES

Besides many uses, algae are also known to create problems for human beings.

1. **Death of aquatic animals and fishes** : Many blue-green algae produce toxins (for details please see separate heading "Toxic algae, toxins and their role") which cause death to domestic animals and fishes. Important among these algae are *Microcystis, Anabaena* and *Aphanizomenon*. Besides death, many harmful effects like loss of weight, weakness, abortion etc. have been reported in cattles. Phycocyanin pigment found in blue-greens is sensitive to light and when they enter in the blood capillaries of cattle, cause an internal burning and peeling of the skin.

The profuse growth of *Aphanizomenon* and *Microcystis aeruginosa* in fish ponds are poisonous to *Crappis perch* and *Gambusia* respectively. Bloom of *Gymnodinium* (Dinophyceae) leads to the death of fish within several square miles of Gulf of Mexico. Algae also cause suffocation of fishes by choking their gils.

2. **Disease and death to human beings** : The indirect consumption of dinoflagellates through fishes has resulted in human death. Dinoflagellates like *Gonyaulax*, when consumed with shell fish, produce several diseases. Reports about the occurrence of paralysis, respiratory failure and death within 2 to 12 hours after consumption of dinoflagellates are available.

3. **Algae and water supply** : Algae interfere and create problems in water supply and storage reservoirs in many ways like a) profuse growth and interference with filtration process, b) production of bad odour by their decomposition, c) imparting objectional tastes, d) formation of slimes in water, e) colouration to the water, f) corrosion of pipes, g) interference in disinfection process by imparting turbidity, h) formation of sediments in service reservoirs etc. Algae causing these problems belong to Cyanophyceae, Chlorophyceae, Bacillariophyceae, Euglenophyceae, Dinophyceae and Chrysophyceae. Common taxa of these classes are *Oscillatoria, Microcystis, Scytonema, Aulosira, Cylindrospermum, Chroococcus, Anabaena* (Cyanophyceae); *Cyclotella, Melosira, Asterionella, Fragilaria, Achnanthes, Navicula, Cymbella* (Bacillariophyceae); *Ankistrodesmus, Chlorella, Micractinium, Netrium, Pediastrum, Stigeoclonium* (Chlorophyceae); *Euglena, Phacus, Trachelomonas* (Euglenophyceae); *Ceratium* (Dinophyceae) and *Dinobryon* (Chrysophyceae).

4. **Algae causing eutrophication and pollution** : Algae are currently considered problem organism in eutrophic waters. The role of algae in causing eutrophication and pollution is well established (dealt separately in this chapter).

5. **Damage to buildings** : During rainy season, on moist wall surface many blue-green algae grow. Gradually these spoil the walls of the buildings. *Scytonema, Tolypothrix, Chlorococcum* are some common algae causing such damage in tropical countries.

6. **Accidents due to blue-green algae** : Due to intensive growth of blue-greens, ground surface becomes slippery by which cattle and human beings get slipped.

7. **Damage to salt by blue-green algae** : Quality of salt in Sambhar lake in Rajasthan is greatly affected by algae. An offensive smell, pink rust colour to the salt are some of the common problems.

8. **Parasitic algae** : Many algal forms are true parasites which cause severe damage to economically important parts of the plants (dealt in detail in this chapter).

PRODUCTIVITY IN FRESHWATER AND MARINE ENVIRONMENT

Production is the weight of new organic matter formed over a period of time including any losses that occurred during that period and **productivity** is the rate of formation of organic matter over a particular period of time such as day or year. Lakes can be classified by the amount of life they can support. This is expressed as productivity (change in biomass over a period of time) or as **standing crop** (amount of biomass present at a specific time). **Oligotrophic** lakes are relatively deep and are characterized by low productivity. These have narrow littoral and sublittoral zone but an extensive profundal area and are not subject to severe oxygen depletion. Diatoms and desmids are commonly found but their standing crop is low. **Eutrophic** lakes are shallow with highly productive littoral and sublittoral region and profundal region is greatly reduced. Here few algal species are present (lower species diversity) but total standing crop of algae is higher. Usually blue-green and green algae dominate in such water.

In a thermally stratified lake, productivity is usually low because the euphotic zone is in the epilimnion and by photosynthesis plants use up all of the available nutrients. If any of the organisms in the epilimnion die, they fall into the hypolimnion and decay. The nutrients released from the decay are then locked in the hypolimnion and are not available for the photosynthesis because there is no exchange of water between the two layers. During an **overturn**, there is an exchange of water. In this process surface water becomes cooler and heavier than the water below it and sinks causing the end of the thermal stratification.

It has been estimated that the net total primary production of plants on the earth is approximately 1.4 X 10^{14} kg dry weight per year. Out of this, about 40% is produced by the marine phytoplankton alone (of which 20-25% is by diatoms only). If the primary production value of inland waters (about 10%) is added then the total value of primary production of algae comes to 50%. It has also been calculated that the prduction of organic carbon is 340 tons per sq km in the water while it is only 160 tons in the cultivated lands. Table 13.1 indicates the values of primary production by phytoplankton in different aquatic systems.

Table 13.1. The primary production in various freshwater ecosystems, with a comparative figure for sublittoral marine algae (After Odum, Wetzel, Likens).

	$gCm^{-2}\ yr^{-1}$
Phytoplankton, marine	
Oceanic	50-100
Coastal	100-200
Upwellings	300
Phytoplankton, freshwater	
Lakes	2-400+
Lake production	
Temperate	2-900
Tropical	30-2000
Oligotrophic	2-180
Mesotrophic	75-300
Eutrophic	300-2000
River production	
Temperate	<1-600
Tropical	1-possibly 1000
Sublittoral Macroalgae, Marine	400-1700

Different methods have been employed to measure productivity values. These are by counting the organisms, by determining the biomass, by measuring the rate of oxygen evolution, pigment content especially chlorophylls, levels of ATP etc. Different important factors which affect and control the productivity of phytoplankton are light, temperature, inorganic nutrients (nitrate, phosphate, silicate etc.), oxygen, organic nutrients and grazing.

Although studies on phytoplankton productivity have attained considerable attention but the role and significance of smaller algae (less than 15 μm size) has not been worked out properly. These **ultraplankton** occur in pelagic (living unattached) waters in lakes along with larger phytoplankton. Smaller flagellates i.e., **microplankton** (less than 10 μm size) are predominant in oceanic biomass and play an important role. These micro- and ultraplankton are relatively more common in oligotrophic lakes than eutrophic ones. Ross and Duthie (1981) found that the photosynthetic efficiency of the ultraplankton was approximately 2.5 times more than that of phytoplankton.

About the benthic macroalgae, the growth in the kelp-beds has been found to increase in the fall months despite the concurrent decrease in temperature and available light. Dissolved and particulate organic matter have been found to influence the production in benthic macroalgae greatly.

(IV) ALGAE, POLLUTION AND THEIR ROLE AS INDICATORS

Algae are the important group of organisms which affect (both positive and negative) the quality of water considerably.When the concentration of naturally occurring substances are increasd or when unnatural synthetic compounds(**xenobiotics**) are released into the environment, pollution is caused. With the origin of pollution, the algal growth may either be inhibited or stimulated and finally algae themselves become the cause of concern.

ORGANIC POLLUTION

As a result of domestic, agricultural and industrial activities, organic substances are released into the environment leading to organic pollution. Palmer(1977) and Round(1981) described the algae and their composition found in different polluted waters. In the most polluted area algae are either absent or are represented by the members of Cyanaphyceae and a few flagellates especially euglenoids. Diatoms and filamentous green algae start appearing with the reduction in pollution level. Sewage discharge has been found to destroy many economically important algae like *Laminaria*(Phaeophyceae). Raw sewage, however, may increae the growth of benthic marine algae. Hydrocarbons including natural petroleum and variety of other chemicals are found in the aquatic environment which are toxic to algae. In the laboratory studies, Gordon and Prouse(1973) showed that levels of hydrocarbons below 30-50 ppm stimulate photosynthesis but progressive inhibition takes place above 50 ppm. Oil spills appear to have little effect on phytoplankton populations which may be attributed to decline in zooplankton abundance.

Most of the information available on algae in relation to polluted waters is confined to water containing treated or untreated domestic sewage and organic wastes. House-hold sewage contains various kinds of organic materials besides phosphates from detergents and other wastes. The growth of *Chlorella* has been found to be enhanced with the addition of sodium triphosphate, an ingredient of a synthetic detergent. The composition and density of algae in the sewage polluted region of the river are different from those present in unpolluted zone. After examining the reports of 165 authors, Palmer(1977) has compiled a list of more than 850 algae commonly found in organic wastes. Some common examples are *Anabaena constricta, Arthrospira jenneri, Oscillatoria* spp., *Phormidium* spp. (Cyanophyceae), *Chlamydomonas reinhardi, Pandorina morum, Eudorina elegans, Actinastrum hantzschi, Ankistrodesmus falcatus, Chlorella* spp., *Coelastrum* spp., *Scenedesmus* spp., *Stigeoclonium tenue* (Chlorophyceae), *Cyclotella meneghiniana, Gomphonema parvulum, Cocconeis placentula, Hantzschia amphioxys, Melosira* spp., *Navicula* spp., *Nitzschia* spp., *Surirella ovata, Synedra* spp. (Bacillariophyceae), *Euglena* spp., *Lepocinclis* spp., *Phacus* spp. (Euglenophyceae) etc. Out of these members *Euglena viridis, Nitzschia palea, Oscillatoria limosa* and *Oscillatoria tenuis* are more likely to be present than any other species when organic pollution exists. Next in order are *Arthrospira jenneri, Stigeoclonium tenue, Euglena gracilis* and *Chlorella vulgaris.*

Some common industrial wastes are organic in nature. In general these wastes tend to stimulate the growth of particular kinds of yeasts, bacteria, molds and Protozoa but are harmful to algae as they increase turbidity in water besides other inhibitary effects.

CULTURAL EUTROPHICATION

In cultural eutrophication the nutrient status of water is increased through human activities. Algal production is enhanced due to increase in available nitrogen, phosphorus and carbon. The increase in nitrogen is mainly through agricultural run-off while domestic sewage (point source pollution) is the main source of phosphates and organic carbon. Phophorus is the most important nutrient in the eutrophication process. In highly eutrophic water, peaks of algal population become less apparent as additional blooms occur throughout the year. The most striking effects of cultural eutrophication are seen in the increased growth of filamentous algae. *Cladophora glomerata* is most common while *Spirogyra, Rhizoclonium, Enteromorpha* (Chlorophyceae), *Thorea, Compsopogon* (Rhodophyceae) are also found.

Eutrophication is limited to lentic (non-flowing) water bodies like reservoirs and lakes. The term pollution is differentiated from eutrophication in the sense that substances added in pollution do not stimulate increased growth of aquatic organisms. Algae are invariably involved in the increased growth by eutrophication. A relatively pure water body has a very little amount of the essential nutrients like nitrogen, phosphorus, carbon compounds etc. Many kinds of algae will grow in abundance if these nutrients are increased. Blue-green algae like *Microcystis, Anabaena* etc. form blooms when gross eutrophication is reached. Besides these, many other algae viz., *Aphanizomenon*, *Oscillatoria*, *Spirulina*, *Cladophora*, *Oedogonium*, *Spirogyra* etc. start growing luxuriently in eutrophic waters. The quantity of algae becomes very large while the diversity is low. An eutrophic lake is characterized by its shallow and extensive littoral zone with plant growth. It is rich in basic nutrients with the average annual concentration of inorganic forms of nitrogen and phosphorus being somewhat greater than 0.300 mg/l and 0.015 mg/l respectively. Water is moderately hard and the alkalinity ranges between 50 to 100 mg/l. The rate of gross primary production generally ranges from 0.5 to 5.0 g dry organic matter/sq m/day during the most fovourable growing season.

METAL AND ACID POLLUTION

Although some algae are remarkably tolerant to heavy matals but heavy metals may be used as algicides. Studies have shown that heavy metals decrease the diversity and productivity of algae. The members of Cyanophyceae and Bacillariophyceae (diatoms) are generally less tolerant than the members of Chlorophyceae. Say and Whitton (1978) listed species of *Netrium*, *Plectonema*, *Nitzschia*, *Hormidium*, *Ulothrix* and *Mougeotia* as typical algal flora of zinc- contaminated streams.

Cassin (1974) reported that *Chlamydomonas acidophila* can grow at pH 2.0. *Chlamydomonas applanata* and *Euglena mutabilis* could be grown at pH 1.3. It has been shown that toxicity of copper and zinc to *Hormidium rivulare* (Cyanophyceae) was less at pH 3.5 than at pH 6.0. Phytoplankton population is gradually changed with the change in pH. Acidification of lakes causes a reduction in species number and diversity though biomass and primary production need not decrease. Many lakes undergoing acidification have shown increase in biomass of benthic and epiphytic algae along with decrease in species diversity which are due to reduced microbial decomposition and invertebrate grazing

THERMAL POLLUTION

Besides electricity generating systems which use water for cooling the condenser and return it at 10-15°C above air temperature, many industries are also responsible for thermal pollution. Phytoplankton diversity generally decreases in water over 30°C with the dominance of blue-greens. Moderate increase in temperature leads to increase in productivity and standing crop of freshwater benthic and epiphytic algae like phytoplankton but the diversity may decrease. Marine macroalgae are also affected by thermal discharges.

ALGAE AS INDICATORS OF POLLUTION

It is now well established that environmental disturbances such as pollution induce changes in the structure and function of biological systems. Changes in water quality exert a selective action on the flora and fauna and the effects produced can be used to establish **biological indices** of water quality. The measurements of physical and chemical criteria indicate only the cause of change in water quality while biological tests depict the effects of the change.

Saprobity system

Kolkwitz and Marsson (1908) forwarded this classical ecological method to describe the pollution oriented changes in the algal composition of a river. They identified the different zones depending upon the degree of pollution and each zone characterised specific algal flora. These zones are oligosaprobic, β-mesosaprobic, α-mesosaprobic and polysaprobic. The characteristic of three distinct zones are:

1. **Polysaprobic** (septic) zone is the segment where there is high molecular decomposible organic matter and deficient oxygen. Algae are rarely found in this zone which may be the species of *Euglena* and *Oscillatoria*.

2. **Mesosaprobic** zone is the segment between polysaprobic and oligosaprobic zone. In this zone decomposible

organic matter is less and oxygen is available which allow a few algae to grow. Algal species belong to *Oscillatoria*, *Phormidium* and *Ulothrix*.

3. **Oligosaprobic** (clean) zone reveals conditions opposite of polysaprobic zone. It is rich in oxygen and a number of algae can grow. Either there is little or no decomposition of organic matter. Algae found in this zone are the species of *Calothrix*, *Batrachospermum*, *Meridion* etc.

Many workers have modified the saprobian zones of Kolkwitz and Marsson. Widely accepted nine modified zones are 1) Coprozoic, 2) α-polysaprobic, 3) β-polysaprobic, 4) γ-polysaprobic, 5) α-mesosaprobic, 6) β-mesosaprobic, 7) γ-mesosaprobic, 8) Oligosaprobic and 9) Katharobic. In Coprozoic zone only bacterial communities are present while in the remaining zones specific algal species are found.

TROPHIC STATE INDICES

It was observed by Thunmark (1945) that lakes receiving domestic sewage have greater number of blue-greens than those of desmids (green algae). Nygaard (1949) carried this idea and proposed 5 indices on the basis of relative dominance of different algal groups. He observed that the members of Cyanophyceae, Euglenophyceae, centric diatoms and Chlorococcales (Chlorophyceae) are characteristic of eutrophic waters while desmids and many pennate diatoms are usually found in oligotrophic waters as they do not tolerate higher nutritional levels.

The five indices of Nygaard and their corresponding ranges are as follows -

	Index	Ratio	Oligotrophic	Eutrophic
1.	**Myxophycean**	Myxophyceae / Desmidiaceae	0.0-0.4	0.1-3.0
2.	**Chlorophycean**	Chlorococcales / Desmidiaceae	0.0-0.7	0.2-9.0
3.	**Bacillariophycean**	Centric diatoms / Pennate diatoms	0.0-0.3	0.0-1.75
4.	**Euglenophycean**	Euglenophyceae / Myxophyceae+Chlorococcales	0.0-0.2	0.0-1.0
5.	**Compund**	Myxophyceae+Chlorococcales+ Centric diatoms+Euglenophyceae / Desmidiaceae	0.01-1.0	1.2-2.5

Palmer's organic pollution Indices

Palmer (1969) analysed reports from 165 authors and developed algal pollution indices for use in rating water samples with high organic pollution. Two lists of organic pollution-tolerant forms were prepared, one containing 20 genera, the other 20 species (Table 13.2, 13.3).

Table 13.2.

Algal genus pollution index (Palmer, 1969)

Genus	Pollution Index
Anacystis	1
Ankistrodesmus	2
Chlamydomonas	4
Chlorella	3
Closterium	1
Cyclotella	1
Euglena	5
Gomphonema	1
Lepocinclis	1
Melosira	1
Micractinium	1
Navicula	3
Nitzschia	3
Oscillatoria	5
Pandorina	1
Phacus	2
Phormidium	1
Scenedesmus	4
Stigeoclonium	2
Synedra	2

Table. 13.3.

Algal species pollution index (Palmer, 1969)

Species	Pollution Index
Ankistrodesmus falcatus	3
Arthrospira jenneri	2
Chlorella vulgaris	2
Cyclotella meneghiniana	2
Euglena gracilis	1
Euglena viridis	6
Gomphonema parvulum	1
Melosira varians	2
Navicula cryptocephala	1
Nitzschia acicularis	1
Nitzschia palea	5
Oscillatoria chlorina	2
Oscillatoria limosa	4
Oscillatoria princeps	1
Oscillatoria putrida	1
Oscillatoria tenuis	4
Pandorina morum	3
Scenedesmus quadricauda	4
Stigeoclonium tenue	3
Synedra ulna	3

Each form was assigned a pollution index number ranging from 1 for moderately tolerant forms to 6 for extremely tolerant forms. In analysing a water sample, any of the 20 genera or species of algae present in concentrations of 50/ml or more are recorded. The pollution index numbers of the algae present are totaled providing a genus score and a species score. Palmer determined that a score of 20 or more for either index can be taken as evidence of high organic pollution while a score of 15 to 19 is taken as probable evidence of high organic pollution. Lower figures suggest that the organic pollution of the sample is not high.

Species Diversity Indices

Diversity is the expression to indicate the relation between number of species and number of individuals in any ecosystem. It has two major components i.e., richness and evenness. Species **richness** is the simple ratio between total species and total number of individuals and **evenness** of species is the distribution of individuals amongst species. There are many methods of measuring species diversity but Shannon and Weaver (1963) formula is widely used. According to which

$$\bar{H} = - \Sigma\, (ni/N) \log (ni/N) \text{ or } \Sigma\, P_i \log P_i$$

where $\bar{H}$ is species diversity, P_i is importance prabability for each species i.e., ni/N, ni is number of individuals in ith sample and N is total of importance value i.e., total number of individuals.

Staub *et al.* (1970) used the diversity to evaluate the effects of industrial wastes on primary planktonic producers in the Mississippi river. They proposed a scale tabulated below.

Species diversity	Condition
3.0-4.5	Slight pollution
2.0-3.0	light pollution
1.0-2.0	moderate pollution
0.0-1.0	heavy pollution

Sequential Comparison Index (SCI)

Cairns et al. (1968) forwarded this index for assessing the quality of water. According to them

$$SCI = \frac{\text{Number of runs}}{\text{Number of organisms}} \times \text{Number of equivalent species}$$

Diatometer, Diatoms and Pollution

Catherwood diatometer developed by Patrick (1977) is used in assessing pollutional level. Glass slides are held upright in racks attached to floats which keeps this simple diatometer in photic zone. It has been found that the species of diatoms present in the water are comparatively more while the individuals of each species are less in unpolluted streams. Similarly the species of diatoms found in mildly polluted waters are more than those present in severely polluted waters while the individuals found in mildly polluted waters are less than those of severely polluted waters. With the increase in pollution load, the species of diatoms decreased along with the increase in the number of individuals.

Diatoms have greatly been used as environmental indicators. Diatoms are usually absent in polysaprobic zones where there is deficiency of oxygen. *Nitzschia palea* and *Gomphonema parvulum* are found in mesosaprobic segment. Due to their easy preservation because of silicified wall and availability of large number of species in different conditions, this group of algae has been studied extensively as indicators.

Chrysophytes as Environmental Indicators

Recently the silica-scale bearing chrysophytes have come into focus as environmental indicators (Kristiansen, 1986; Saha and Wujek, 1990). They have varied and often very specific ecological requirements. In his system, Sladecek (1973) enumerated 8 species and placed them in saprobic categories, viz., *Mallomonas acaroides* (ß-mesosaprobic), *M. akrokomos* (oligosaprobic), *M. coronifera* (oligosaprobic), *M. caudata* (oligosaprobic), *M. tonsurata* (ß-mesosaprobic), *Synura petersenii* (ß-mesosaprobic), *S. spinosa* (ß-meso-oligosaprobic) and *S. uvella* (ß-meso- oligosaprobic).

Kristiansen (1981) summarized the indicator value of some silica-scaled species. *Synura sphagnicola* and *Mallomonas paludosa* are indicators of acid, somewhat humic water. *M. heterospina* is one of the few species found in watering ponds for cattle. Except for this and a few others, the presence of silica-scaled chrysophytes indicates that there is no heavy pollution. Saha and Wujek (1990) observed the rich chrysophyte assemblage from eutrophic waters of north-east India which contradicts the earlier idea that these silica-scaled chrysophytes are the special indicators of oligotrophy.

Algae as indicators of high acidity : *Chlamydomonas, Cladophora, Desmidium, Mougeotia, Ulothrix zonata, Xanthidium antilopaeum* (Chlorophyceae), *Diatoma vulgare, Eunotia lunaris, Frustulia* *megaliesmontana, Navicula roteana, N. bryophila, N. viridis, Pinnularia acoricola, Tabellaria flocculosa* (Bacillariophyceae), *Euglena hiemalis, E. mutabilis, E. stellata, Lepocinclis ovum* (Euglenophyceae), *Dinobryon, Ochromonas* (Chrysophyceae).

Algae as indicators of marine and estuarine pollution : *Calothrix confervicola, Spirulina subsalsa* (Cyanophyceae), *Chlorella variegata, Enteromorpha intestinalis, Acrochaetium virgatulum, Erythrotrichia carnea, Porphyra leucosticta* (Rhodophyceae).

Algae indicating industrial wastes

Distillery wastes : *Chlamydobotrys, Chlorogonium euchlorum, Chlorobrachis gracillima* (Chlorophyceae).

Oil wastes : *Ankistrodesmus, Chlamydomonas, Closterium, Gonium* (Chlorophyceae), *Amphora ovalis, Asterionella, Cyclotella, Fragilaria, Gomphonema herculaneum, Melosira varians, Navicula radiosa, Synedra acus* (Bacillariophyceae), *Lyngbya, Oscillatoria* (Cyanophyceae), *Euglena, Trachelomonas* (Euglenophyceae).

Hydrogen sulfide wastes : *Achnanthes affinis, Cymbella ventricosa, Hantzschia amphioxys, Navicula minima, Neidium bisulcatum, Nitzschia ignorata* (Bacillariophyceae).

Iron wastes : *Chlorella variegata* (Chlorophyceae), *Gomphonema acuminatum, Pinnularia microstauron, Stauroneis phoenicenteron, Surirella linearis* (Bacillariophyceae), *Trachelomonas hispida* (Euglenophyceae).

Chromium wastes : *Tetraspora, Closterium acerosum, Stigeoclonium tenue* (Chlorophyceae), *Navicula atomus, Nitzschia linearis* (Bacillariophyceae), *Euglena oxyuris, E. sociabilis, E. stellata* (Euglenophyceae).

Salt brine (mainly Nacl) : *Anacystis, Calothrix, Lyngbya aestuarii, Phormidium tenue, Trichodesmium* (Cyanophyceae), *Amphiprora paludosa, Amphora ovalis, Chaetoceros elmorei, Diatoma elongatum, Diploneis elliptica, Gyrosigma attenuatum, Melosira arenaria, Meridion circulare, Navicula salinarum, Nitzschia frustulum* (Bacillariophyceae), *Chaetomorpha, Dunaliella salina, Pediastrum simplex, Scenedesmus bijugatus* (Chlorophyceae), *Euglena, Trachelomonas* (Euglenophyceae).

Copper wastes : *Calothrix braunii, Symploca erecta* (Cyanophyceae), *Scenedesmus obliquus, Chlorococcum botryoides, Stigeoclonium tenue* (Chlorophyceae), *Asterionella formosa, Cymbella naviculiformis* (Bacillariophyceae).

Phenolic wastes : *Ceratoneis arcus, Cyclotella Kuetzingiana, Fragilaria virescens, Pinnularia borealis* (Bacillariophyceae).

Paper mill wastes : *Pediastrum, Ulothrix, Pandorina* (Chlorophyceae), *Cocconeis diminuta, Gomphonema herculaneum, Navicula radiosa, Synedra pulchella* (Bacillariophyceae).

Algae found in organic wastes and eutrophic waters have already been discussed in earlier part of this chapter.

(V) ALGAL BLOOMS

When the planktonic algae grow in sufficient number and accumulate or cover the water surface, this condition is known as **algal bloom** or **water bloom.** This gives a distinct colour to the entire body of water which depends upon the colour of bloom-forming algae. Bloom is usually formed by single alga or by a few algae in some instances. It may occur in freshwater, inland salt lakes and sea.

Due to favourable conditions (rich nutrients) a small initial number of cells of a species grows exponentially for several weeks which lead to bloom formation. Majority of bloom-forming algae belong to Cyanophyceae but the members of different classes also form blooms. It may be temporary or permanent in nature. The thickness of water bloom also varies due to change in environmental and seasonal conditions. *Microcystis* bloom may reach upto 8 cm in thickness. With the advent of rainy season there is dilution in the thickness of algal bloom. Gradually during winter and summer months again bloom-forming algae multiply rapidly.

Algae causing water blooms : *Microcystis aeruginosa, M. viridis, M. flos-aquae, Coelosphaerium kuetzingianum, Anabaenopsis elenkinii, Aphanizomenon flos-aquae, Nostoc linckia, Nodularia spumigena, Trichodesmium erythraeum, Lyngbya limnetica, L. hieronymusii, Oscillatoria prolifica, O. planktonica, O. rubescens, Arthrospira platensis, Spirulina gomontiana, Gloeotrichia echinulata, Anabaena circinalis, A. microspora, A. spiroides, A. flos-aquae* (Cyanophyceae), *Chlamydomonas, Pandorina, Volvox, Golenkinia, Scenedesmus, Botryococcus, Cosmarium, Zygnema, Mougeotia, Oedogonium* (Chlorophyceae), *Fragilaria, Asterionella, Surirella, Tabellaria* (Bacillariophyceae), *Synura* (Chrysophyceae), *Ceratium, Gonyaulax, Gymnodinium* (Dinophyceae), *Hornellia* (Cryptophyceae), *Prymnesium* (Prymnesiophyceae), *Euglena* (Euglenophyceae) are the common bloom -forming algae.

Throughout the centuries there have been records of algal blooms in Europe. In the Old Testament there is an account of a river changing to a blood red colour. A permanent bloom of *Oscillatoria, Chroococcus, Synechocystis, Merismopedia* and *Anabaena* can be observed in Lake Magadi of Kenya which is a saline lake (sodium concentration 11.4, pH 10). A bloom of *Trichodesmium* was observed occupying about 52,000 km^2 area off the coast of Australia. In India bloom formation is a common phenomenon in the temple ponds, lakes and other lentic watersystem. *Microcystis aeruginosa* and *M. flos-aquae* are the two blue-green algae which are most commonly involved in the bloom formation. Some other algae associated with bloom formation in.the aquatic systems of India are *Arthrospira, Spirulina, Oscillatoria, Anabaenopsis, Raphidiopsis* (Cyanophyceae), *Pandorina, Gonium, Volvox, Cosmarium, Staurastrum, Botryococcus* (Chlorophyceae), *Nitzschia, Melosira, Navicula, Cyclotella, Surirella* (Bacillariophyceae), *Ceratium, Peridinium* (Dinophyceae) and *Euglena* (Euglenophyceae). The members of blue-green algae usually form permanent blooms. A permanent bloom of *Anabaenopsis* is seen in Sambhar salt lake (salt concentration 14-17%). Due to their capabilities to tolerate high salt concentration, blue-green algae form water blooms in the saline lakes of different types. In the esturine regions of rivers in South Indian, *Anabaenopsis arnoldii* v. *indica* forms bloom. In Indian Ocean, Arabian sea and Bay of Bengal, water bloom is caused by *Trichodesmium erythraeum* and *T. thiebautii*. In these areas *Trichodesmium* has been observed to be present for more than a century.

Red Tide (A single species bloom) : *Gymnodinium* and *Gonyaulax* (dinoflagellates) are responsible for large marine fish kills or paralytic shellfish poisoning. When they are present in large number (up to 6 x 10^7 cells/litre) they may change the water colour to red or yellow to brown. They account for more than 99% of the organisms present during that period. The development of red tide takes place with increase in nutrient levels along with drop in coastal salinity. These tides are commonly found off the Gulf Coast of Florida (California), in New England and are also reported from South Africa, Peru, Japan and India. There are views that such single species blooms must result from the release of substances that are either self-stimulatory or are toxic to other phytoplankton.

Factors of bloom formation : A large number of factors have been found to be associated with the formation of water blooms. There has been the records of some correlation between algal blooms and water temperature, water movements, light and inorganic nutrients (mainly phosphorus and nitrogen). Major elements and heavy metals like zinc play an important role in the bloom formation. Blooms of dinoflagellates are related to cobalamine being washed out in the sea from the soils which possess sufficient amount of this vitamine. As the planktonic blue-green algae have floating device (gas vocuoles), blooms of the members of this class are mostly of permanent nature.

Prymnesium (Prymenesiophyceae) is a marine alga which is not found in waters with temperature greater than 30°C. The production of bloom of this alga is dependent upon the presence of vitamin B_{12} and thiamine.

Mechanism : Swingle (1966) forwarded a mechanism of algal bloom causing fish mortalities. According to this, a type of phytoplankton becomes dominant and rises to surface as a dense scum under certain conditions (Fig. 13.6). These algae like *Microcystis* and *Anabaena* ususally have gas vacuoles which

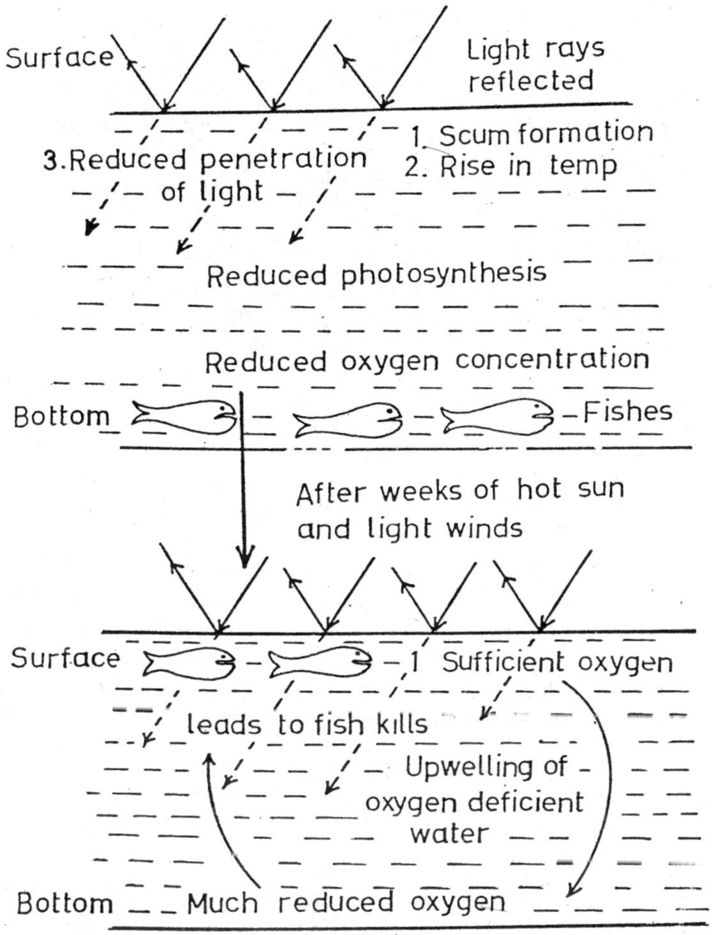

Fig.13.6. Mechanism of fish kill by algal bloom. (After Swingle, 1966).

expand in warm water and rise to the surface. The dense plankton in surface layers absorbs and reflects the sunlight causing an inadequate penetration of light for photosynthesis. Due to fall in light intensity, oxygen concentration decreases to 1 ppm at depths. After few weeks when there is no oxygen or insufficient oxygen in the greater depth, fish moves into the upper water where oxygen is sufficient. But when the oxygen deficient water upwells to the surface, fish of all types are trapped and killed.

Uses : Blooms of Cyanophyceae provide the major source of food for flocks of flamingoes. It is supposed that pink colouration on their feathers is due to carotenoids of the blue-green algae.

Water blooms composed of nitrogen-fixing blue-greens are common in many countries. These help in recuperation of fertility in soils by the fixation of nitrogen. To suppress the growth of submerged weeds, bloom formation is induced sometimes. Urea or superphosphate is added in higher concentration or cow dung may also be used for the enhanced growth of algae. Certain rice fields in California were free of mosquito larvae due to the blooms of *Anabaena unispora*, *A. variabills* and *Aulosira implexa*. The toxins produced by these algae kill larvae of *Culex tarcalis*.

Harmful effects : Many water bloom causing algae produce toxins and other chemicals which are harmful to various aquatic fauna and fishes. Fish mortality is also caused due to depletion in dissolved oxygen in water. Presence of water blooms in the reservoirs impart unpleasant taste and smell to drinking water. They also choke up slow sand filters.

Control : The application of copper sulphate or chlorine (as bleaching powder or calcium hypochlorite) is the easy and most common method. As the higher concentration of chemicals (**algicides**) may be lethal to fish and cattles, it is applied in low concentration. Blooms are also removed by mechanical means, by covering the water with floating plants or with parts of plants like banana leaves.

Biological method to control the algal blooms is also used. Cyanophages (virus infecting the members of Cyanophyceae), bacterial species and the members of fungi known to infect algae are used for controlling the blooms. Cyanophages have been recorded on different genera of blue-green algae which are specific to hosts. The pathogenic fungi mainly belong to order Chytridales (Phycomycetes) and Blastocladiales. A number of gram negative bacteria have been reported which cause lysis of blue-green algae.The algal bloom of *Microcystis* can be destroyed by using a specific bacterium (CP-1). The destruction of algal blooms can easily be obtained if the specific bacteria or bacteriophage able to lyse the concerned algae are known. The biological control of algal blooms is much more economical, easier and advisable than the chemical control.

(VI) TOXIC ALGAE, TOXINS AND THEIR ROLE

Toxins, produced by many freshwater and marine algae, are when ingested by man, fish or other animals can cause detrimental or even lethal effects. A number of blue-greens, prymnesiophytes, dinoflagellates and some sea-weeds are well known for their toxic effects (Fig. 13.7). Since 1878 there has been reports regarding algae as causative agents for animal toxicity. The most common clinical manifestations are gastrointestinal, neuromuscular, hepatic, cardiovascular and respiratory disorders. The gastrointestinal symptoms include disturbance of intestinal motility, anorexia, extreme salivation and abdominal distention. The respiratory disorders include sneezing, coughing, dyspnoea, tachypnoea and cyanosis resulting in death sometimes. Hepatic damages normally result in ascites and jaundice.

Toxic cyanophytes : Most of the toxic freshwater algae belong to the class Cyanophyceae along with some marine species. The blooms of Cyanophyceae have long been associated with animal poisoning. Three confirmed toxin-producer species are *Microcystis aeruginosa, Anabaena flos-aquae, Aphanizomenon flos-*

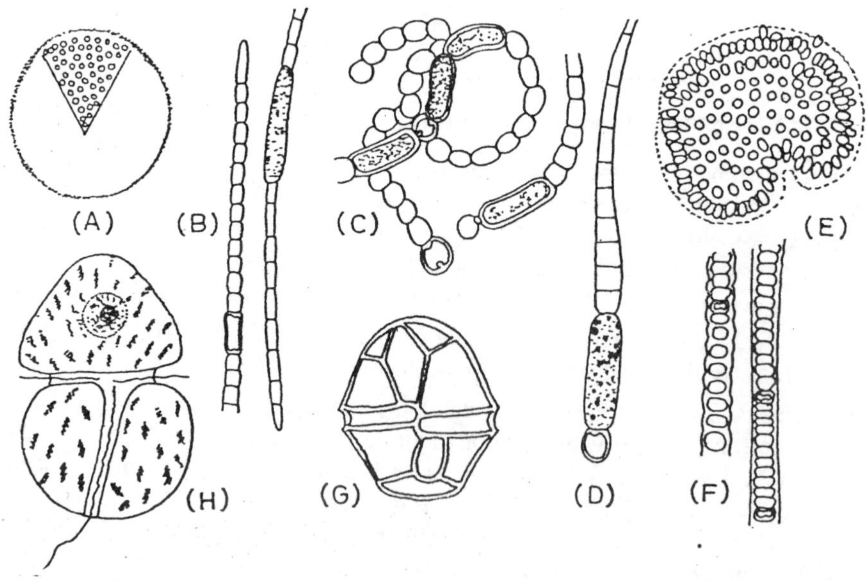

Fig.13.7. Common examples of toxin-producing algae. (A) A portion of globular colony of *Microcystis flos-aquae.* (B) *Aphanizomenon flos-aquae.* (C) *Anabaena flos-aquae.* (D) *Gloeotrichia echinulata.* (E) *Coelosphaerium* sp. (F) *Nodularia spumigena.* (G) *Gonyaulax* sp. (H) *Gymnodinium* sp.

aquae. *Aphanizomenon* is least frequently reported, *Microcystis* is the most widespread and the *Anabaena* blooms have the most dramatic effects (Carmichael, 1981). Some other common toxic blue-green algae out of more than 50 recorded genera are *Coelosphaerium, Gloeotrichia, Lyngbya, Nodularia* and *Nostoc*. These members are the causal organisms for the majority of reported alga- caused deaths of fish, livestock, waterfowl and humans. All of the reported cases of blue-green algal toxicity have occurred during the summer months when the concentration of cyanophytes are sufficient to form blooms. *Schizothrix calcicola* and *Oscillatoria* species found in drinking waters are capable of inducing nonmortal gastrointestinal disorders in humans. Similarly, a diarrhea-producing toxin was separated from the lethal toxin of *Microcystis aeruginosa*. *Nostoc rivulare* has also been reported for carcinogenic activity.

Anabaena toxins are termed as **anatoxins** which appear to be of three types - alkaloids, peptides and pteridines. The alkaloids are the most acutely toxic which produce neurological symptoms within minutes of ingestion whereas the peptides are slower requiring one to several hours for action which produce liver necrosis. The toxin from *Anabaena* was named **very fast death factor (VFDF)** to distinguish it from the **FDF** of *Microcystis*. VFDF differs from FDF in that it is an **exotoxin** which is excreted by *Anabaena* into the surrounding water. *Anabaena* VFDF has recently been renamed as **anatoxin A**. Chemically it was found to be water soluble as well as heat, light and alkaline labile. It has been hypothesized that there are atleast three other toxins (**anatoxins B, C** and **D**) produced by *Anabaena flos-aquae*. *Microcystis* toxins are mainly peptides and *M. aeruginosa* is probably the one which causes the most harm. The deaths of many poultry and cattle have occurred as results of blooms of this alga. The alga is found most frequently in shallow freshwater lakes and ponds. Two toxins isolated from this taxa are slow death factor (SDF) and fast death factor (FDF). **Microcystin** was found to be a nonvolatile substance that was irreversibly adsorbed on activated carbon and water soluble.

Aphanizomenon toxins are closely related to the saxitonins produced by the dinoflagellates. The toxins produced by *Aphanizomenon flos-aquae* have been implicated in the deaths of livestock and fish for a long time .Like the toxins of other blue-green algae the production of *Aphanizomenon* toxin can be activated or inhibited by a variety of environmental parameters. It was originally classified as another VFDF, even though it differed chemically from the *Anabaena* toxin because it also acts rapidly.The symptoms and cause of death, when injected into mice, are similar to other blue-green toxins. Mice exhibit irregular breathing, gaping mouth, spasms, loss of coordination, violent tremors and death due to respiratory paralysis.

Toxins have also been reported from marine blue-green algae. The marine toxic blue-green algae mainly belong to Oscillatoriaceae. Toxins found in marine *Lyngbya* spp. may be responsible for a **contact dermatitis** among swimmers. Another toxin has been isolated from the nondermatitis - producing type of *Lyngbya majuscula*. This toxin has been named **lyngbyatoxin A** which is a nitrogenous toxin that is optically active and nonbasic.

Toxic prymnesiophyte : *Prymnesium parvum* (Prymnesiophyceae) produces a potent exotoxin which causes extensive fish mortalities in marine habitat and brackish water conditions in many countries. The toxin is most effective in aquatic gill-breathing animals like fish and molluscs . As it is capable of causing mass mortality in fish it is a serious problem in aquaculture. The toxic components are a mixture of proteo-phospholipids. They get activated by cationic cofactors, sodium, magnesium or calcium ions and act by changing the permeability of the gill membranes. If fishes are removed promptly from such toxin solutions, the gill damage is repaired within hours.

The evidence shows that several toxic components are present in *P. parvum*. These toxins are responsible for causing three separate physiological responses (ichthyotoxic activity, hemolytic activity and antispasmodic effects) on smooth muscle. Salinity is a critical factor for the growth and toxin production by this taxa. Many other physico-chemical factors like inorganic phosphate, pH,light etc. are found to be linked with *Prymnesium* toxicity. As in the case of blue-green and dinoflagellate toxins, there is no simple numerical correlation between the number of algal cells and the amount of toxins produced.

Toxic dinoflagellates : Paralytic shellfish poisoning (PSP) or **neurotoxic shellfish poisoning** (NSP) in humans, fish and sea birds are caused due to toxins produced by monospecific blooms of dinoflagellates. Dinoflagellate **red tides** occur in coastal areas and estuaries of both temperate and tropical seas. About 20 dinoflagellates out of more than 1000 species are known to produce PSP toxins. PSP is potentially fatal in humans. *Protogonyaulax* (= *Gonyaulax*) toxin may produce fish kills as the toxins are concentrated in the filter-feeding zooplankton.

Steidinger (1973) divided the toxic blooms of dinoflagellates into three types: 1) Blooms that kill fish but few invertebrates (*Ptychodiscus brevis*, the Florida red tide organism), 2) Blooms that kill primary invertebrates (*Gonyaulax* spp.), and 3) Blooms that kill few marine organisms but the toxins are concentrated within the digestive glands or siphons of filter-feeding bivalve molluscans (mussels, oysters, clams etc.) causing PSP.

The most notorious PSP-causing dinoflagellate on the Pacific Coast is *Gonyaulax catenella*. The toxin produced by this species (**saxitoxin**, a neurotoxin) is 100,000 time more potent than cocaine (Steidinger and Joyce, 1973). Twelve closely related water soluble toxins are implicated in PSP, of which the best known is saxitoxin. All are potent neurotoxin which inhibit nerve transmission by blocking sodium channels in the nerve axons. *Gonyaulax monilata*, a chain-forming species has also been associated with fish kills.

In the west coast of Florida, the most spectacular red tides occur which are caused mainly by *Ptychodiscus* (= *Gymnodinium*) *brevis*, a NSP agent. Massive fish mortalities may result from these tides although the toxin was found to be milder than saxitoxin. The lipid - soluble NSP toxin has a different structure and action from the PSP toxins. Numbness accompanied by food poisoning may occur in humans after consumption of those shellfish which have accumulated the NSP toxin but fatalities and paralysis have not been recorded.

Another species of dinoflagellates (*Prorocentrum minimum* = *Exuviella lebouriae*) found in Japan has been shown to be hepatotoxic and nephrotoxic in animals. The benthic dinoflagellates *Gambierdiscus toxicus* is responsible for the production of **ciguatera toxin**. By the consumption of the visceral organs of some coral-reef fish, ciguatera poisoning is caused. Symptoms are muscular weakness, nausea and cramps but mortality is low. Benthic *Prorocentrum lima* has also been implicated in ciguatera. *Noctiluca miliaris* red tide along the coast of Madras (India), Japan has led to the poisoning of many fish and invertebrates.

Most of the toxic dinoflagellates are marine and they are generally within the range of 40 to 60 •m in diameter. Blooms are usually dependent upon an annual cycle of environmental parameters to produce the suitable temperature, salinity, illumination, pH, concentration of trace elements and growth factors. Occurrence of red tides essentially on all shores of North America have been correlated with the surfacing of nutrients by upwelling or by tidal turbulence, their competitive advantage over coastal diatoms and addition of humic substances etc.

Toxic seaweeds : Secondary metabolites produced by many toxic seaweeds act as defences against herbivores and microbial pathogens. Such compounds may be toxic to humans and many are used in medicine. They are found in the Chlorophyceae, Phaeophyceae and Rhodophyceae and include terpenoids, halomethanes, polyphenols, acetylenes and alkaloids. Algae producing these substances are almost restricted to tropical waters. According to Fenical (1980) most seaweeds consumed by humans do not produce toxic compounds or contain relatively low amounts which are readily metabolized.

Chlorophyceae : Only in rare instances the Chlorophyceae have been associated with toxicity. The marine benthic green alga *Caulerpa* produces two toxic substances (**caulerpin** and **caulerpicin**). Both of these compounds have demonstrated toxicity in mice. It is interesting to know that *Caulerpa* is probably the most popular edible alga in the Philippines but becomes toxic during rainy months which is believed to be due to agitation of the plant. The human physiological symptoms associated with caulerpicin ingestion include numbness and a cold sensation of the extremities, rapid and difficult breathing, slight depression and eventually loss of balance.

The organic extract from the green alga *Chaetomorpha minima* has experimentally shown hemolytic activity and fish toxicity. Three distinct **hemolysins** have been isolated from *Ulva pertusa*. The hemolysins did not produce developmental modification but they induced toxicity.

Phaeophyceae : A general feature of the Phaeophyceae is the production of phenolic compounds which are usually polymerization products of phloroglucinol. They are antibacterial and also cause feeding avoidance by invertebrates. Many tropical species of Dictyotaceae and Cystoseiraceae also produce a variety of terpenoids which may reach upto 5% of the dry weight in some species of *Dictyota*.

Rhodophyceae : The widest variety of secondary metabolites are produced by the Rhodophyceae. *Asparagopsis taxiformis*, considered as a delicacy by native Hawaiians, produces the carcinogen bromoform, halogenated acetates, acrylates and ketones. *Liagora farinosa* (Nemaliales) produces an acetylenic ichthyotoxic lipid. *Laurencia obtusa* (Ceramiales) is avoided by most herbivores including the urchin *Diadema* and it produces brominated compounds showing toxicity to fertilized sea urchin eggs. The carrageenans (rhodophycean polysaccharide) have capability to induce serious physiological effects which include the induction of acute and chronic inflammatory responses, strong immunosuppression etc.

(VII) FOSSIL ALGAE

Because of delicate, soft and readily decomposiable structures, majority of algae are not easily preserved as fossils. However, they have one feature in common, namely the formation of resistant skeletal material or superficial precipitation. These are either the carbonate of calcium or magnesium (Chlorophyta, Rhodophyta and coccoliths of certain Prymnesiophyceae) or silica (diatoms and silicoflagellates). The preservation of cell walls of non-mineralized species is rare e.g., *Pediastrum* and akinetes of Cyanophyceae. The impressions of algae are even more rare.

The best known fossil algae are among the microalgae including blue-green algae, diatoms and coccolithophores. The blue-green algae are among the most ancient known for atleast 2 billion and possibly even 3 billion years. Coccoliths, scales from some Prymnesiophyceae, dinoflagellates and diatoms are more recent (no older than 200 million years old). The largest number of fossils are from Cretaceous period though they are common in different geological periods. Some fossil members are remarkably similar to present day genera and are often described as species in these extant genera. Schopf (1970) assigned the period for different algal groups (Table 13.4).

Table 13.4. Dates of early fossil records in various algal groups

Algal Group	Geological Time
Cyanophyta	Precambrian
Chlorophyta	Precambrian
Rhodophyta	Cambrian
Phaeophyta	Late Precambrian
Charophyta	Silurian
Prymnesiophyceae	Triassic
Xanthophyceae, Euglenophyceae, Bacillariophyceae and Chrysophyceae	Cretaceous

In macroalgae, the best fossils are found in those groups of algae on which lime deposits are present. Fleshy genera lacking this firm outer coating and microalgae without firm cell coverings preserve poorly. The process of fossilization is continuous and can be observed today in sediments being formed at the bottom of lakes and oceans which may be used to study the more recent changes in algal floras.

The algae are preserved as impressions, carbon films, casts, molds and petrifications. Out of these types, petrifications are most important and reveal structural details. The common fossil members of different divisions of algae are as follows :

Cyanophyta : Schopf (1970) and his associates have described a number of Precambrian algae and designated the Precambrian as the age of blue-green algae. *Archaeosphaeroides barbertonensis* was considered to be the most ancient recognisable algae which showed similarity to extant coccoid blue-green algae (Fig. 13.8 A). The blue-green algae are the oldest group of algae with definite fossil remains in the form of **stromatolites** dating back almost 3000 million years. Stromatolites are rocklike formations of limestone and other materials trapped in layers by members of blue-green algae like *Schizothrix calcicola* either in freshwater or marine habitat. The members of blue-green algae have persisted from the Precambrian to the present with little modification.

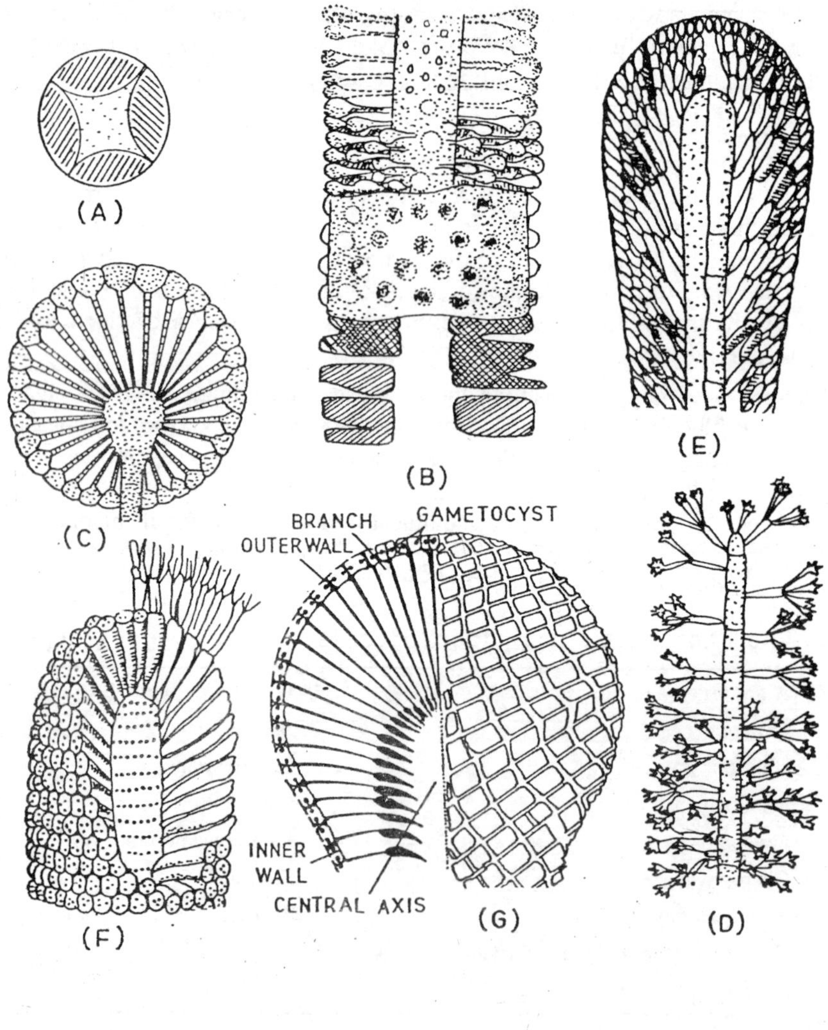

Fig.13.8. Representatives of fossil algae. (A) Outline of *Archaeosphaeroides barbertonensis.* (B) *Rhabdoporella.* (C) *Cyclocrinus.* (D) *Primicorallina.* (E) *Paleodoscycladus.* (F) *Triploporella.* (G) *Ischadites.*

The repeated lime deposition of *Aphanothece, Gloeocapsa, Microcystis, Rivularia* appear as biscuit-shaped calcareous masses. *Plectonema* may precipitate crystals in the mucilage between the cells of filaments and can deposit lime as in *Lyngbya, Schizothrix*. Mucilage of blue-greens may be replaced by a mass of crystal in some species. In later geological periods also there are records of structures like blue-green algae. *Girvanella* (= *Symploca*) is bilieved to be a calcareous cyanophycean fossil reported from the Cambrian to Cretaceous. *Marpolia* is very similar to the modern *Schizothrix*. *Gloeocapsomorpha* is found in Ordovician deposits which bears considerable resemblance to *Gloeocapsa*. *Oscillatoriformis*, a genus from Middle Devonian may be compared with living genus *Oscillatoria*. *Langiella scourfieldii* (Stigonemataceae) was found to be well preserved in Devonian. *Aphanocapsa* has been recorded from Permocarboniferous beds of Rewa (M.P.), *Synechocystis, Scytonema* from kashmir, *Palaeonostoc* from Ganga valley as fossil algae.

Chlorophyta : The members of Codiaceae and Dasycladaceae of the order Siphonales are lime depositors and represent best known fossils amongst Chlorophyta. The fossil species of *Halimeda* and *Codium* (Codiaceae) are found in Tertiary and Quarternary deposits. *Dimorphosiphon rectanulare* from Ordovician of Norway is probably oldest known Codiaceae representing calcified form of branched tubular cells with no crosswalls. *Palaeoporella* (Codiaceae) is also observed during Paleozoic. *Codium*, an important constituent of coral reef is found in reef deposits mainly in Indo-Pacific region. The members of Dasycladaceae were common in the Permian and Jurassic. The most simple member is probably *Rhabdoporella* (Fig. 13.8 B) which has a central siphon surrounded by simple, club-shaped branches showing carbonate depositions. In *Cyclocrinus* branches originate from clavate head (Fig. 13.8 C). *Primicorallina* shows scattered branches (Fig. 13.8 D). In *Palaeodascycladus*, the ultimate branches form a pseudoepidermis showing increased complexity (Fig. 13.8 E).Gametangial branches can also be recognised in many fossil Dasycladaceae. *Triploporella* (Fig. 13.8 F) obtained from Mexico represents a very complex Cretaceous Dasycladacean form. *Ischadites* (Dasycladaceae) also represents a complex structure (Fig 13.8 G).

Due to the delicate nature, the reports of flagellated members are lacking. However, there is one record of *Chlamydomonas* like alga *Gloeocystis* from upper Jurassic. *Phacotus*, a flagellated form is known from Tertiary. In 'brown coal' deposits of Germany the chaetophoralean algae *Phycopeltis* has been found. The zygospores of desmid have been recorded from Devonian, Jurassic and Cretaceous periods. The desmid *Oocardium* secretes a basal mucilage stalk around which calcium carbonate is deposited (Fig. 13.9 D,E). *Botryococcus*, a hydrocarbon oil rich form was abundant in Permian seas and was obtained in "Boghead coal" from Scotland. The genus was often associated with desmids and *Pediastrum* and also occurs in interglacial and post-glacial peat deposits. *Pediastrum* is found as fossil in oil drillings and well borings as its cell wall is resistant to decay. *Paleoclosterium leptum*, a freshwater desmid occurs in Middle Devonian strata of New york.

Since Ordovician and Silurian, fossil records of *Chara* and other members of Charales are available. The main organs of Charales which get fossilized are oogonia known as **gyrogonites** and nodal regions of stems. Gyrogonites (Fig. 13.9 B,C) is incomplete oogonium which reveals inner calcareous part enclosing cutilized membrane and egg. Due to lime deposition around oogonial wall only, coronal cells are lost. Oogonial wall consists of five spiral tube cells. Fossil genus *Clavator* shows oogonium surrounded by a circle of elongated cells (Fig. 13.9 A). Two form genera (*Lagynophora* and *Palaeonitella*), similar to present day Charales, have been reported from the Paleozoic period. Mesozoic and Cenozoic genera of Charales are *Perimnesta, Nitella, Nitellites, Tolypella, Kosmogyra, Çhara* etc. Paleozoic genera are *Actistochara Paleochara, Sycidium* and *Trochiliscus*. Gyrogonites of *Chara, Nitellites* have been reported from Andhra Pradesh and Madhya Pradesh.

Bacillariophyta : *Pyxidicula* (Fig. 13.9 F) is considered to be oldest fossil diatom recorded from Jurassic. In general diatoms seem to exhibit maximum occurrence during Tertiary. Out of about 200 genera of diatoms, 70 are recorded as fossils. Both freshwater and marine diatoms are well preserved in large amounts as **diatomaceous earth** or **Kieselguhr** or **diatomite.** It may be thin (Saharan diatomites) or form

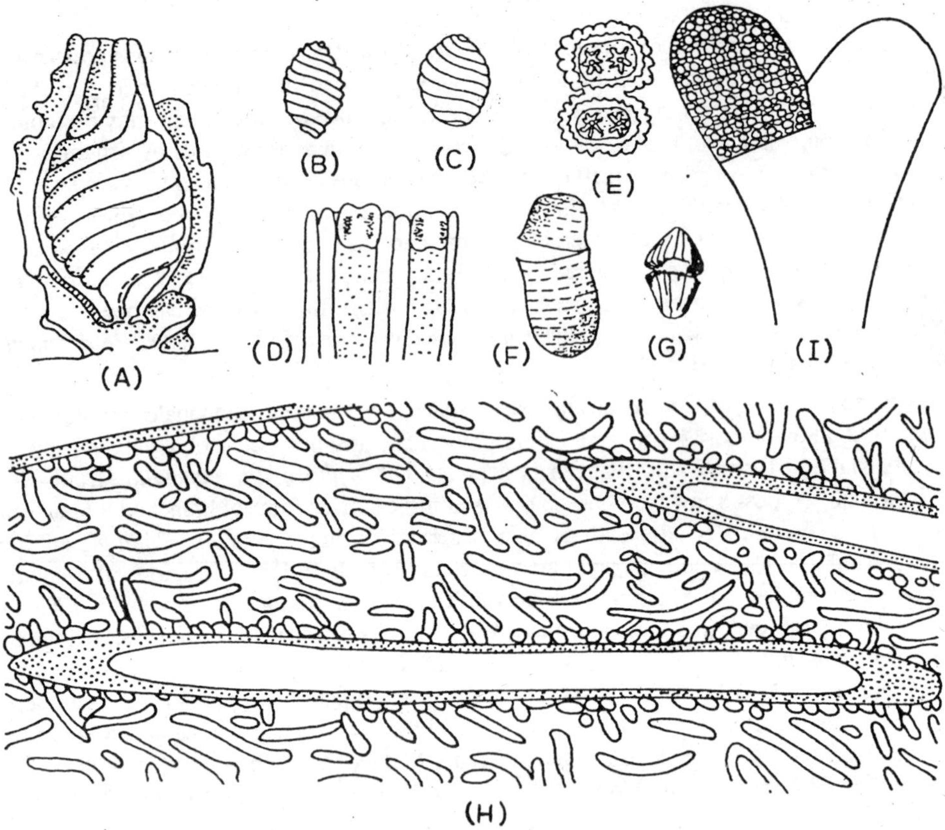

Fig.13.9. Representatives of fossil algae and their organs. (A) Internal view of the oogonium of *Clavator reidi* surrounded by the utricle. (B-C) Gyrogonites of *C.reidi.* (D) *Oocardium* showing two cells on mucilage stalk surrounded by columns of lime. (E) The same seen from above. (F) *Pyxidicula.* (G) Fossil dinoflagellate. (H) *Prototaxites southworthii* in longitudinal section. (I) *Protosalvinia furcata* showing bifurcation.

more than 100 metre thick beds (about 425 meteres in Lompoc, Colifornia). Majority of the fossil diatoms belong to Centrales suggesting that these are primitive than Pennate diatoms. *Scyphasphaera* and *Thorosphaera* have been reported from early Paleozoic age. 15 diatom species are reported from Pleistocene deposits of Kashmir. From Laradura, Kashmir, 4 species of *Cyclotella*, 4 species of *Melosira, Stephanodiscus, Eunotia, Fragilaria kashmeriensis, Diploneis ovalis, Gyrosigma acuminatum, Cymbella affinis* etc. have been reported. Fossil diatoms have also been observed in Andaman-Nicobar islands and Orissa.

Dinophyta : Due to calcareous and siliceous inclusions in cell walls (mainly in cysts), the members of this divisions (Fig. 13.9 G) are suitable for fossilization. Well preservation of the cysts help in placing them to proper systematic position. *Biecherlla, Cacisphaerella* reported from Jurassic of France, *Eodinia* and *Lithodinia* represent calcareous impregnations. From Miocene of Algeria, *Bicarnelleem* has been observed. Many naked and thecate dinoflagellates have been described from Cretaceous of victoria, Australia. *Peridinians* have been obtained from the chalk deposits of Germany and France.

Phaeophyta : The members of this division are supposed to have appeared in Precambrian period. The most striking example is *Prototaxites* (Fig. 13.9 H) which appeared in Silurian and Devonian. It has a stout trunk like axis having about 3 feet diameter with a system of small and large tubes. Some of the specimens are upto 2.2-2.3 meters long and the structure resembles stipe of kelps. Few recent studies have disclosed that this does not belong to algae as there are specific septal pores in it. Cuticle and tetrads of spores withir these tubes have been seen. Like *Prototaxites*, the flat leaf like structure *Nematothallus* are also observed. *Thamnocladus* is supposed to be fucoid or dictyotan in nature and has been observed from Devonian of New York. Since the existence of *Prototaxites* many brown algae like *Fucus, Ectocarpus, Dictyota, Padina, Cutleria* etc. are recognised.

Rhodophyta : Impressive accumulations of several thousand feet thick deposits of *Lithothamnion* and other coralline algae are found in Eocene and Miocene of Cenozoic. Almost all the fossil Rhodophyta belong to Corallinaceae because of their ability to deposit lime. Coralline fossils marked their presence in Cretaceous of Mesozoic. In the formation of coral reefs, the encrusting Corallinaceae (Nullipores) are very important as they act as cementing organisms. The most important genera forming coral reef in the Indian ocean are *Porolithon, Gonilithon* and *Lithophyllum.*

The calcareous genus *Lilkophyllum, Melobesia* have been found in fossiliferous rocks. A completely extinct group Solenoporaceae with main genus *Solenopora* has been described. As only vegetative remains and occasional tetraspores enclosed in sunken tetrasporangia have been observed, it is difficult to place them with Rhodophyta conclusively. However, they resemble present day *Lomentaria.* In India *Lithothamnion, Melobesia, Distichoplax* were reported from Assam, *Amphiora* and *Corallina* from Andaman, *Lithothamnion* from Surat, *Archaeolithothamnion, Lithophyllum* and *Petrophyton* from Pondicherry.

Phillips et al. (1972) reported several species of the algal genus *Protosalvinia* (Fig. 13.9 I) from the upper Devonian. The members have frequently bifurcate thalli with tetraspores located at their apices within depressions. It shows phaeophycean affinities also.

Other algae : Silicious chrysophytes and the coccolithophores are abundant in certain Mesozoic and Tertiary rocks. These remains are in the form of silicious scales, statospores (Chrysophyceae) and as calcareous structure called coccoliths (Prymnesiophyceae) which constitute the cell coverings. Crystals of calcium carbonate are also deposited on the filaments of *Vaucheria.*

Uncertain algal groups : Some fossils are considered algae of uncertain systematic position. During Precambrian and upper Cambrian and from calcareous beds, algae with large concentric structures upto 2 feet in diameter (*Cryptozoon*) has been reported. Many other members are *Newlandia, Camasia, Collenia* etc.

Uses : The importance of the study of algal fossils are as follows 1) The use of diatomaceous earth is diverse which has been discussed earlier (please see economic importance of algae). 2) The deposits of algal fossil indicate the process of evolution of the aquatic system. Their composition depict the past environment. It also helps in geographical mapping. 3) In locating petroleum bearing oil rocks, the members of Coccolithophorides are important. 4) In various parts of the world limestone rocks are used as building stones which include *Lithothamnion, Lithakalk.* As the bricks of these limestones are very light, they are used for high structures of building. 5) Due to their specificity, the members of Chrysophyceae and Bacillariophyceae indicate the possible changes in the ecological conditions. As in three different strata of same area of Europe, three different species of *Melosira* indicating different conditions are found. *Melosira italica* represents cold water and presence of poor nutrients, *M. ambigua* is an indicator of warm water and rich nutrients and *M. granulata* represents less warm water and not rich nutrients. 6) In tracing the evolution, fossil algae are very much useful. Fossil studies indicated that blue-greens were present in early and middle Precambrian. Eucaryotes developed during later part of late Precambrian. Red algae were clearly present in Cambrian. Codiaceae and Dasycladaceae (green algae) were abundant in Ordovician while Charales appeared in Silurian. The members of Euglenophyceae, Xanthophyceae, Chrysophyceae appeared during Createaceous and diatoms in Jurassic. This shows possible line of evolution.

(VIII) ALGAL CYTOLOGY AND GENETICS

Procaryotic algae reveal primitive cytological features i.e., presence of incipient nucleus, absence of bounding membranes and genetic apparatus (nucleus) while in eucaryotic algae these structures are well developed.

Nucleus : The nucleus is frequently visible in algal cells appearing as a greyish body often enclosing a darker region, the **nucleolus**. Genetic information (**genome**) is contained principally in nucleus but additional genetic informations are found in the chloroplast and mitochondria. In Cryptophyceae, **nucleomorph** is found which is supposed to have subsidiary genome. In both pro- and eucaryotic algae the genetic information is stored in the triplet sequences of nucleotide bases on the double helix of nucleotide chains of DNA molecules.

In procaryotic algae the DNA is present in a ring-like configuration in the centre of the cell and is not contained within a membrane system. Process of transcription and translation occur in a tightly coupled manner. In eucaryotic algae chromosomes are present in which DNA contains nuclear proteins with small amounts of RNA in a complex termed as **chromatin**. The nuclear envelopes are double membraned and are continuous with cell ER. Nuclei of Dinophyceae, Euglenophyceae and Cryptophyceae possess unique features of the nucleus, nuclear division and cytokinesis (Dodge, 1973). They have chromosomes which remain condensed during all cell phases. The retention of the nuclear membrance during nuclear division (Dinophyceae and Euglenophyceae but not Cryptophyceae) and persistence of the nucleolus (endosome) during division (Euglenophyceae) are some of the other unique features. These types of nuclei are termed as **mesocaryotic**.

The nuclear envelopes are about 7-8 nm thick and the membranes are separated by a perinuclear cavity. Many pores are present in the nuclear envelope which may be arranged in specific patterns. Nucleus contains the ground substance (**nucleoplasm**) and one or more nucleoli in addition to chromosomes. Nucleoli are rich in RNA and are intimately associated with the specific region of the chromosomal DNA.

The majority of eucaryotic algae possess uninucleate cells while in the members of Chlorococcales, siphonaceous members of Chlorophyceae, Xanthophyceae (e.g., *Vaucheria*), vegetative cells of most Florideophycidae (Rhodophyceae), the multinucleate cells are of normal occurrence. Algal nuclei and chromosomes are usually small except a few like Dinophyceae, Euglenophyceae. It varies from less than a micron in some algae to 80 µm in the primary nucleus of *Acetabularia*. The earlier informations on algal chromosomes was reviewed by Godward (1966). Much of the cytological informations and karyological studies on Indian algae have been carried out by Sarma and his team of workers. Four types of chromosomes were categorised by Sarma (1983) after their extensive studies.

1. Chromosomes appreciably long with localised centromeres (members of Cladophorales, Oedogoniales, Charales)
2. Chromosomes extremely small without any indication of centromeric position (many species of Volvocales, Chlorococcales, Chaetophorales, Siphonales, many desmids and some filamentous Conjugales)
3. Polycentric chromosomes (several *Spirogyra* species and some desmids)
4. Chromosomes fairly large without any evidence of centromere and no association of spindle during mitosis (members of Dinophyceae and Euglenophyceae)

Inspite of many difficulties in determining the chromosome numbers of algae, much data have been accumulated on this aspect. In the Volvocales haploid numbers vary from 5 in some *Volvox* species through 12, 13, 14, 15 and 16 to 22. In the Ulotrichales the chromosome counts tend to be 5 or multiples of 5 or 8 and its multiples. Polyploidy is a common feature is Cladophorales where the basic haploid number is 12. In many desmids chromosome number goes up to 112, 182 and even 592. Cytological analysis of various taxa of *stigeoclonium* (Chaetophorales) revealed a continous series of chromosome numbers ranging from 5 to 20 except 15, 17, 18 and 19. The lowest number obtained in *Oedogonium* was 9. The basic chromosome numbers for *Nitella* and *Chara* were 3 and 7 respectively. In Phaeophyceae, the Ectocarpales often have 8 chromosomes, the Cutleriales 24, the Sphacelariales and Dictyotales 16 and in the Fucales this number was 32. In Dinophyceae, the chromosome numbers range from 56 (*Gymnodinium indicum*) to 220 (*Woloszynskia hiemale*). The basic chromosome counts in *Euglena limnophila*, *E. gracilis* and *E. spirogyra* were found to be 16, 95 and 182 ± 5 respectively. In Ceramiales (Rhodophyceae) this number from n = 2 to n = 90 have been recorded.

Nuclear division : Vegetative growth of eucaryotic algae is achieved by nuclear division and cytokinesis (the division of cytoplasm).

Mitosis : Nuclear division in algae (except mesocaryotic forms) shows considerable similarity to that in animals, although the kinetochore and centrioles are not found in algae. In Ceramiales (Rhodophyceae), **polar rings** arise during early prophase and remain throughout mitosis. As flagella are absent in red algae, these are not regarded as true centrioles. In the Xanthophyceae (*Vaucheria*), very long interzonal spindle and an intact nuclear membrane during mitosis have been recorded. Its mitotic feature is more typical of phycomycetous fungi than of other algae. In Dinophyceae, the nucleolus and nuclear membrane persists during mitosis. The chromosomes appear to be attached to the nuclear membrane. Chromosomes remain condensed throughout the cell cycle. Chromosomes are devoid of centromeric localisation. Occasional unequal distribution of chromosomes at anaphase was observed in *Gymnodinium*. Cytoplasmic microtubules (**spindles**) are not involved in the separation of chromosomes. The spindle of Phaeophyceae is intranuclear with extranuclear centriole. The formation of spindle microtubules prior to prophase and the breakdown of the nuclear envelope are the common features in Cryptophyceae and Prymnesiophyceae. Leedale (1982) observed many unusual features in Euglenophyceae. The nuclear membrane remains intact and the nucleolus divides into two. During metaphase some chromatids may reach the poles while others remain at the equator. No spindle and centromere was detected. In Chlorophyceae, the spindle is both nuclear and cytoplasmic while in *Oedogonium*, it is entirely intranuclear. A complex kinetochore has been found in *Oedogonium*.

Meiosis : Very few detailed studies have been undertaken which revealed its similarity with other organisms. Due to absence of centromeres in *Spirogyra crassa*, movement of the chromatids and terminalization of chiasmata do not involve centromeres. At metaphase I, the four visible chromatids are held together by a sticky matrix.

Cytokinesis : After the completion of nuclear division, cytokinesis normally occurs either by growth of a new wall or by constriction of the cell by means of a furrow. Three types of cytokinesis have been described in blue-green algae : a **constrictive** type, a **septum forming** type and a **budding** type. The most primitive form of eucaryotic algal cytokinesis i.e., longitudinal cleavage is found in flagellates. The **phragmoplast** type of cytokinesis (a coalescence of dictyosome vesicles containing wall components and between the spindle microtubules) is characteristic of higher plants and is found in some Chlorophyceae, all the members of Charales and Prasinophyceae. In Rhodophyceae there is incomplete separation of daughter cells and cytoplasmic continuity. After this there is development of pit plug. Cell plate develops centrifugally in Phaeophyceae and gets thickened with fibrillar material through the action of Golgi vesicles.

Certain conflicting problems of algal taxonomy have very successfully been solved by cytological findings. Besides classification, it also helps in tracing evolutionary relationships.

Genetic analysis : Pascher (1918) was able to obtain mature heterozygotes in crosses between *Chara ceratophyllum* (female) and *Chara foetida* (male). For the genetical studies he recommended *Chara* and *Oedogonium* as experimental materials. Many papers were published subsequently on Mendelian inheritance in *Chlamydomonas*. Many workers started working seriously on algal genetics after 1950.

There is no sexual reproduction (syngamy and meiosis) in blue-green algae but a type of parasexuality, known as **genetic recombination** has been observed for the first time in *Anacystis nidulans* by kumar (1962). Shestakov and Khyen (1970) demonstrated **genetic transformation** (a mode of gene transfer leading to genetic recombination) for antibiotic resistance in *A. nudulans*. Genetic recombination has also been reported in *Cylindrospermum majus* (Singh and Sinha, 1965) and *Anabaena doliolum*. Mutations have been induced by using penicillin, erythromycin, polymyxin B, sulphanilamide, nitrosomethyl urea, X-rays, UV etc. Mutants may be grouped as **resistant mutants** (clonally pure mutants resistant to antibiotics, toxic chemicals etc.), **morphological and developmental mutants** (mutants induced by diethyl sulphate, ethyl methanosulphonate etc.) and **pigment mutants** (mutants with altered or changed proportions of pigments).

In green algae, genetical studies have mainly been done on *Chlamydomonas reinhardtii* because of many reasons: 1) They are unicellular, 2) can be cultured in continuous or synchronous cultures, 3) mutations develop spontaneously at low frequencies and can be induced by mutagens, 4) sexuality can be induced by nitrogen deficiency and 5) life cycle is completed within short period (about 2 months). Three different genetic systems (Mendelian, non-Mendelian assigned to chloroplast DNA, and mitochondrial DNA system) have been found in *C. reinhardtii*. A number of mutants like acetate requiring mutants, pigment deficient mutants, phototactic mutants etc. have been studied. Chloroplast DNA does not carry all the necessary genes for plastid structure and function while nuclear genes are known to affect chloroplast DNA polymerase and photosynthetic capacity of the chloroplast. A good amount of work has been done on *C. moewusii* and *C. eugametos* also.

Small nucleate portion of *Acetabularia* is capable of regenerating a complete plant and an anucleate portion can continue to grow and develop a sterile reproductive structure. Interspecific grafts in *Acetabularia* has shown that nucleus produces the morphogenetic substances for cap formation. Although chloroplasts are partly dependent on the nuclear genome but they have more genetic autonomy than other algae as *Acetabularia* can grow and differentiate in absence of nucleus. Experiments have shown that the divisions of the primary nucleus do not commence until the cap has reached a certain size i.e., nuclear division is dependent on some nutritional/cytoplasmic factors.

A convincing evidence has been found for a chromosomal sex-determination in British *Laminaria* species (Phaeophyceae) in which female gametophyte having a large X chromosome pairs with a small Y chromosome of male gemetophyte. The genetical studies of *Gracilaria* (Rhodophyceae) revealed that most of the mutant phenotypes result from the action of single gene exhibiting Mendelian inheritance, although many pigmentation genes revealed non-Mendelian cytoplasmic inheritance. Due to lack of sexual reproduction, genetic analysis in Euglenophyceae is not possible. Isolated mutant strains of *Euglena* have been found to be useful in photosynthetic studies. The chloroplasts and proplastids are semiautonomous as they need products of nuclear genome for their synthesis.

(IX) ALGAE IN BIOTECHNOLOGY

The controlled or deleberate application of simple and biological agents (living or dead) in technically useful operations is known as **biotechnology**. During the last couple of years, both basic and applied aspects of biotechnology have developed rapidly. It may involve bacteria, algae, fungi, yeasts, cells of the higher plants or animals. These technologies are utilised in several areas like chemical industry, food industry (mass production of yeast, algae and bacteria), pharmaceutical industry, environmental protection and abatement of pollution, agricultural productivity etc.

Biotechnologists are actively engaged in improving the efficiency of microorganisms to fix atmospheric nitrogen and to improve agricultural productivity. The fixation of atmospheric nitrogen by procaryotic organisms is termed as **diazotrophy**. The nitrogen-fixing blue-green algae can be broadly classified into three types : 1) The most important group contains heterocystous filamentous forms which can fix nitrogen both aerobically and anaerobically. 2) Certain nonheterocystous filamentous forms like *Oscillatoria* which can fix it only under anaerobic conditions. 3) Unicellular forms like *Aphanothece*, *Gloeocapsa* which can fix nitrogen both under aerobic and anaerobic conditions.

The enzyme **nitrogenase**, responsible for nitrogen fixation, is sensitive to oxygen. Although a large number of blue-green algae exhibit this phenomenon, the common examples are *Aulosira, Tolypothrix, Nostoc, Anabaena, Mastigocladus* and *Cylindrospermum*. Enzyme nitrogenase is a complex of iron, sulphur and molybdenum. It catalyzes the reduction of N_2 to NH_3. Protein I and protein II are the two subunits of nitrogenase. Protein I contains iron, sulphur and molybdenum while protein II has only iron and sulphur. Both the proteins are sensitive to oxygen. The activity of nitrogenase requires high amounts of ATP, anaerobic condition, magnesium ion and an electron donor (ferredoxin).

The heterocystous species contribute greatly to the fertility of rice fields in tropical countries. Nitrogen fixaton in temperate regions are also exhibited mainly by cyanophycean members rather than by bacteria. The free living diazotrophic Cyanophyceae are the largest contributors of the process of biological nitrogen fixation, the second most important biological process on this planet. Many blue-greens are found is symbiotic association with other plants and the nitrogen fixed by them is largely responsible for such partnership.

The nitrogen-fixing ability was first related to the presence of specialized non-oxygen evolving cells (**heterocysts**). Dinitrogen is first reduced to ammonia by the enzyme nitrogenase (Fig. 13.10). The enzyme

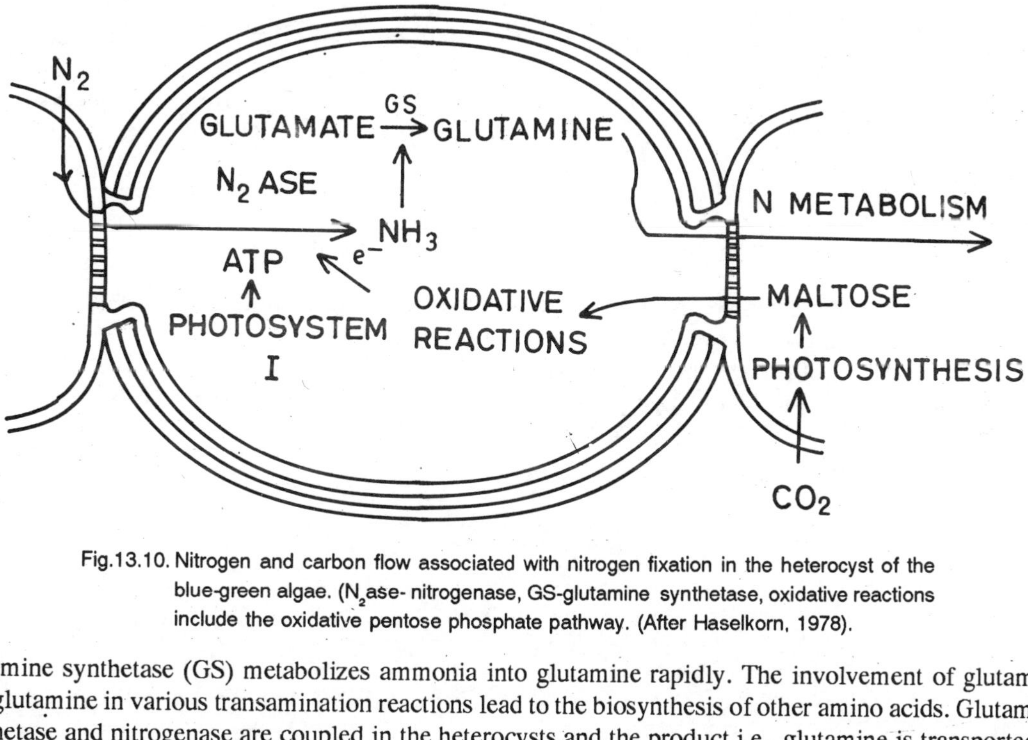

Fig.13.10. Nitrogen and carbon flow associated with nitrogen fixation in the heterocyst of the blue-green algae. (N_2ase- nitrogenase, GS-glutamine synthetase, oxidative reactions include the oxidative pentose phosphate pathway. (After Haselkorn, 1978).

glutamine synthetase (GS) metabolizes ammonia into glutamine rapidly. The involvement of glutamate and glutamine in various transamination reactions lead to the biosynthesis of other amino acids. Glutamine synthetase and nitrogenase are coupled in the heterocysts and the product i.e., glutamine is transported to the nearby vegetative cell. The electron required by nitrogenase is provided by hydrogen gas in the light. It means that hydrogen donates its electron to photosystem I in heterocysts. The transfer of electron is accomplished by a specific transfer proteins (**ferrodoxins** or **flavodoxins**, Fig. 13.11). ATP serves as energy source. The consumption of ATP molecules is 12 to 15 per nitrogen molecule.

The study of nitrogen fixation gens (**nif**) was first carried out in *Klebsiella pneumoniae*. Although the first record came from the mutants of nitrogen-fixing bacteria which was isolated and biochemically characterized in 1959, the organisation of nif-genes has also been studied from many other free-living diazotrophs including a few blue-green algae. Successful attempts are made in transferring nitrogen fixation genes to *Klebsiella pneumoniae*. Later on they could able to transfer these genes to *Escherichia*

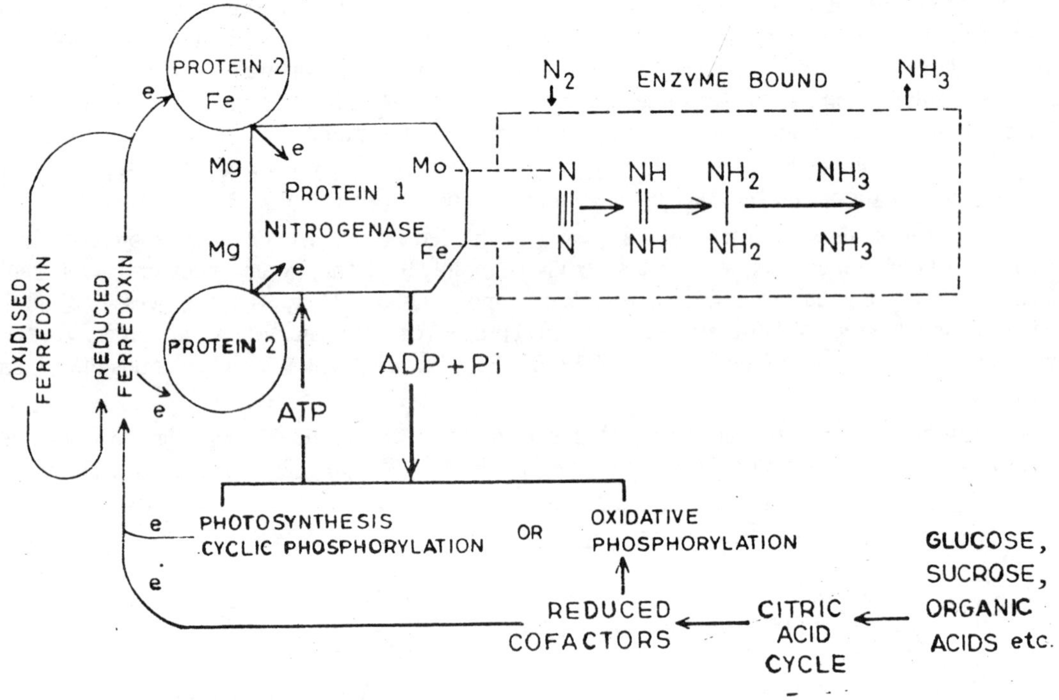

Fig.13.11. A generalized scheme to show the action of nitrogenase.

coli. More than a dozen genes are involved in the assembly of nitrogen fixation apparatus. The regulation of nif-genes is a complex process. Ammonia and nitrogenous compounds like nitrates and amino acids inhibit the expression of nitrogenase activity. The transfer of nif-genes from one species to other may be done by any of the three methods of gene transfer : conjugation, transduction and transformation.

REFERENCES

Ahbel, D. E., Alexander, A.H. and Klein, M.L. (1980) Protothecal olecranon bursitis - a case and review of the literature. *J. Bone. Jt. Surg. Am.* 62 : 835-836.

Asmund, B. and Kristiansen, J. (1986) The genus *Mallomonas* (Chrysophyceae). A taxonomic survey based on the ultrastructure of silica scales and bristles. *Opera Bot.* 85 : 1-128.

Atkinson, A.W. Jr., Gunning, B.E.S. and John, P.C.C. (1972). Sporopollenin in the cell wall of *Chlorella* and other algae: ultrastructure, chemistry and incorporation of ^{14}C acetate, studied in synchronous cultures. *Planta* 107 : 1-32.

Bazin, M.J. (1968) Sexuality in a blue green alga : genetic recombination in *Anacystis nidulans. Nature* 218 : 282-283.

Berdach, J.T. (1977) In situ preservation of the transverse flagellum of *Peridinium cinctum* (Dinophyceae) for scanning electron microscopy. *J. Phycol.* 13 : 243-251.

Beskow, R. (1978) Silicosis in diatomaceous earth factory workers in Sweden. *Scand. J. Resp. Dis.* 59 : 216-221.

Bold, H.C. and Wynne, M.J. (1985) *Introduction to the Algae. Structure and Reproduction.* 2nd ed. Printice-Hall, Englewood Cliffs, New Jersey. 720 pp.

Bouck, G.B. (1971) The structure, origin, isolation and composition of the tubular mastigonemes of the *Ochromonas* flagellum. *J. Cell Biol.* 50 : 362-384.

Bouck, G.B. and Sweeney, B.M. (1966) The fine structure and ontogeny of trichocysts in marine dinoflagellates. *Protoplasm* 61 : 205-223.

Bronchart, R. and Demoulin, V. (1977) Unusual mitosis in the red alga *Porphyridium Purpureum. Nature* 268 : 80-81.

Brook, A. J. (1980) Barium accumulation by desmids of the genus *Closterium* (Zygnemaphyceae). *Br. Phycol. J* . 1 : 262-264.

Cairns, J. Jr., Albaugh, D.W., Busey, F. and Chanay, M.D. (1968) The sequential comparison index - a simplified method for nonbiologists to estimate relative difference in biological diversity in stream pollution studies. *J. Wat. Pollut. Control Fed.* 40 : 1607-1613.

Carmichael, W.W. (ed.) (1981) *Water Environment. Algal Toxins and Health.* Environmental Science Research Vol. 20 : International Conference on Toxic Algae. Plenum Press, New York.

Carmichael, W.W. and Gorham, P.R. (1981) The mosaic nature of toxic blooms of Cyanobacteria. In *Water Environment. Algal Toxins and Health.* (Carmichael, W.W., ed). Environmental Science Research Vol. 20, pp. 161-172. International Conference on Toxic algae. Plenum Press, New York.

Cassin, P.E. (1974) Isolation, growth and physiology of acidophilic Chalamydomonads. *J. Phycol.* 10 : 439-447.

Caubergs, R. and de Greef, J. (1971) Pigments studies in the genus *Vaucheria. Bot. Mag. Tokyo* 84 : 222-230.

Chapmann, V. J. and Chapman, D.J. (1973) *The Algae.* 2nd ed. Macmillan, New York. 497 pp.

Chapman, V.J. and Chapman, D.J. (1980) *Seaweeds and Their uses.* 3rd ed. Chapman & Hall, London. 334 pp.

Christensen, T. (1962) Alger. In *Systematisk Botanik*, Vol. 2, No. 2 (Bocher, T.W., Lange, M. and Sorensen, T., eds.) Munksgaard, Copenhagen. 178 pp.

Clayton, M.N. (1974) Studies in the development, life history and taxonomy of the Ectocarpales in southern Australia. *Aust. J. Bot.* 22: 743-813.

Coute, A. (1971) Sur le cycle morphologique du *Liagora tetrasporifera* compare a celui du *Liagora distenta* (Rhodophycees, Nemalionales, Helminthocladiacees). C.R. Acad. Sci. (Paris). 273 D : 626-629.

Crossett, R.N., Drew, E.A. and Larkum, A.W.D. (1965) Chromatic adaptation in benthic marine algae. *Nature* 207 : 547-548.

Dodge, J.D. (1969) A review of the fine structure of algal eyespots. *Brit. Phycol. J.* 4 : 199-210.

Dodge, J.D. (1973) *The fine structure of Algal cells.* Academic Press, London 261 pp.

Dodge, J.D. (1974) Fine structure and phylogeny in the algae. *Sci. Prog. (oxford)* 61 : 257-274.

Dodge, J.D. (1984) The functional and phylogenetic significance of dinoflagellate eyespots. *Biosystems* 16 : 259-267.

Dodge, J.D. and Crawford, R.M. (1968) Fine structure of the dinoflagellate *Amphidinium carteri* Hulbert. *Protistologica* 4 : 231-242.

Drew, K.M. (1955) Life histories in the algae with special reference to the Chlorophyta, Phaeophyta and Rhodophyta. *Biol. Rev.* 30 : 348-390.

Evans, L.V., Callow, J.A. and Callow, M.E. (1973) Structural and physiological studies of the parasitic red algae *Holmsella. New Phytol.* 72 : 393-402.

Fagerberg, W.R. and Dawes, C.J. (1973) An electron microscopic study of the sporophytic and gametophytic plants of *Padina vickersiae Hoyt. J. Phycol.* 9 : 199-204.

Fan, K.C. (1961) Studies on *Hypneocolax* with a discussion on the origin of parasitic red algae. *Nova Hedw.* 3 : 119-128.

Faust, M.A. and Gantt, E. (1973) Effect of light intensity and glycerol on the growth, pigment composition and ultrastructure of *Chroomonas* sp. *J. phycol.* 9 : 489-495.

Feldmann, J.(1952) Les cycles de reproduction des algues et leurrapports avec la phylogenie. *Rev. Cytol. Biol. Veg.* 13: 1-49.

Feldmann, J. and Feldmann, G. (1958) Recherches sur quelques Floridees parasites. *Rev. Gen. Bot.* 65 : 49-128.

Fenical, W. (1980) Distributional and taxonomic features of toxin producing macroalgae. In *Pacific Seaweed Aquaculture* (Abbott, I.A., Foster, S. and Eklund, L. F., eds.) pp. 144-151. Institute of Marine Resources, University of California, La Jolla.

Fogg, G.E.(1951) *Ann. Bot. Lond. N.S.* 15: 23-36.

Fogg, G.E., Stewart, W.D.P., Fay, P. and Walsby, A.E. (1973) *The Blue-green Algae.* Academic Press, London. 459 pp.

Fott, B. (1962) Toxonomy of *Mallomonas* based on electron micrographs of scales. *Preslia* 34 : 69-84.

Fott, B. (1971) Algenkunde. 2nd ed. Gustav Fischer, Jena. 581 pp.

Fraser, T.W. and Gunning, B.E.S. (1969) The ultrastructure of the plasmodesmata in the filamentous green alga, *Bulbochaete hiloensis* (Nordst.) Tiffany. *Planta* 88 : 244-254.

Frederick, J.F. (1951) Preliminary studies on the synthesis of polysaccharides in the algae. *Physiol. Plant.* 4 : 621-626.

Fritsch, F.E. (1935) *The Structure and Reproduction of the algae.* Vol. I. Cambridge University Press, Cambridge. 791 pp.

Fritsch, F.E. (1945) *The Structure and Reproduction of the Algae.* Vol. II. Cambridge University Press, Cambridge. 939 pp.

Fritsch, F.E. (1951) The heterocyst : a botanical enigma. *Proc. Linn. Soc. Lond.* 162 : 194-211.

Gallois, R.W. (1976) Coccolith blooms in the Kimmeridge clay and origin of North Sea oil. Nature 259 : 473-475.

Gantt, E. (1969) Properties and ultrastructure of phycoerythrin from *Porphyridium cruentum. Plant Physiol.* 44 : 1629-1638.

Gantt, E. (1971) Micromorphology of the periplast of *Chroomonas* sp. (Cryptophyceae). *J. Phycol.* 7 : 177-184.

Gantt, E. and Conti, S.F. (1965) The ultrastructure of *Porphyridium cruentum*. *J. Cell Biol.* 26 : 365-381.

Gillott, M.A. and Gibbs, S.P. (1980) The cryptomonad nucleomorph : its ultrastructure and evolutionary significance. *J. Phycol.* 16 : 558-568.

Godward, M.B.E. (1966) *The Chromosomes of the Algae*. Edward Arnold, London 212 pp.

Gordon, R. and Drum, R. (1970) A capillarity mechanism for diatom gliding locomotion. *Proc. Natl. Acad. Sci. USA* 67 : 338-345.

Gordon, D.C. Jr. and Prouse, N.J. (1973) The effects of three oils on marine phytoplankton photosynthesis. *Mar. Biol.* 22 : 329-333.

Grambast, L.J. (1974) Phylogeny of the Charophyta. *Taxon* 23 : 463-481.

Griffiths, D.W. (1970) The Pyrenoid. *Bot. Rev.* 36 : 29-58.

Halfen, L.N. (1973) Gliding motility of *Oscillatoria* : ultrastructural and chemical characterization of the fibrillar layer. *J. Phycol.* 9 : 248-253.

Harper, M.A. (1977) Movements. In *The Biology of Diatoms* (Werner, D., ed.). pp. 224-249. Blackwell Scientific, Oxford.

Harper, M.A. and Harper, J.F. (1967) Measurements of diatom adhesion and their relationship with movement. *Br. Phycol. J.* 3 : 195-207.

Haupt, W. and Schonbohm, E. (1970) Light oriented chloroplast movements. In *Photobiology of Microorganisms* (Halldal, P., ed.) pp. 283-307. Wiley Interscience, London.

Hendey, N.I. (1964) *An introductory Account of the smaller Algae of British Coastal Waters. Part. V. Bacillariophyceae (Diatoms)*. Ministry of Agric. Fisheries and Food, Fishery Investigation Ser.IV. 317 pp.

Hibberd, D.J. (1976) The ultrastructure and taxonomy of the Chrysophyceae and Prymnesiophyceae (Haptophyceae). A survey with some new observations on the ultrastructure of the Chrysophyceae. *Bot. J. Linn. Soc.* 72 : 55-80.

Hibberd, D.J. and Leedale, G.F. (1971) Cytology and ultrastructue of the Xanthophyceae. II. The zoospore and vegetative cell of coccoid forms, with special reference to *Ophiocytium majus* Naegeli. *Br. Phycol. J.* 6 : 1-23.

Hopkins, J.M. (1970) *J. Cell. Sci.* 823-839.

Hovasse, R., Mignot, J.P. and Joyon, L. (1967) Nouvelles observations sur les trichocystes des Cryptomonadines et les "R bodies" des particules kappa de *Paramecium aurelia* Killer. *Protistologica* 3 : 241-255.

3 : 241-255.

Jarosch, R. (1960) *Phyton (Buenos Aires)*. 15 : 43-66.

Jarosch, R. (1962) In *The physiology and biochemistry of algae* (Lewin, R.A., ed.) pp. 573-581. New York and London.

Kamat, N.D. and Patel, M.Z. (1973) *Botanique* 4 : 101.

Kersey, Y.M. and Wessels, N.K. (1976) Localization of actin filaments in internodal cells of characean algae. *J. Cell Biol.* 68 : 264-275.

Kolkwitz, R. and Marsson, M. (1908) Okologie der pflanzlichen Saprobien. *Ber. dt. bot. Ges.* 26 : 505-519.

Kristiansen, J. (1981) Distribution problems in the Synuraceae. *Verh. int. Verein. theor. angew. Limnol.* 21 : 1444-1448.

Kristiansen, J.(1986) Silica-scale bearing chrysophytes as Environmental Indicators. *Br. Phycol. J.* 21: 425-436.

Kumar, H.D. (1962) Apparent genetic recombination in a blue-green alga. *Nature*, Lond. 196 : 1121-1122.

Kylin, H. (1933) Lunds Univ. Arsskr, N.F. 29 : (2) No. 7.

Lang, N.J. and Fay, P (1971) The heterocysts of blue-green algae. II. Details of ultrastructure. *Proc. R. Soc. Lond.* B 178 : 193-203.

Lazaroff, N. (1973) Photomorphogenesis and Nostocaceen development. In *The Biology of Blue-green Algae* (Carr, N.G. and Whitton, B.A., eds.). University of California Press. Berkeley. 676 pp.

Leadbeater, B.S.C. (1970) Preliminary observations on differences of scale morphology at various stages in the life cycle of *'Apistonema - Syracosphaera' sensu* von Stosch. *Br. Phycol. J.* 5 : 57-69.

Leadbeater, B.S.C.(1971) Observations by means of cine photography on the behaviour of the haptonema in plankton flagellates of the class Haptophyceae. *J. Mar. Biol. Assoc. U.K.* 51: 207-217.

Leadbeater, B.S.C. (1986) Scale-case construction in *Synura petersenii* (Chrysophyceae). In *Chrysophytes - Aspects and Problems* (Kristiansen, J and Anderson, R.A.,eds.) pp. 121-131. Cambridge University Press.

Lee, R.E. (1971) The pit connections of some lower red algae; Ultrastructure and phylogenetic significance. *Br. Phycol. J.* 6 : 29-38.

Lee, R.E. (1978) Formation of scales in *Paraphysomonas vestita* and the inhibition of growth by germanium dioxide. *J. Protozool.* 25 : 163-166.

Lee, R.E. (1980) *Phycology*. Cambridge University Press, Cambridge. 478 pp.

Leedale, G.F. (1967) *Euglenoid Flagellates*. Printice Hall, Inc., Englewood Cliffs, N.J. 242 pp.

Leedale, G.F. (1982) Ultrastructue. In *The Biology of Euglena* Vol.III, Physiology (Buetow, D.E., ed.) pp. 1-27. Academic Press, New York.

Lewin, R.A. (1977) *Prochloron* - type genus of the Prochlorophyta. *Phycologia* 16 : 217.

Lewin, R.A. and Gibbs, S.P. (1982) Algae of uncertain position : introduction and bibliography. In *Selected Papers in Phycology II* (Rosowski, J.R. and Parker, B.C., eds.) pp. 659-662. Phycological Society of America, Allen Press, Lawrence, kansas.

Linnaeus, C. (1754) *Genera plantarum*., Ed. 5. Holmiae.

Lucas, I.A.N. (1970) Observations on the fine structure of the Cryptophyceae. I. The genus. *Cryptomonas. J. Phycol.* 60 : 30-38.

Magne, F. (1964) Recherches caryologiques chez les Floridees (Rhodophycees). Cahiers Biol. Mar. 5 : 461-671.

Marchant, H.J. (1972) Pyrenoids of *Vaucheria woroniniana* Heering. *Br. Phycol. J.* 7 : 81-84.

Metting, B. (1981) The systematics and ecology of soil algae. *Bot. Rev.* 47 : 195-312.

Migaki, G., Garner, F.M. and Imes, G.D. (1969) Bovine Protothecosis. *Pathol. Vet.* 444-453.

Mignot, J.-P. (1967) Structure et ultrastructure de quelques Euglenomonadines. *Protistologica* 2 : 51-113.

Mignot, J.-P. and Brugerolle, G. (1982) Scale formation in chrysomonad flagellates. *J. Ultrastruct. Res.* 81 : 13-26.

Mix, M. (1969) Zur Feinstruktur der Zellwande in der Gattung *Closterium* (Desmidiaceae) unter besonderer Beruck-sichtigung des Porensystems. *Arch. Mikrobiol.* 68 : 306-325.

Mix, M. (1975) Die Feinstruktur der Zellwande der Conjugaten und ihre systematische Bedeutung. Beih. *Nova Hedw.* 42 : 179-194.

Moikeha, S.N. and Chu, G.W. (1971) Dermatitis producing alga *Lyngbya majuscula* Gomont in Hawaii. II. Biological properties of the toxic factor. *J. Phycol.* 7 : 8-13.

Morse, D.E., Morse, A., Duncan, H. and Trench, R.K. (1981). Algal tumors in the Caribbean Octocorallian *Gorgonia ventalina*. II. Biochemical characterization of the algae, and first epidemiological observations. *Bull. Mar. Sci.* 31 : 399-409.

Nultsch, W. (1956) Studien uber die Phototaxis der Diatomeen. *Arch. Protistenk* 101 : 1-68.

Nultsch, W. (1974) Movements. In *Algal Physiology and Biochemistry* (Parker, W.D.P., ed.). University of California Press, Berkeley. 989 pp.

Nygaard, G. (1949) Hydrobiological studies on some Danish ponds and lakes. Part II : The quotient hypothesis and some new or little known phytoplankton organisms. *Kgl. Dan. Vid. Selsk. Biol. Skr.* 7 : 1-293.

Orkwizewski, K.G. and Kaney, A.R. (1974) Genetic transformation of the blue-green bacterium, *Anacystis nidulans, Arch. Mikrobiol.* 98 : 31-37.

Ott, D.W. and Brown, R.M., Jr. (1974) Developmental cytology of the genus *Vaucheria*. I. Organization of the vegetative filament. *Br. Phycol. J.* 9 : 111-126.

Palmer, C.M. (1969) A composite rating of algae tolerating organic pollution. *J. Phycol.* 5 : 78-82.

Palmer, C.M. (1977) *Algae and water Pollution : An Illustrated Manual on the Identification, Significance and Control of Algae in Water Supplies and Polluted Water.* EPA-600/9-77-036. US Environmental Protection Agency, Cincinnati, U.S.A.

Pankratz, H.S. and Bowen, C.C. (1963) Cytology of blue-green algae. I. The cells of *Symploca muscorum. Am. J. Bot.* 50 : 387 - 399.

Papenfuss, G.F. (1966) A review of the present system of classification of the Florideophycidae. *Phycologia* 5 : 247-255.

Parker, S.P. (ed.) (1982) *Synopsis and Classification of living Organisms,* Vols 1 & 2. Mc Graw-Hill, New York.

Pascher, A. (1918) Uber die Beziehung der Reduktionsteilung zur Mendelschen Spaltung. *Ber. Dtsch. Bot. Ges.* 36 : 163 - 168.

Pascher, A. (1931) Systematische Uebersicht uber die mit Flagellaten in Zusammenhang stehenden Algenreihen, etc. *Beih. Bot. Centralbl.* 48 : 317 - 332.

Patrick, R. (1977) *In The Biology of Diatoms.* pp. 284-332. Blackwell, London.

Patrick, R. and Reimer, C.W. (1966) *The Diatoms of the United States exclusive of Alaska and Hawaii. Vol. I. Fragilariaceae, Eunotiaceae, Achnanthaceae, Naviculaceae.* Monogr. Acad. Nat. Sci. Phila. 13. 688 pp.

Phillips, T.L., Miklas, K.J. and Andrews, H.N. (1972) Morphology and vertical distribution of *Protosalvinia (Foerstia)* from the New Albany shale (Upper Devonian). In *Advances in Paleozoic Botany* (Streel, M., Bonamo, P.M. and Fairon - Demaret, M., eds.) pp. 171 - 196. Elsevier, New York.

Pickett - Heaps, J.D. (1970) Some ultrastructural features of *Volvox* with particular reference to the phenomenon of inversion. *Planta* 90 : 174 -190.

Pickett - Heaps, J.D. (1974) Cell division in *Bulbochaete*. II. Hair cell formation. *J. Phycol.* 10 : 148 - 164.

Porter, J. and Jost, M. (1976) Physiological effects of the presence and absence of gas vacuoles in the blue-green alga *Microcystis aeruginosa* kuetz. emend. Elenkin. *Arch. Microbiol.* 110 : 225 - 231.

Prescott, G.W. (1969) *The Algae : A Review.* Thomas Nelson and sons Ltd. London. 436 pp.

Puiseux - Dao, S. (1970) *Acetabularia* and Cell Biology. Logos Press, 162 pp.

Roberts, K., Gurney-Smith, M. and Hills, G.J. (1972) Structue, composition and morphogenesis of the cell wall of *Chlamydomonas reinhardii*. I. Ultrastructue and preliminary chemical analyses. *J. Ultrastr. Res.* 40 : 599 - 613.

Ross, P.E. and Duthie, H.C. (1981) Ultraplankton biomass, productivity and efficiency in Lac Matamee : a Precambrian Shield Lake. *J. Phycol.* 17 : 181 - 186.

Round, F.E. (1963) The taxonomy of the Chlorophyta. *Br. Phycol. Bull.* 2 : 224 - 235.

Round, F.E. (1971) The taxonomy of the Chlorophyta. II. *Br. Phycol. J.* 6 : 235 - 264.

Round, F.E. (1973) *The Biology of the Algae.* St. Martin's Press, New York. 278 pp.

Round, F.E. (1981) Morphology and phyletic relationships of the silicified algae and the archetypal diatom - monophyly of polyphyly. In *Silicon and Siliceous Structures in Biological Systems* (Simpson, T.L. and Volcani, B.E., eds.) pp. 97 - 128. Springer-Verlag, New York.

Sager, R. and Palde, G.E. (1957). *J. Biophys. Biochem. Cytol.* 3 : 463 - 487.

Saha, L.C. and Wujek, D.E. (1990) Scale-bearing chrysophytes from tropical Northeast India. *Nord. J. Bot.* 10 : 343-354.

Santore, U.J. (1977) Scanning electron microscopy and comparative micromorphology of the periplast of *Hemiselmis rufescens, Chroomonas* sp. and members of the genus *Cryptomonas* (Cryptophyceae). *Brit. Phycol. J.* 12 : 255 - 270.

Sandgren, C.D. (1981) Characteristics of sexual and asexual resting cyst (statospore) formation in *Dinobryon cylindricum* Imhof (Chrysophyta). *J. Phycol.* 17 : 199 - 210.

Sarma, Y.S.R.K. (1983) Contribution to the study of algae with particular reference to their karyology. pp. 1 - 11. In *Sixth Botanical Conference, Bhuwaneshwar,* India.

Say, P.J. and Whitton, B.A. (1978) Chemistry and benthic algae of a zinc polluted stream in the northern Pennines. *Brit. Phycol. J.* 13 : 206.

Schopf, J.W. (1970) Precambrian micro-organisms and evolutionary events prior to the origin of vascular plants. *Biol. Rev.* 45 : 319 - 352.

Sepsenwol, S. (1973) Leucoplast of the Cryptomonad *Chilomonas paramecium. Exp. Cell Res.* 76 : 395 - 409.

Shannon, C.E. and Weaver, W. (1963) *The Mathematical Theory of communication.* University of Illinois Press. 117 pp.

Shestakov, S.V. and Khyen, N.G. (1970) Evidence for genetic transformation in blue- green algae *Anacystis nidulans. Molec. gen. Genet.* 107 : 372 - 375.

Shukla, A.C (1967) Effect of algal hormones on stomatal and epidermal development in rice leaves. *Hydrobiologia* 30 : 221 - 224.

Silva, P.C. (1982) Thallobionta. In *Synopsis and Classification of Living Organisms,* Vol. 1 (Parker, S.P., ed.) pp. 59-60. McGraw-Hill, New York.

Simon, R.D. (1973) Measurement of the cyanophycin granule polypeptide contained in the blue-green alga *Anabaena cylindrica. J. Bacteriol.* 114 : 1213-1216.

Simonsen, R. (1972) Ideas for a more natural system of the centric diatoms. Beih. *Nova Hedw.* 39 : 37 - 54.

Singh, R.N. (1961) *Role of Blue-green Algae in nitrogen economy of Indian Agriculture.* ICAR Monograph, New Delhi. 175 pp.

Singh, R.N. and Sinha, R. (1965) Genetic recombination in blue-green algae *Cylindrospermum majus. Nature* 207 : 782.

Sladecek, V. (1973) System of water quality from a biological point of view. *Arch. Hydrobiol., Beih. Ergeb. Limnol.* 7 : 1 - 128.

Smith, G.M. (1950) *The Fresh-water Algae of the United States.* 2nd ed. McGraw-Hill, New York. 719 pp.

Smith, G.M. (1955) *Cryptogamic Botany.* Vol. I. McGraw Hill, New York. 546 pp.

South, G.R. and Whittick, A. (1987) *Introduction to Phycology.* Blackwell Scientific Publications. 341 pp.

Stanier, R.Y., Kunisawa, R., Mandel, M. and Cohen - Bazire, G. (1971) Purification and properties of unicellular blue-green algae (order Chroococcales). *Bacteriol. Rev.* 35 : 171 - 205.

Stanier, R.Y. and Van Niel, C.B. (1962) *Arch. Mikrobiol.* 42 : 17 - 35.

Staub, R., Appling, J.W., Hofstetter, A.M. and Haas, I.J. (1970) The effects of industrial wastes of Memphis and Shelby County on primary planktonic producers. *Bioscience* 20 : 905 - 912.

Steidinger, K.A. (1973) Phytoplankton ecology : a conceptual review based on eastern Gulf of Mexico Research. *CRC Crit. Rev. Microbiol.* 3 : 49 - 68.

Steidinger, K.A. and Joyce, E.A. Jr. (1973) Florida red tides. *Fla. Dept. Natl. Resourc. Mar. Res. Lab. Educ. Ser.* 17 . 26 pp.

Stewart, K.D. and Mattox, K.R. (1975) Comparative Cytology, evolution and classification of the green algae with some consideration of the origin of other organisms with chlorophylls a and b. *Bot. Rev.* 41 : 104 - 135.

Swingle, H.S. (1966) In *FAO Warm-Water Pond Fish Symposium*, Rome.

Tamiya, H. (1960) Role of algae as food. In *Proceedings of Symposium Algology*. pp. 379-389. ICAR, New Delhi.

Taylor, F.J.R. (1975) Non-helical transverse flagella in dinoflagellates. *Phycologia* 14 : 45 - 47.

Taylor, D.L. and Lee, C.C. (1971) A new cryptomonad from Antarctica : *Cryptomonas cryophila* sp. nov. *Arch. Mikrobiol.* 75 : 260 - 280.

Thunmark, S. (1945) Zur Soziologie des Susswasser planktons. Eine methodologischokologische Studie. *Folia Limnol. Scandin.* 3:66 pp.

Tischer, I. (1957) Untersuchungen uber die granularen Eihschlusse und das Reduktions-Oxydations-Vermogen der Cyanophyceen. *Arch. Mikrobiol.* 27 : 400 - 428.

Trainor, F.R. (1978) *Introductory Phycology*. John Wiley, New York. 525 pp.

von Stosch, H.A. (1967) Haptophyceae. In : Vegetative Forplanzung, Parthenogenese und Apogamie bei Algen (Ruhland, W., ed.) *Encyclopedia of Plant Physiology* 18 : 646 - 656.

Volcani, B.E. (1981) Cell wall formation in diatoms : morphogenesis and biochemistry. In *Silicon and Siliceous Structures in Biological Systems* (Simpson, T.L. and Volcani, B.E., eds.) pp. 157 - 200. Springer - Verlag, New York.

Walsby, A.E. (1973) Gas vacuoles. In *The Biology of the Blue-Green Algae* (Carr, N.G. and Whitton, B.A., eds.) pp. 340 - 352. Blackwell, Oxford.

Wang, W.S. and Tischer, R.G. (1973) study of the extracellular polysaccharides produced by a blue-green alga, *Anabaena flos-aquae* A - 37. *Arch. Mikrobiol.* 91 : 77 - 81.

Whittle, S.J. and Casselton, P.J. (1969) The chloroplast pigments of some green and yellow-green algae. *Br. Phycol. J.* 4 : 55 - 64.

Wik -Sjostedt, A. (1970) Cytogenetic investigations in *Cladophora*. *Hereditas* 66 : 233 - 262.

Wildman, R.B., Loescher, J.H. and Winger, C.L. (1975) Development and germination of akinetes of *Aphanizomenon flos-aquae*. *J. Phycol.* 11 : 96 - 104.

Wujek, D.E. (1978) Ultrastructure of flagellated chrysophytes. III. *Mallomonas caudata*. - *Trans. Kansas Acad. Sci.* 81 : 327 -336.

Wujek, D.E. and Kristiansen, J. (1978) Observations on bristle - and scale production in *Mallomonas caudata* (Chrysophyceae). *Arch. Protistenk.* 120 : 213 - 221.

GLOSSARY

Acropetal : toward the apex.

Acronematic : flagellum that is smooth and terminated with a fine fibril.

Adelphoparasites : parasites that are closely related to their host.

Agar : gelling compound (polysaccharide) derived from certain red algal walls. It is used in the preparation of solidified culture media.

Akinete: thick walled resting spore.

Algicide : one of many chemicals or agents used to kill algae, especially troublesome forms.

Alginate : salt form of alginic acid such as with calcium or barium ions.

Allelopathy : harmful direct or indirect effect of one organism on another by production of a chemical which escapes into environment.

Alloparasites : parasites not closely related to their host.

Amphiesma : outer covering of a dinophycean cell comprised of only membranes in naked species or thecal plates enclosed in membranes in armored species.

Amylum star : star-shaped aggregate of cells filled with starch that forms new plants in Charophyta.

Androsporangium : sporangium that forms androspores.

Androspore : spore that forms a dwarf male filament in the Oedogoniales (Chlorophyceae).

Anisogamy : sexual fusion between flagellated gametes of distinctly different sizes.

Antapical : opposite from the apex.

Antheridium : reproductive structure in which sperms are produced.

Antherozoid : a flagellated motile male gamete.

Apical growth : growth by means of apical cell.

Aplanosporangium : a cell or sporangium containing aplanospores.

Aplanospore : non-motile spore in which the spore wall is not fused to the vegetative wall of the cell which producted it.

Apochlorotic : colourless.

Aquaculture : the science of growing aquatic organisms in the laboratory or field under artificial conditions, often utilizing human waste products in liquid form for nutrients.

Archegonium : multicellular female gametangium producing egg and consisting of neck and venter. It includes sterile cells in addition to fertile egg.

Areolae : cavity like depressions covered by a perforated membrane found in diatoms.

Autocolony : A newly formed coenobium with the same form as the parent colony.

Autospore : aplanospore with the same shape as the parent cell.

Autotroph : Not needing an external source of organic compounds as an energy source. Energy is obtained from light or inorganic chemical reactions.

Auxiliary cell : cell that receives the diploid nucleus from fertilization in the Rhodophyceae.

Auxospore : the zygote of diatoms.

Auxotrophic : requiring an exogenous source of one or more vitamins for growth.

Axenic culture : pure culture.

Axoneme : the 9+2 complex of microtubules contained within the flagellum and enclosed by a surrounding membrane.

Basal body : bottom part of a flagellum beneath the transition zone. Basal bodies divide to perpetuate flagella.

Basipetal : toward the base.

Benthos (Benthic) : bottom-dwelling nonplanktonic organisms attached to or resting on the substrate.

Bilaterally symmetrical : symmetrical about a line.

Binary fission : division into two products.

Bioluminescence : emission of light by a living organism.

Biomass : the amount of living matter in a habitat at any one time.

Bisporangium : a sporangium whose contents divide into two spores.

Blade : the broad membranous distal portions of kelp plants or foliaceous algae like *Ulva.*

Bloom : heavy growth of planktonic algae in a body of water.

Brackish : saline water with a salinity less than that of seawater.

Bristle : long, thin cytoplasmic projection through pectic layer of certain algae.

Brittlewort : a name applied to calcareous members of the Charophyta.

Calcification : deposition of calcium carbonate usually in association with smaller amounts of other carbonates.

Calciphilic : organisms well adapted to calcium rich water.

Calciphobe : organism which grows poorly in areas high in calcium salts.

Calcite : rhombohedral crystals of calcium carbonate.

Callose : polysaccharide associated with pores in sieve cells.

Canal : a tube-like area connecting the anterior invagination or reservoir to the surrounding medium in euglenoids.

Capitulum : a cell in the antheridium of Charophyta from which the antheridial filaments arise.

Carotene : oxygen-free, unsaturated, hydrocarbon carotenoid.

Carpogonium : female gametangium in the Rhodophyceae.

Carposporangium : carpospore-producing sporangium derived directly or indirectly from the zygote in the Rhodophyceae.

Carpospore : usually diploid spore produced by the carposporangium in the Rhodophyceae.

Carposporophyte : usually diploid generation in the Rhodophyceae derived from the zygote and which forms the carpospores.

Carrageenin : red algal polysaccharide similar to agar but requires higher concentration to form gel.

Caulerpicin : a toxin produced by *Caulerpa.*

Cellulose : polysaccharide composed of ß-1,4 linked glucose molecules that forms the main skeletal framework of most algal cell walls.

central nodule : a wall swelling that divides the raphe into two parts in the Bacillariophyceae.

Centrosome : polar body at each pole of dividing cell. Spindle is seen to diverge from these bodies.

Chemotaxis : movement of an organism toward or away from a chemical stimulus.

Chitin : polysaccharide made up of repeating units of N-acetylglucoseamine.

Chlorophyll : fat-soluble, green, porphyrin-type pigment.

Chloroplast : plastid with chlorophyll.

Chloroplast endoplasmic reticulum : one or two membranes surrounding the chloroplast envelope. Ribosomes are usually attached to the outside of the outer membrane.

Chromatic adaptation : the ability to change the relative amounts of the various photosynthetic pigments to reduce or increase the absorption of light and in response to light quality.

Chromatophore or **Chromoplast** : a chloroplast with some other colour than green.

Chrysolaminarin or **Leucosin** : a liquid polysaccharide storage product composed principally of ß-1,3 linked residues of glucose.

Cilia : flagella.

Cingulum or **Girdle** : transverse furrow in the Dinophyceae containing the transverse flagellum.

Circadian rhythm : repeated sequence of events that occur at about 24-hour intervals.

Clone : group of individuals produced asexually from single individual.

Coccolith : calcareous, minute and ornamented scale on cell surface of a coccolithophore.

Coccolithophores : unicellular organisms of the order Coccosphaerales (Prymnesiophyceae) which bear variously sculptured calcareous scales (coccoliths) upon their cell surface.

***Codiolum* stage** : a unicellular diploid stage in some green algae formed by enlargement of a zygote which remains photosynthetic often elongate or club-shaped.

Coenobium : colony of algal cells in a specific arrangement and number that does not increase once mature.

Coenocyte : large multinucleate cell without cross walls except where reproductive bodies are formed.

Compensation depth : depth in a body of water where photosynthetic production is equal to respiration for a species.

Conceptacle : nearly spherical cavity containing reproductive structures such as in Fucales.

***Conchocelis* stage** : a filamentous, microscopic phase in the life cycle of some Bangiales (Rhodophyceae).

Conchospore : a spore produced and released singly by a conchosporangium.

Conjugation : connection of cell by papillae and tubes through which amoeboid gametes move or in which they conjugate.

Connecting band : part of the girdle in the Bacillariophyceae.

Contractile vacuole : vacuole fed by smaller vesicles that expels water and solutes rhythmically to the outside of the cell.

Coralline : calcified algae.

Corona : 5 or 10 cells present at the apices of oogonia of Charophyta.

Corps de Maupas : a vesicular body in the Cryptophyceae used in the digestion of unwanted cell components.

Cortex : outer part of an algal thallus.

Cosmopoliton : found in all or most part of the world.

Costa : an elongated solid thickening of the valve of a diatom frustule.

Cruciate : having shape of a cross.

Cryptostomata : flask-like opening in the thallus with hairs.

Cyanophhage : virus in the cells of the Cyanophyceae.

Cyanophycin granule : proteinaceous food reserve found in granular form in blue-green algae.

Cyst : a thick-walled dormant cell.

Cystocarp : the carposporophyte and surrounding gametophytic tissue (pericarp) in Rhodophyceae.

Cytokinesis : process of division of the cytoplasm.

Delophycean : the macroscopic or conspicuous phase in the life history of an alga.

Dendroid : a type of nonmotile colony which produces mucilage in one area and usually forms a stalk.

Detritus : particulate organic matter.

Diastole : filling of a contractile vacuole.

Diatomaceous earth : a deposit consisting of the remains of the silicified frustules of the Bacillariophyceae.

Dichotomy : division of a thallus into two equal branches.

Dictyosome ; stack of vesicles in a Golgi apparatus.

Diffuse growth : generalized and no localized growth.

Dimorphic : where gametophytic and sporophytic generations are morphologically distinct.

Dioecious : an organism that has male and female gametes borne on separate plants.

Diplobiontic : having two free-living phases in the life history of an organism.

Discobolocyst : type of projectile in the Chrysophyceae.

Distal : opposite of basal.

Distichous : arranged in two equal rows.

Diurnal : daily periodicity.

Dorsiventral : flattened in one plane.

Dulse : a preparation of *Rhodymenia* (red algae) used as food.

Dystrophy : condition in which water has a high content of organic material (humus) and brown colour.

Ecad (or **Ecotype** or **Ecophene**) : a form of a plant species produced in response to a particular habitat. The modification is not heritable.

Ecdysis : a process in which the thecal wall layer of a dinoflagellate cell is shed.

Echinate : spiny.

Ectoderm : outer protoplasm next to the plasma membrane in the Charophyta.

Ectoparasite : a parasite living on the outside of its host.

Egg : large nonmotile female gamete.

Ejectosome : type of projectile found in the Cryptophyceae and Prasinophyceae.

Electron-dense or **Electron-opaque** : term used to describe a material that absorbs electrons and appears dark in electron micrographs.

Electron-transparent : term used to describe a material that does not absorb electrons and appears light in electron micrographs.

Endemic : occurring in a specific area.

Endocytosis : phagotrophic engulfment of particulate matter.

Endoderm : inner protoplasm next to vacuole in the Charophyta capable of cytoplasmic streaming.

Endolithic : living inside rock.

Endosome : nucleolus in the Euglenophyceae.

Endospore : asexual spore in the Cyanophyceae formed by the internal division of a cell.

Endosymbiotic : an organism living inside a host in a symbiosis.

Endozoic : living inside an animal.

Enucleate : without a nucleus.

Epicone : part of a cell above the girdle in the Dinophyceae.

Epidermis : outer layer of cells.

Epipelic : growing on mud or sand.

Epiphyte : plant living on another plant.

Epitheca : larger of the two halves of a diatom frustule.

Epizoic : living on an animal.

Estuary : the mouth of a river where freshwater and seawater mix and tidal effects are evident.

Eucaryotic : a cell having membrane-bound organelles like nucleus and mitochondria.

Euphotic zone : upper layer of water receiving sufficient light for photosynthesis. It lies above compensation depth.

Eurysaline or **Euryhaline** : tolerant of a wide salinity range.

Eutrophic : a stagnant aquatic system that receives large amounts of nutrients usually resulting in a large growth of algae.

Evection : the process by which a lateral branch and the main axis shift relative to each other resulting into dichotomous appearance.

Exochite : the outermost layer of the oogonium in the Fucales.

Exospore : spores of blue-green algae abstricted from the protoplast of the parental cell.

Exotoxin : toxin secreted into the medium.

Extinct : not living today.

Facultative : occassional and can live under differnet conditions of life.

Facultative heterotroph : an autotroph which can live as heterotroph.

False branching : breakage of a trichome through a sheath giving the appearance of branch in some Cyanophyceae.

Filament : one or more trichomes embedded in a sheath in the Cyanophyceae.

Flagellar swelling : a thickening near the flagellar base of euglenoids.

Floridean starch : storage product in red algae composed of α -1,4 and α-1,6 linked glucose residues.

Floridеoside : primary product of photosynthesis in the Rhodophyceae.

Foliose : leaf- like.

Fragmentation : type of asexual reproduction where a thallus breaks into two or more parts, each of which forms new thallus.

Free-nuclear division : repeated mitosis without intervening cytokinesis.

Frustule : the silicious wall of a diatom.

Fucoidin : polysaccharide found in the cell wall and mucilage of the Phaeophyceae composed of sulfated fucose units.

Fucosan vesicles : small, refractive vesicles present in the cells of brown algae containing fucosan.

Funori : commercially produced phycocolloid of some Rhodophyceae.

Furrow : instricting cell membrane during cell division, usually occurs at the equator of a cell.

Fusiform : spindle- shaped.

Gametangium : structure forming gametes.

Gamete : cell capable of fusion with another to form a zygote.

Gametophore : a branch bearing a gametangium or gametangia.

Gametophyte : plant generation that forms the gametes, usually haploid.

Gamone : a chemical involved in the attraction of one gamete to another (androgamone and gynogamone).

Gas vacuole : a collection of gas vesicles.

Gas vesicle : hollow cylindrical gas-filled structures in the Cyanophyceae.

Gelatinous : having a jelly-like consistency.

Generative auxiliary cell : cell forming the gonimoblast filaments in the Rhodophyceae.

Genophore : a gene-containing body, such as the discrete ringlike structure found in some algal chloroplasts.

Girdle : the transvers groove containing the transverse flagellum in the Dinophyceae.

Girdle band : part of a diatom wall where the theca overlap.

Girdle lamella : lamella encircling the periphery of a chloroplast lying just within the chloroplast envelope.

Gliding : active movement of an organism in contact with a solid substrate where there is no visible organ responsible for the movement.

Globule : male reproductive structure in the Charophyta.

Glomerule : cluster of branches at node as in *Batrachospermum.*

Glycoside : a polysaccharide composed of glucose.

Gonidium : asexual reproductive nonmotile cell that divides to form a daughter colony as in *Volvox.*

Gonimoblast : filaments formed from zygote in red algae which bear carposporangia.

Grana : stacks of thylakoid within a chloroplast.

Gynandrosporous : species of *Oedogonium* in which the androspore develops on the female filament.

Gyrogonite : fossilized oospore of Charophyceae.

Habitat : natural place of growth of a plant.

Haematochrome : orange or red oily pigment like carotene present in eyespot and in some algae.

Hairs : elongate, colourless, unicellular or multicellular structure.

Haplobiontic : with a single free-living phase in the life history of an organism.

Haploid : a single basic set of Chromosomes Characteristic of the species.

Haplostichous : type of construction in which the thallus is composed of free or consolidated filaments but lacking a true parenchymatous organization.

Hapteron : multicellular attaching organs of the kelps.

Haptonema : appendage that arises between flagella in the Prymnesiophyceae.

Hermaphrodite or **Homothallic** : producing both male and female gametangia on the same thallus.

Heterocyst : thick-walled, enlarged, usually translucent cell of the Cyanophyceae which are the site of nitrogen fixation.

Heterodynamic flagella : flagella having independent patterns of beating and not correlated with each other.

Heterokont : flagella of unequal length.

Heteromorphic alternation of generation : having haploid and diploid generations of differeent morphology.

Heteroplastidy : having two kinds of plastids (chloroplasts and starch-storing leucoplasts).

Heterothallic : producing male and female gametangia on different plants.

Heterotrichous : differentiation of plant into an erect and prostrate system.

Heterotrophy : the ability to take up and use organic compounds in the dark as energy sources.

Histone : basic protein.

Holdfast : an attaching cell or cells.

Holocarpic : conversion of the entire thallus into a reproductive structure.

Hologamy : fusion of mature individuals directly as gametes.

Holoplanktonic : an organism that spends its entire existence as free-floating individuals.

Holozoic : feeding like animals by ingesting solid food.

Homothallic : producing male and female plants on the same plant.

Hormocyst : thick-walled hormogonium or multicellular akinete found in some Cyanophyceae.

Hormogonium : short pieces of trichome in blue-green algae which get detached from the parent filament and move away by gliding. It develops into new filament subsequently.

Hypha : the elongate cells found in the medulla of kelps and fucoids.

Hypnospore : aplanospore with greatly thickened cell wall.

Hypocone : lower part of the cell in the Dinophyceae usually having a longitudinal sulcus.

Hypolimnion : water beneath the thermocline in thermally stratified water bodies.

Hypotheca : smaller half of a diatom frustule.

Ichthytoxin : toxin that kills gill-bearing animals

Idioandrosporous : species of *Oedogonium* in which the androspore is produced by a special filament other than the female.

Intercalary bands : bands between the valve and girdle band in the Bacillariophyceae.

Internode : part of axis between nodes.

Intertidal : occurring between the low and high-trade marks.

Inversion : phenomenon in some Chlorophyceae where a colony turns itself inside out through a pore.

In vitro : describes laboratory experiments using cell-free extracts or organs or organelles that have been removed from a living organism.

In vivo : describes laboratory experiments with cells, tissues, organs or whole organisms.

Isoenzyme : enzymes having the same function but of a somewhat different structure.

Isogamy : fusion of similar gametes.

Isokont : cell with flagella of the same length.

Isomorphic alternation of generation : generations that are morphologically alike.

Isthmus : the equatorial region connecting the semicells of desmids.

Juvenile : not well developed or fully mature.

Juxtaposed : side by side.

Karyogamy : fusion of nuclei.

Kelp : common name for some of the larger Phaeophyceae.

Kambu : a vegetable made from Laminariales (Phaeophyceae) in Japan.

Lacustrine : growing in lakes.

Lamina : blade.

Laminarin : food storage polysaccharide in the Phaeophyceae composed principally of ß-1,3 linked glucose residues.

Laminate : flat.

Laver : vegetable made from dried *Porphyra* (Rhodophyceae).

Lentic : related to stagnant systems like pond, lake etc.

Leucoplast : colourless plastid usually having a large number of starch grains and few thylakoids.

Leucosin or **Chrysolaminarin** : food storage polysaccharide of golden-brown algae composed mostly of ß-1,3 linked glucose residues.

Lithophyte : plant growing on rock.

Littoral : near the shore.

Littoral zone : zone from the water's edge to a water depth of about 6 m or the maximum depth of rooted vegetation, if exists.

Loculus : hexagonal chamber in the wall of the Bacillariophyceae.

Lorica : a nonliving investment of protoplasts having openings for organelles of locomotion.

Lotic : related to flowing systems like rivers or streams.

Luciferin : compound responsible for bioluminescence.

Lysosome : membrane- bounded cytoplasmic particle containing destructive enzymes.

Macrandrous : species of *Oedogonium* where male and female filaments are approximately similar in diameter.

Macroplankton : plankton larger than 75 µm.

Mannitol : a sugar alcohol reserve food in Phaeophyceae.

Marl : deposits of calcium and magnesium carbonate.

Mastigonemes : fine hair- like appendages of flagella.

Medulla : inner part of algal thallus usually composed of packed colourless filaments.

Meiosporangium : structure where spores are produced by meiosis.

Meristem : dividing tissue that forms new cells.

Meristoderm : a superficial layer of dividing cells covering the thalli of Laminariales.

Meroplanktonic : refers to organisms that spend a part of their life cycle as planktonic and a part as benthic form.

Mesochite : the middle layer in the oogonial wall of fucalean algae.

Mesocaryotic or **Dinocaryotic** : a group of organisms (Dinophyceae) supposed to be in between Procaryotes and Eucaryotes.

Metachromatic granule : protoplasmic body containing stored polyphosphate in the Cyanophyceae.

Microaerophilic : with small amounts of oxygen.

Micrometer (µm) : 10^{-6} meter. 1 µm equals one micron.

Microtubule : submicroscopic tubule in the protoplasm.

Moniliform : arranged like string of beads.

Monopodial : method of development in which the primary axis is maintained as the main line of growth and secondary laterals or offshoots are produced from it.

Monosporangium : a sporangium that produces a monospore in the Rhodophyceae.

Monospore : nonflagellate asexual spore which germinates to form parent in the Rhodophyceae.

Monostromatic : thallus with one-cell layer thick.

Muciferous body : organelle containing mucilage usually found in the outer protoplasm of a cell in euglenoids and dinoflagellates. It discharges mucilage usually explosively.

Mucopeptide : polysaccharide of the walls of the Cyanophyceae composed of sugars and aminoacids.

Multiaxial : an axis composed of many elongeted filaments.

Multiseriate : with more than one row of cells.

Myxophycean starch : storage polysaccharide of the Cyanophyceae similar to glycogen.

Nannandrous : species of Oedogonium in which the male filament is a dwarf epiphyte on the female.

Nannoplankton or **Nanoplankton** : plankton with dimension of 2.20 µm.

Nanometer (nm) : 10^{-9} meter, 1nm equals 10 angstrom units.

Necridium or **Separation disc** : a dead cell between two hormogonia in filamentous blue-green algae.

Nemathecium : wart- like area on the surface of a thallus containing reproductive organs in Rhodophyceae.

Nematocyst or **Cnidocyst** : a complex ejectile produced in a few dinoflagellates.

Neritic region : ocean water over the bottom that extends from the high-tide mark to a depth of 200 m.

Neuston : community of organisms living at the water - atmosphere interface.

Nitrogen fixation : the intracellular fixation of nitrogen gas from the atmosphere to ammonia in

Cyanophyceae or bacteria.

Node : part of thallus that bears branches.

Nori or Laver : vegetable prepared from dried *Porphyra* (Rhodophyceae) in Japan.

Nucleolar organizing chromosome : chromosome concerned with formation of nucleolus.

Nucule : female reproductive structure in the Charophyceae.

Nucleomorph : a structure in the periplastidal space of cryptomonads with suggested role of subsidiary genome.

Nurse cell : cell laden with reserve food and borne near carpogonium in red algae.

Obligate : essential or necessary, not facultative.

Oceanic region : open seas beyond 200 m bottom depth.

Oligotrophic : term describing a lentic aquatic system low in nutrients.

Ontogeny : development of organism.

Ooblast : a filament produced by the zygote that fuses with an auxiliary cell in the Rhodophyceae.

Oogamy : fusion of a large nonmotile egg with a small notile sperm.

Oogonium : the female gametangium that produces an egg or eggs.

Oospore or **Zygospore** : thick- walled zygote with food reserves.

Organelle : a membrane- bounded part of a cell.

Osmotroph : organisms which absorb nutrients in soluble form through plasma membrane.

Ostiole : an opening or pore of conceptacle of Fucales.

Ovum or **Egg** : nonmotile large female gamete.

Palmella stage : nonmotile cells embedded in an amorphous gelatinous matrix.

Pantonematic or **Tinsel flagellum** : flagellum with hairs attached to the surface.

Paraflagellar body : photoreceptor in the Euglenophyceae consisting of a crystalline swelling in one of the flagella.

Paramylon : storage polysaccharide composed of ß-1,3 linked glucose molecules.

Paraphysis : sterile filament among sporangia or gametangia.

Parasite : heterotrophic organism that derives nutrients from a living host.

Parietal : peripheral.

Parthenogenetic : germination of an egg without fertilization to form a new plant.

Pedicel : supporting structure of a reproductive tissue.

Peduncle : a cytoplasmic extention used as a stalk or means of affixing the cell.

Pelagic : living in the open ocean i.e., living at or near the surface of the open sea.

Pellicle : proteinaceous outer covering in the Euglenophyceae.

Pennate : describes diatoms in Pennales showing ornamentation arranged on either side of a central line (bilateral symmetry).

Peptidoglycan : polysaccharides composed of sugars and amino acids in the walls of the Cyanophyceae.

Pericarp : a sterile covering around carposporophyte in the Florideophycidae.

Pericentral cell : a small cell formed around a central axis.

Periclinal : parallel to the circumference of the surface.

Periphyton : organism attached to submerged vegetation.

Perizonium : an outerlying membrane surrounding the auxospore and derived from the fertilization membrane.

Phagotrophic or **Holozoic** : ingesting solid food particles into a food vesicle for digestion.

Phialopore : hole in an inverted daughter colony in the Volvocales (green algae).

Photic or **Euphotic zone** : depth of water above the compensation depth.

Photoautotrophic : autotroph which derives its energy from light.

Photoauxotrophic : photosynthetic but requiring vitamins.

Photoheterotroph : organism capable of using organic compounds as a source of carbon in the light but not in the dark.

Photoreceptor : the part of the cell that receives the stimulus in phototaxis, usually a dense area in a flagellar swelling.

Phototaxis : movement of a whole organism toward (positive) or away from (negative) light.

Phragmoplast : wall formation by the coalescence of Golgi vesicle between spindle microtubules.

Phycobiliprotein or **Phycobilin** : water-soluble blue-green or red pigment in Cyanophyceae, Rhodophyceae and Cryptophyceae.

Phycobilisome : an aggregation of phycobiliproteins on the surface of a thylakoid.

Phycobiont : algal partner in a lichen.

Phycocyanin : blue biliprotein.

Phycoerythrin : red biliprotein.

Phycology or **Algology** : the study of algae.

Phylogeny : evolutionary development of a species.

Physiological anisogamy : isogamy with physiological differences between the fusing isogametes.

Physode : vesicle containing phaeophycean tannins in the Phaeophyceae.

Phytochrome : chemical regulating photoperiod.

Phytoplankton : plants that float along with water current.

Pit - connection : cytoplasmic strand connecting two adjoining cells of Rhodophyceae through pit in their respective cell wall.

Placental cell : cell resulting from the fusion of an auxiliary cell and nearby cells that give rise to carpospores.

Placoderm desmid : desmids having two semicells frequently joined by narrow isthmus. Pores are normally present in cell walls.

Plakea : flat plate of cells in the Volvocales (Chlcrophyceae).

Plankton : organisms that float in water.

Planogamete : motile gamete.

Planozygote : motile zygote.

Plasmodesmata : cytoplasmic connection between adjoining cells.

Plastid : double- membraned organelle usually containing the photosynthetic apparatus.

Pleomorphic : having more than one shape.

Plurilocular : having many small chambers and locules.

Polar nodule : wall swelling near the end of a cell in the diatoms.

Polyglucan granule : protoplasmic structure containing the storage product in the Cyanophyceae.

Polyhedral body : polygonal structure in Cyanophyceae associated with DNA microfibrils.

Polyphosphate body : protoplasmic structure containing stored phosphate in the Cyanophyceae.

Polyploidy : cells having three, four or more times the haploid number of chromosomes.

Polysiphonous : thallus having vertical files of parallel cells.

Polystichous : type of construction in which the thallus is composed of true parenchymatous organization (found in some Phaeophyceae).

Proboscis : microtubule in the anterior part of a sperm, probably associated with chemotactic response.

Procarp : multicellular female sex organ of certain red algae.

Procaryotic cell : a type of cell lacking membrane-bounded organelles.

Productivity : change in biomass per unit time.

Profundal zone : part of a lake beneath the compensation depth.

Progressive cleavage : gradual cytokinesis of multinucleate protoplasm ultimately into uninucleate segments.

Prapagule : a multicellular structure meant for asexual reproduction.

Proplastid : a small plastid that usually matures to a chloroplast or amyloplast.

Protonema : the product of the germination of the zygote in Charophyta.

Proximal : opposite of distal, nearest to the point of origin or attachment.

Pseudocilia : nonfunctional flagella.

Pseudoparenchyma : densely packed filaments resembling parenchyma tissue.

Pseudoraphe : unornamented area on a theca of the Bacillariophyceae where a raphe would occur.

Puncta : opening in the frustule in the Bacillariophyceae, either a pore or a loculus.

Pusule : a structure in the Dinophyceae that is associated with the expulsion of excess protoplasmic water.

Pyriform : pear- shaped.

Pyrenoid : proteinaceous area of the chloroplast associated with the formation of storage product.

Quadrate : squarish, four-sided or rectangular.

Radial : occurring around a central point.

Raphe : longitudinal slit in the valve of some diatoms.

Receptacle : a fertile area containing conceptacles which bear gametangia or sporangia.

Red tide : water with a large number of dinoflagellates or other organisms that colour the water red.

Replicate : septum showing two circular infoldings, one on either side.

Reservoir : large empty space at the bottom of the canal in some flagellates.

Resting spore : thick-walled cell resistant to unfavourable environmental conditions.

Reticulate : net like.

Rhizoid : rootlike filament without vascular tissue.

Rhizoplast : a rootlike striated band arising form the basal body of a flagellum and usually directed to the nuclear envelope.

Rhizopodium : long delicate cytoplasmic protrusion for locomotion.

Rib : extension of the cell wall.

Saprobic : pertaining to organisms that feed on decaying organic matter.

Saprophyte : heterotrophic organism living on dead material.

Saxitoxin : a neurotoxin produced by *Gonyaulax catenella.*

Scalariform conjugation : sexual reproduction in Conjugales in which the connected filaments and their conjugation canals give a ladderlike appearance.

Scintillon : particle associated with bioluminescence.

Seaweed : a macrophytic marine alga.

Segregative cell division : type of cell division found in Siphonales (Chlorophyceae) in which the protoplast is cleaved into spherical masses which then expand and function as independent entities.

Seirospore : a spore produced by a seirosporangium.

Semicell : one half of desmidiacean cell.

Separation disc or **Necridium** : a cell that dies in a trichome of the Cyanophyceae resulting in the separation of a hormogonium.

Septum : cross wall.

Seston : particulate matter in free water.

Seta : a stiff hair or bristle or a hollow projection of the diatom frustule extending beyond the valve margin.

Sheath : extracellular mucilage.

Shield cell : a wall cell of the charophycean antheridium.

Silica deposition vesicle : vesicle in which silica is deposited.

Silicalemma : membrane of a silica deposition vesicle.

Siphonaceous or **Siphonous** or **Coenocytic cell** : large multinucleate cell without cross walls except at the base of reproductive bodies.

Slime layer : extracellular mucilage.

Somatic : vegetative.

Sorus : cluster of reproductive bodies.

Sperm : male gamete.

Spermatangium : male gametangium forming one spermatium in the Rhodophyceae.

Spermatogenesis : formation of sperm.

Spermocarp : zygote with an enclosing layer of cells in *Oedogonium* (Chlorophyceae).

Sporangium : spore- producing organ.

Spore : cell which germinates to form a new individual without fusion.

Sporogenesis : process of spore production.

Sporophyll : a fertile, spore-producing blade.

Sporophyte : diploid plant that forms spores.

Standing crop : the amount of biomass of plants present at a specific time.

Starch : storage polysaccharide composed of α-1,4 and α-1,6 linked glucose residues.

Statospore or Cyst : resting spore of some Chrysophyceae and Xanthophyceae.

Stellate : star-shaped.

Stephanokont : cell with a ring of flagella at one pole.

Stichonematic : type of flagellum with a single row of hairs (mastigonemes).

Stigma or **Eyespot** : group of pigmented lipid bodies that are associated with phototaxis.

Stipe : organ between a holdfast and a blade.

Stipulode : a single- celled process subtending branchlets in some genera of Charales.

Stoneworts : calcified members of Charophyta.

Striae : row of punctae (pores or loculi) in the diatoms.

Stromatolite : rocklike deposition of carbonates and trapped sediments formed by the Cyanophyceae.

Sublittoral zone : zone in a lake from the end of rooted vegetation (about 6 m) to the compensation depth. It varies from the lowest low-tide mark to 200 m depth in the sea.

Suffultory cell : cell to which the dwarf male filament attaches in *Oedogonium* (Chlorophyceae).

Sulcus : longitudinal groove in the hypocone of Dinophyceae.

Supporting cell : a cell that bears the carpogonial branch in some Rhodophyceae.

Supralittoral zone : zone above the standing water mark in lakes and above the high-tide mark in the ocean which receives splash during windy periods.

Symbiosis : two organisms living together to the mutual benefit of each.

Syngamy : fusion of gametes.

Synzoospore : multiflagellate, multinucleate compound asexual spore found in *Vaucheria*.

Systole : contraction of a contractile vacuole.

Taxon : a taxonomic group.

Terrestrial : growing on soil.

Tetrasporangium : a sporangium producing four tetraspores usually by meiosis.

Tetraspore : spore formed in a tetrasporangium, usually by meiosis.

Tetrasporophyte : usually diploid plant forming tetraspores in the Rhodophyceae.

Thallophyta : plant kingdom lacking root, stem and leaves.

Theca : outer covering of the Dinophyceae.

Thecal plate : a plate in a vesicle under the plasmalemma in the Dinophyceae.

Thermal stratification : phenomenon in which the water is progressively colder with depth resulting in no interchange of water between the bottom and top.

Thermocline : layer in a thermally stratified lake where the temperature changes suddenly with depth.

Thermophile : organism which lives at high temperature.

Thylakoid : membrane- bound sac or photosynthetic lamella.

Tinsel flagellum : flagellum with hairs.

Trabeculae : wall ingrowths in some coenocytic green algae.

Transition zone : the region between stipe and blade where active cell division is localized i.e., intercalary meristem, found in Laminariales (Phaeophyceae).

Trellisoid : type of uniformly arranged ornamentation without reference to a point or line in the Bacillariophyceae.

Trichoblast : uniseriate, usually colourless hair bearing sex organs in the Rhodophyceae.

Trichocyst : projectile in the Dinophyceae and Rhaphidophyceae.

Trichogyne : long, colourless, apical portion of a carpogonium that receives the spermatium in the Rhodophyceae.

Trichome : a row of cells without the sheath in the Cyanophyceae.

Trichothallic growth : growth in which the site of active cell division is at the base of a filament or group of filaments.

Tube cell : the helical cells enveloping charophycean oogonia.

Trumpet hyphae : long, tubular, achlorophyllous cells of medulla found in Laminariales (Phaeophyceae). These help in conduction of water and nutrients.

Unarmored : lacking a cell covering composed of polysaccharide plates or valves (Dinophyceae).

Unduliseptate : septum with wavy margins.

Uniaxial : having a main axis consisting of a single row of usually large cells.

Unilocular sporangium : sporangium composed of a single cell producing zoospores usually by meiosis.

Uniseriate : having a single row of cells.

Unisexual : having only one type of gametangium formed on one plant.

Upwelling : an area of the ocean where nutrient-rich bottom water rises to the surface.

Utricle : inflated branchlet as in *Codium*.

Vacuole : small, usually spherical space within cell, bounded by membrane and containing sap.

Valve : part of the cell wall in diatoms. A valve and a connecting band make up a theca.

Vegetative : formation of plant body lacking reproductive structures.

Velum : wall extension over a loculus in the Bacillariophyceae.

Vesicle : membrane-bound organelle.

Volutin granule : protoplasmic body containing stored polyphosphate.

Water bloom : luxurient growth of one or a few species of algae in water often imparting colour to water.

Whiplash flagellum : flagellum without hairs on its surface.

Xanthophyll : a carotenoid composed of an oxygenated hydrocarbon.

Xylan : polysaccharide composed of xylose sugar residues.

Zonate : division of tetrasporangium in three parallel planes so that tetraspores appear arranged in linear row.

Zoochlorella : a symbiotic green-pigmented algal cell (Chlorophyceae) living inside tissue of an animal host.

Zoosporangium : sporangium that forms zoospores.

Zoospore : motile flagellated asexual cell.

Zoosporogenesis : formation of zoospores.

Zooxanthella : a symbiotic golden-pigmented algal cell living in the tissue of an animal host.

Zygospore : a thick-walled zygote or resting spore.

Zygote : product of the fusion of two gametes.

Zygotic meiosis : meiosis occurring during germination of zygote.

AUTHOR INDEX

SUBJECT INDEX

TAXONOMIC INDEX